AF531362

UNDERSTANDING MOLECULAR BIOLOGY

By
Dr. Rajiv Tyagi
Department of Zoology
M.M. College
Modi Nagar (U.P.)
(India)

DISCOVERY PUBLISHING HOUSE PVT. LTD.
NEW DELHI-110 002

First Published-2009

ISBN 978-81-8356-455-7

Published by:

DISCOVERY PUBLISHING HOUSE PVT. LTD.
4831/24, Ansari Road, Prahlad Street,
Darya Ganj, New Delhi-110002 (India)
Phone: 23279245 • Fax: 91-11-23253475
E-mail: dphbooks@rediffmail.com
dphtemp@indiatimes.com
Website: www.discoverypublishinghouse.com

Printed at:
Sachin Printers
Delhi

Preface

The present title "Understanding Molecular Biology" has been written for those students interested in careers in diverse fields of biological sciences. It provides a structured approach to learning by covering all the important topics in a uniform, systematic format. The book has been comprehensively designed incorporating recent advances in this fast moving field. It also provides accessible information on molecular biology in compact form for undergraduate students in biology and related life sciences. It is intelligible to the educated layman, though it deals with some complex ideas. It is an adequate text for all the requirements of students in this area. In addition, busy lecturers who require a quick reference compendium will find it useful, particularly for tutional planning. Simple, yet hopefully clear figures and tables are provided throughout the book.

The over-riding goal of this book, and indeed of the whole *Understanding series*, is to present the essential information concering molecular biology in a compact, readily accessible form which leads itself to student learning and revision. The convergence of various approaches has generated a rich panorama of detail, the significance of which we are still attempting to unraval. The present text has been written as an introduction to this rapidly growing field.

To make the work more comprehensive and informative, the author has consulted many authoritative books, research journals, abstracts, monographs etc., so there can be no claim to originality except in the manner of treatment.

The author expresses his thanks to his friends and colleagues whose continue inspirations have initiated him to bring out this book.

The author expresses his gratitude to Mr. Wasan and staff of M/s Discovery Publishing House Pvt. Ltd. for their whole hearted co-operation in the publication of this book.

In the mean time, the author will remain sincerely responsible for any shortcomings of the book and be grateful to the readers for their suggestions and constructive criticism for the continuous betterment of the book. He takes this opportunity to appeal to the readers to send their suggestions straightaway to his Publisher.

Author

CONTENTS

1

INTRODUCTION

In 1985, scientists discovered that the high-altitude ozone layer, which shields organisms from harmful ultraviolet radiation, had thinned greatly over Antarctica. Ozone (O_3) is produced by the action of sunlight on atmospheric oxygen (O_2) in tropical regions and is transported to high latitudes, where it is destroyed. Chlorine compounds, produced mainly by humans, appear to be the main cause of the unusually high rates of ozone destruction. Ozone is now seriously depleted at very high latitudes, where conditions favour its destruction.

Because ultraviolet radiation damages DNA, it increases the incidence of mutation and cancer. A 1 percent decrease in atmospheric ozone is estimated to result in a 6 percent increase in the incidence of skin cancer. Average ozone concentrations have decreased about 3 percent, enough to have caused a 20 percent increase in skin cancers at midlatitudes. Skin cancer, which is sometimes lethal, is a serious human health problem.

The modification of Earth's atmosphere by human activity is just the latest of a series of atmospheric changes caused by organisms. The most dramatic of these changes was the generation of O_2 by photosynthetic prokaryotes 2.5 billion years ago. Organisms have also changed Earth's waters and soils. Conditions on Earth, in turn, have influenced the evolution of life. In this chapter we examine how conditions on Earth 3.8 billion years ago influenced where and how life evolved and how life in turn, during the early phases of its evolution, changed Earth.

The phenomenon of reproduction distinguishes living organisms from nonliving and was central to discussions at many points in Parts One and Two of this book. How could something that is self-replicating have existed before there were cells? What was the earliest self-replicating unit, and how did it reproduce? Remarkably, scientific evidence is good enough that we can suggest answers to these questions.

Earth's oldest rocks contain no fossils, showing that for at least several hundreds of thousands years after its formation, Earth had no life. Today we are confident that all organisms come from other organisms. The first life, however, must have come from nonliving matter. How did this happen? Under what conditions did life originate on Earth?

Is Life Forming from Nonlife Today?

Until the nineteenth century, people believed in *spontaneous generation*—the regular formation of living organisms from nonliving matter. Flies and maggots were believed to arise from rotting meat and barnyard manure, eels and fish from sea mud, and frogs and mice from moist soil. For more than 2,000 years, belief in spontaneous generation was accepted by most scholars, clergy, and plain folk, although attempts to disprove it had been made.

In 1862, the great French scientist Louis Pasteur performed a series of meticulous experiments showing that microorganisms come only from other microorganisms and that a genuinely sterile solution remains lifeless indefinitely unless contaminated by living creatures. His most elegant experiment relied on swan-necked flasks that were open to the air. Pasteur filled the flasks with nutrient medium, heated them to kill any microorganisms present, then cooled them slowly. No new growth appeared in these flasks. The shape of the necks kept any new organisms or minute dust particles from falling into the medium. However, in flasks without the narrow swan necks, microorganisms entered and grew rapidly.

As a result of Pasteur's experiments, similar experiments by other scientists, and other discoveries of nineteenth-century science, most people accepted that all life comes from existing life. These experiments suggested that life was not forming from nonliving matter under the current conditions on Earth, but they shed no light on when or how life did form on Earth 3.8 billion years ago.

How Did Life Evolve from Nonlife?

To understand how life arose, we need to know the physical conditions that prevailed on Earth before there was life. And we need to know when life arose. The early conditions on Earth determined which types of chemical reactions could take place. If we know what types of reactions took place, we can suggest how those reactions could have led to the appearance of life.

For its First Billion Years, no Life Existed on Earth

How did the universe form? Scientists now believe that between 10 billion and 20 billion years ago there was a mighty explosion. The matter of the universe, which had been highly concentrated, began to spread apart rapidly. Eventually clouds of gases collapsed on themselves through gravitational attraction, forming the galaxies-great clusters of hundreds of billions of stars.

Somewhat less than 5 billion years ago, toward the outer edge of our galaxy (the Milky Way), our solar system (the sun, Earth, and our

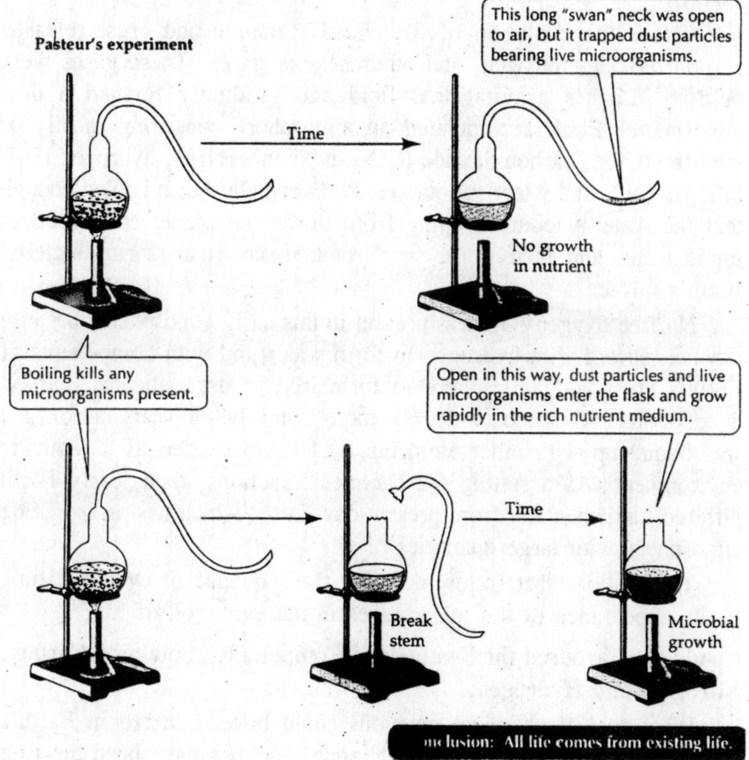

Fig. 1.1. Experiments disproved the spontaneous generation of life. Louis Pasteur's classic experiments showed that a genuinely sterile solution remains lifeless indefinitely; only "contamination" by living organisms could cause life to appear in the flasks.

sister planets) took form. Earth probably formed by gravitational attraction of rocks of various sizes. As Earth grew by this process over millions of years, the weight of the outer layers compressed the interior of the planet. The resulting pressures, combined with the energy from radioactive decay, heated the interior until it melted. Within this viscous liquid, the heavier elements settled to produce a fluid iron and nickel core with a radius of approximately 3,700 km that persists to this day. Around the core lies a mantle of dense silicate materials that is 3,000 km thick. Over the mantle is a lighter crust, more than 40 km thick under the continents but as little as 5 km thick in some places under the oceans.

Conditions on Earth 3.8 Billion Years Ago Differed from Those of Today

Before the evolution of life, Earth's mantle and crust released carbon dioxide, nitrogen, and other heavier gases. These gases were held by Earth's gravitational field and gradually formed a new atmosphere. Earth accumulated an atmosphere consisting mostly of methane (CH_4), carbon dioxide (CO_2), ammonia (NH_3), hydrogen (H_2), nitrogen (N_2), and water vapour (H_2O). Eventually Earth cooled enough that the water vapour escaping from inside the planet condensed to liquid water and formed the seas. Violent electrical storms battered Earth's surface.

No free oxygen (O_2) was present in this early atmosphere, because oxygen reacted with hydrogen to form water and with components of Earth's crust and atmosphere to form iron oxides, silicates, carbon dioxide, and carbon monoxide. For more than 2 billion years, all oxygen was bound up with other elements, and Earth existed as a reducing environment. As a setting for chemical reactions, then, early Earth differed fundamentally from present-day Earth, which has an oxidizing atmosphere with large quantities of O_2.

The sections that follow describe the sequence of events leading to the appearance of the main stages in the evolution of life.

Conditions Favoured the Synthesis of Compounds Containing Carbon, Nitrogen, and Hydrogen

What sort of chemical reactions could have occurred in Earth's early reducing environment? Could such reactions have been the first step toward the origin of life? To investigate these questions, in the 1950s Stanley Miller studied chemical reactions proceeding under conditions that resembled those believed to exist on Earth 4 billion years ago.

Within a closed system of glass tubes, he established a reducing atmosphere of hydrogen, ammonia, methane gas, and water vapour. Through these gases, he passed a spark to simulate lightning. Within a few hours the system was found to contain numerous simple organic compounds, including hydrogen cyanide, formaldehyde, cyanogen, acetaldehyde, cyanoacetylene, and propionaldehyde. In water, these compounds dissolved and were rapidly converted into amino acids, simple acids, purines, and pyrimidines—the building blocks of life.

The same or similar compounds are produced under a variety of conditions, provided that free oxygen is absent—that is, if the environment is a reducing one. Thus, once Earth cooled enough for water to condense and form oceans, molecules of many kinds formed, and they probably accumulated until they reached relatively high concentrations. From such a prebiotic ("before life") molecular soup, life emerged from nonliving matter.

Polymerization Provided Macromolecules with Diverse Properties

The next stage in the sequence leading to life was the generation of large molecules by *polymerization* of small molecules. The important large molecules of which organisms are composed (polysaccharides, proteins, and nucleic acids) are polymers formed by the combination of subunits called monomers. The polymerization reactions that generate these large molecules belong to a class of reactions called *condensations*, or *dehydrations*. During a condensation reaction, water forms. Large molecules are assembled through repeated condensations of monomers, and each condensation reaction requires energy. In the prebiotic soup, polymers that formed faster or were more stable would have come to predominate.

Some polymers can direct the synthesis of molecules identical to themselves. Which of the molecules on prebiotic Earth were most likely to reproduce themselves? Nucleic acids, the basis of today's genetic code, are clearly capable of self-copying, and the purine and pyrimidine constituents of nucleotides are formed under conditions similar to those believed to have prevailed on early Earth.

Sugars, another basic component of living cells, are generated by the polymerization of formaldehyde, another molecule formed in Miller's experiments. Solutions of formaldehyde spontaneously polymerize to form several different five- and six-carbon sugars, but these sugars are unstable in aqueous solutions and break down to alcohols and carboxylic acids. However, three and four-carbon sugars are very stable and can accumulate for hundreds of years in aqueous solutions.

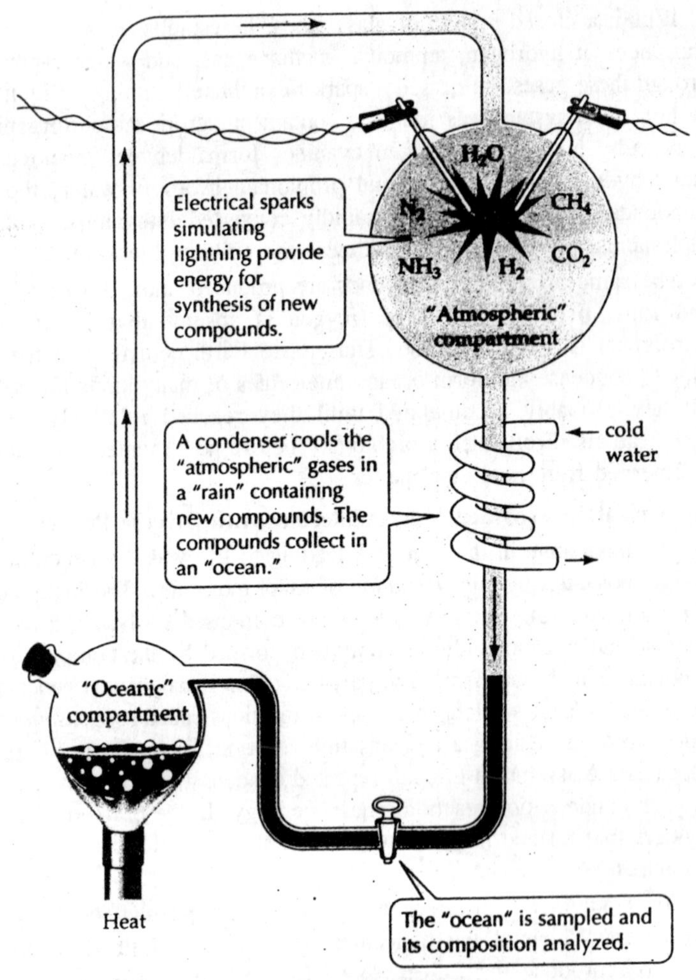

Fig. 1.2. Synthesis of molecules in an experimental atmosphere.

Over millions of years, these organic molecules would have accumulated in the oceans. They would have reached even higher concentrations in drying ponds. High concentrations of polymers, in turn, would have stimulated further polymerization and chemical reactions leading to the synthesis of sugars, which were important as components of nucleotides and for energy storage in early cells.

Phosphate-based Polymerizations Occurred in the Prebiotic Soup

Monomers formed under prebiotic conditions can polymerize by phosphorylation (the addition of a phosphate group). Phosphorylated monomers are stable enough to accumulate in solution but reactive enough to polymerize further. The first biologically active polymers of nucleic acid bases may have been compounds formed from three- and four-carbon sugars. They could have self-replicated by forming complementary double-stranded molecules, just as pentose-based nucleic acid polymers do today. At some point, by as yet unidentified processes, polymers based on small sugars were replaced by those based on larger sugars, such as ribose, the sugar in the nucleotides of RNA.

RNA was Probably the First Biological Catalyst

The enzymes that control the types and rates of reactions within organisms are proteins. Proteins are synthesized by a process that begins with transcription of information from DNA to an RNA molecule that has a base sequence complementary to that of one strand of the DNA. This information is then translated into mRNA and is eventually used to synthesize a specific polypeptide from amino acids. Amino acids are brought to a ribosome by specific tRNA molecules and are attached sequentially to the growing polypeptide. But how could such a system have evolved if protein catalysts needed nucleic acids for their formation but nucleic acids needed proteins to catalyze their own replication? Which came first, if both are necessary?

The inability to solve this dilemma held up research on the origin of life for several decades. The discovery that provided a solution came in 1981 from researchers working with the unicellular protist *Tetrahymena thermophila*. These scientists were studying the excision of introns and the splicing together of exons. They found entirely contrary to expectation—that excision is catalyzed in the absence of enzymes. The intron itself a 400-nucleotide sequence of RNA-carries out the excision and splicing. In addition, *Tetrahymena* ribosomes, which contain several molecules of RNA and a variety of proteins, have a catalytic RNA that operates in protein synthesis. RNAs that catalyze chemical reactions, called *ribozymes*, have now been found in many organisms.

The current system of macromolecular synthesis (DNA → RNA → protein) probably evolved gradually from much simpler processes. Biochemists believe that the first information-carrying molecules were short strands of RNA that replicated themselves without the help of enzymes. Evidence that RNAs can replicate themselves came first from experiments conducted by Manfred Eigen in the late 1970s. Eigen added RNA molecules to solutions containing monomers for making more RNA and found that sequences of five to ten nucleotides formed.

If he added a simple inorganic molecule such as zinc, much longer sequences were copied.

Although these experiments showed that RNA could replicate itself, they did not demonstrate that RNA could catalyze the synthesis of other molecules as true enzymes do. Within two years, however, studies of a tRNA-processing enzyme that contains RNA and a protein (ribonuclease P) showed that the RNA alone can cut the pre-tRNA molecule at the correct spot, whereas the protein cannot.

Many scientists believe that the first genetic code was based on RNA that catalyzed both its own replication and other chemical reactions. In such conditions a high concentration of RNA would have been needed, so that it could participate in many different chemical reactions. The accumulated products of RNA-catalyzed reactions could then participate in other reactions to form structures—for example, the production and accumulation of lipidlike molecules to form cell membranes, and the synthesis of proteins. However, after proteins evolved, they eventually took over most enzymatic functions because they are better catalysts than RNA and are capable of more diverse specificities.

In order to replicate, different RNAs would have competed with one another for monomers. Some RNA molecules would have been better at replicating in certain environments because their base sequences produced the most stable configurations (folded structures) under the particular conditions of temperature and salinity. With their higher rates of replication and greater stability, these molecules would have come to dominate the populations, of RNA in the corresponding environments.

In the precellular world, RNA that contained both genetic information and catalytic capacity evolved. Molecules with these capacities have been produced in the laboratory by investigators who started with completely random-sequence RNA. These ribozymes ligate (bind together) two RNA molecules in a reaction similar to that employed by the enzymes that synthesize RNA.

The investigators simulated the "evolution" of RNA molecules by selecting in test tubes for ribozymes with high ligation ability. By this method, they produced ribozymes with reaction rates 7 million times faster than the uncatalyzed reaction rate. These experiments suggest that ribozymes could have evolved rapidly when conditions on Earth became suitable for the formation of nucleic acids. But ribozymes in solution, even active ones, do not constitute life. Life required a barrier

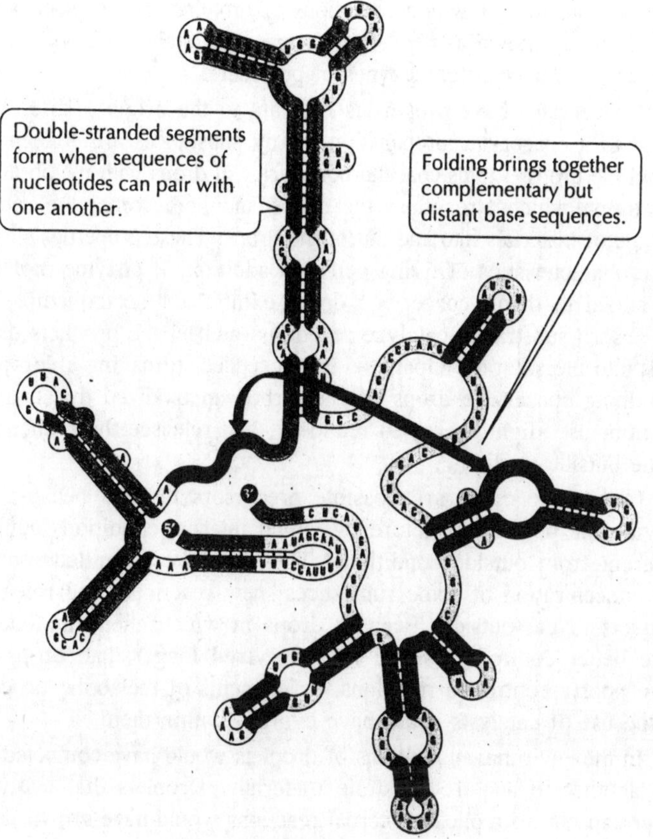

Fig. 1.3. A ribozyme from a protist.

that permitted the homeostatic control of internal conditions; life could not appear until cells enclosed in membranes evolved. The prebiotic RNA world still lacked cells.

Membranes Permitted a Stable Internal Environment to Form

The way in which ribozyme-based systems in solution might have evolved into cells was first suggested in the 1920s by the experiments of the Russian scientist Alexander Oparin, who spent much of his career studying complex solutions. Oparin observed that if he shook a mixture of a large protein and a polysaccharide, the drops that formed were divided into two "phases": an interior separated from but in contact with an exterior. These interiors, which were primarily protein and polysaccharide, with some water, were surrounded by an aqueous solution containing low concentrations of proteins and polysaccharides. These drops, known as *coacervates*, are quite stable and will form in solutions of many different types of polymers.

Coacervates have properties relevant to the origin of life. When added to a coacervate preparation, many substances are concentrated within the drops. Lipids coat the boundaries of drops with membranelike structures, which strengthen the drops and help contol the rates of passage of materials into and out of the drops. These properties simulate some characteristics of living cells. In addition, if enzyme molecules are added to them, coacervate drops exhibit a "metabolism": They can absorb substrates, catalyze reactions, and let the products diffuse back into the solution. Oparin even succeeded in making chlorophyll-containing coacervate drops that absorbed an oxidized dye from the solution, used light energy to reduce it, and released the reduced dye to the outside medium.

Coacervate drops are possible precursors to cells because they provide the physical structure by which internal conditions could be different from outside conditions. This structure permitted retention and concentration of some substances, new reactions, and release of products to the outside. Because drops in which chemical reactions were better controlled would have survived longer than drops with more poorly controlled reactions, refinements of metabolic processes by the use of catalysts could have evolved within them.

In the environment, billions of droplets would have competed with one another to acquire substrate materials. Droplets that evolved a better capacity to replicafe internal reactions would have grown larger. But when coacervates become large, they are less stable, and physical agitation breaks them apart into smaller droplets. During this early form of reproduction, they would have passed on at least rough copies of their molecules to daughter droplets. At some point, such droplets may have accumulated enough RNA to display the key property we associate with living cells—the use of energy to maintain homeostasis.

DNA Evolved from An RNA Template

If the first cells used RNA as their hereditary molecule, then RNA must have provided the template for the synthesis of DNA. In solution, DNA is less stable than RNA. Therefore, DNA probably did not evolve until RNA-based life became contained in membrane-enclosed cells where water concentrations were lower than in the surrounding environment. Because DNA is a more stable storage location for genetic information than is RNA, once the appropriate environments were available, DNA probably evolved rapidly, replacing RNA as the genetic code for most organisms. But by then, RNAs had assumed their current roles as intermediaries in the translation of genetic information into proteins.

Early Anaerobic Cells

Earth's atmosphere lacked the gas O_2 for about a billion years after life evolved. Therefore, the earliest organisms must have processed energy chemically without using O_2 as an electron acceptor. No traces of the metabolic pathways of those early organisms have been preserved, but some living anaerobic prokaryotes still employ similar pathways. By studying them, we can make reasoned guesses about the metabolism of ancient prokaryotes.

Three major types of anaerobic photosynthetic prokaryotes live today in sediments that lack oxygen: green sulfur bacteria, purple sulfur bacteria, and purple nonsulfur bacteria. In fact, oxygen is poisonous to them, as it must have been to organisms that evolved in a reducing atmosphere. These bacteria contain types of chlorophyll, called bacteriochlorophyll *a, c,* and *d.* Anaerobic photosynthetic bacteria also contain red and yellow carotenoids. These pigments absorb wavelengths of light that are not absorbed by chlorophyll and pass the absorbed energy along to chlorophyll for conversion into chemical energy.

The photosynthetic system of these bacteria is embedded in membrane complexes that also possess the electron transport chains and enzymes by which captured solar energy is converted to chemical energy and used to generate ATP. Photosynthetic prokaryotes similar in their metabolism to today's forms were so abundant about 3.4 billion years ago that their partly decomposed fossil remains formed extensive deposits of carbon, resembling the coal produced by vascular plant fossils 3 billion years later.

To reduce carbon dioxide (CO_2), a photosynthetic cell needs a source of electrons (hydrogen atoms). Many photosynthetic prokaryotes use light energy to generate ATP and NADPH + H^+. The particular waste product that they liberate depends on the source of hydrogen atoms they use. The green and purple sulfur photosynthetic bacteria obtain their hydrogen atoms from hydrogen sulfide (H_2S) and generate sulfur as a waste product. The purple nonsulfur bacteria typically obtain hydrogen atoms from organic compounds such as ethanol, lactic acid, or pyruvic acid, or directly from hydrogen gas (H_2).

In some environments today, the H_2 is produced by other prokaryotes as the end product of their fermentations. Under the anaerobic conditions of early Earth, hydrogen sulfide and other compounds containing hydrogen were more abundant than they are now. Today O_2 directly oxidizes H_2S, H_2, pyruvic acid, and similar compounds into water, carbon dioxide, and sulfur oxides.

Aerobic Photosynthesis is the Source of Atmospheric O_2

The evolution of aerobic photosynthesis, slightly more than 3 billion years ago, changed the course of evolution and changed Earth. The key change was the acquisition of the ability to use water as the source of hydrogen: 2 H_2O - 4 H + O_2. The chemical splitting of H_2O produced O_2 as a waste product and made available hydrogen atoms for reducing CO_2. This ability appeared first in certain sulfur bacteria that evolved into cyanobacteria. Remains of these bacteria are abundantly preserved in fossil *stromatolites*. Cyanobacteria are still forming stromatolites in a few very salty places on Earth. The ability to split water was doubtless the cause of the extraordinary success of the cyanobacteria. The O_2 they liberated opened the way for the evolution of oxidation reactions as the energy source for the synthesis of ATP. Their success made possible the evolution of the full respiratory chain of reactions now carried out by all aerobic cells. The evolution of life irrevocably changed the nature of our planet. Life created the O_2 of our atmosphere, and it removed most of the carbon dioxide from the atmosphere by transferring it to ocean sediments. Oxygen was poisonous to the anaerobic organisms that lived on Earth, but the prokaryotes that evolved a tolerance to it were able to live successfully in environments empty of other, organisms. Also, aerobic (oxygenated) metabolism could proceed much more rapidly and efficiently than the anaerobic metabolism that had dominated life until then.

Eukaryotic Cells: Increasing Complexity

The generation of atmospheric O_2 by cyanobacteria probably set the stage for the evolution of eukaryotes. Small, prokaryotic cells can obtain enough oxygen by simple diffusion even when oxygen concentrations are very low. Larger cells, however, have a lower surface area-to-volume ratio and need higher environmental O_2 concentrations to meet their needs.

About 1.5 billion years ago oxygen concentrations became high enough for larger cells to flourish and diversify. Some of these cells grew big enough to consume smaller cells; they were the first predators. Most of the prey died after they were engulfed, but some of them survived and eventually evolved into the mitochondria and chloroplasts that now maintain mutually beneficial relationships with their host.

Chromosomes Organize Large Amounts of DNA

All early organisms were haploid. They reproduced by dividing, and each division was accompanied by a duplication of their circular chromosomes. The single circular DNA molecule of prokaryotes fits into a small cell, and it can be duplicated rapidly. Over time, the amount of DNA in the cells of early organisms increased, as it proved advantageous to encode information for the enzymatic catalysis of new reactions and to pass this information to progeny cells.

But replicating a long DNA molecule would take many hours if it could be started only at one spot. Replication can proceed much more rapidly if genes are separated into chromosomes, each of which has multiple initiation sites for duplication. Division of the genome into chromosomes also makes it less likely that the DNA molecules will become tangled when they divide. For these reasons, biologists believe that chromosomes evolved soon after the appearance of eukaryotes, about 2.5 billion years ago.

Variable Offspring are Advantageous in Variable Environments

All early organisms reproduced by simple fission, and each division was accompanied by a duplication of their chromosomes. Occasionally and accidentally, the duplication of DNA was not accompanied by a cell division, and the result was a diploid cell. A diploid cell has certain advantages. It can repair more kinds of chromosome damage than a haploid cell because the undamaged duplicate copy can guide the repair. But even if repair was not possible, a duplicate copy of the genome offered assurance that the life of the cell could continue. Today, chromosome breakage can be repaired only by diploid cells. Because chromosome damage is especially likely in stressful environments, diploidy is advantageous in those conditions. Many present-day unicellular organisms that live in a haploid state most of the time produce diploid cells when they are environmentally stressed. Diploid cells also have protection from point mutations, most of which alter only one copy of a gene. The unaltered copy on the homologous chromosome continues to function normally.

A diploid cell can also result if two haploid cells fuse. The chromosomes of two haploid cells are unlikely to have the same damages. Thus, a diploid cell formed by acquiring chromosomes from another cell would be able to repair all its damaged chromosomes. This ability may have provided the original advantage of sexual recombination, but we can only speculate. Whatever its original advantages, sex appeared early during the evolution of life, and nearly all living organisms engage in sexual recombination at least occasionally during their life cycles. Those that do not are believed to have evolved recently from sexual ancestors; they are destined to have short evolutionary lives.

Sex is maintained in multicellular organisms today by a benefit different from the ones that initially favoured it. Whereas the original benefits accrued directly to the individuals accepting genetic material, sex is maintained because of its effects on offspring. Whereas in unicellular organisms genes are not exchanged during reproduction (cell division), in multicellular organisms genes are nearly always exchanged at the time of reproduction, which is the only time during the life cycle when a single-cell stage exists. The offspring of sexual multicellular organisms are genetically highly variable. In an environment that varies in space and time, the genetically uniform offspring of an asexual organism are less likely to find suitable conditions than are the variable offspring of a sexual organism.

Matter, Time, and Inevitable Life

Scientists have been able to gather information that provides many insights into the origin of life on Earth. In laboratory experiments they have studied chemical reactions under conditions similar to those believed to have prevailed on early Earth. Under laboratory conditions, they have witnessed the "evolution" of ribozymes with catalytic activity from random-sequence RNA. Taken together, this information suggests that the evolution of life as we know it was highly probable under the conditions that prevailed on Earth 3.8 billion years ago. The molecules on which life is based form readily under such conditions, as does the organization of those molecules into larger units. Much remains to be learned about the early evolution of the complex metabolism of living organisms, but techniques are available to help us.

If the origin of life was almost inevitable, why is new life not still being assembled from nonliving matter on today's Earth? The reason is that simple biological molecules released into today's environment are quickly consumed by existing life. They cannot accumulate to the densities that characterized the prebiotic "soup," even in anaerobic environments. In aerobic environments these molecules are quickly oxidized to other forms. They would not accumulate even if they were not consumed. Life was generated from nonlife on Earth, but it was an event of the remote past. Once life had evolved, it prevented the formation of other life from nonlife.

2

Molecular Origin of Life

One of the oldest arguments in the history of biology concerned the basic nature of living things. Many biologists in the past thought that the substances composing living things are fundamentally different from those in nonliving things. They thought there is a special life chemistry, governed by its own set of laws different from those governing nonliving matter. They also proposed that there is a special force or energy, "the vital force," that is unique to life. Despite all efforts to detect it, however, no such vital force has ever been found, and chemical laws have been found to operate the same whether the chemical processes occur in living or nonliving things.

Today, biologists recognize that while all matter–living and nonliving—consists of atoms and molecules, many of the molecules in living organisms are larger and more complex than those commonly found in nonliving systems. As we will see, these relatively large and complex molecules are arranged into highly organized systems such as membranes and cells. Because of the high degree of organization among their molecules, living things have properties not found in nonliving systems.

In order to be able to explore the properties of living things, we need to understand some characteristics of the substances that constitute them.

Elements and Compounds

All matter is composed of basic substances called *elements*. An element cannot be broken down into simpler units by chemical reactions; it contains only one kind of atom. An *atom* is the smallest characteristic unit of an element.

Atoms of most elements do not occur in isolation. Usually they are combined with other atoms to form *molecules*. A molecule is a combination of like or different atoms. For example, we do not find individual oxygen atoms in the atmosphere. Rather, oxygen in air is in the form of oxygen molecules, each of which contains two oxygen atoms.

There are ninety-two naturally occurring elements and a number of artificial elements that can be made in laboratories. Each element has a symbol consisting of a one- or two-letter abbreviation of its English or Latin name. For example, the symbols for hydrogen (H) and oxygen (O) are derived from their English names, while the symbols for sodium (Na) and potassium (K) are derived from their Latin names, *natrium* and *kalium*.

A *compound* is a substance that can be split into two or more elements. Water is a compound because it can be split into its components, hydrogen and oxygen. The *formula* of a compound gives information about the kinds and numbers of atoms that make up each molecule of that compound. A formula contains the symbols (abbreviations) of the kinds of atoms in each molecule and subscripts that indicate the number of each kind of atom in the molecule. For example, the formula for water, H_2O, indicates that a water molecule contains two hydrogen atoms and one oxygen atom; and a molecule of the sugar glucose, $C_6H_{12}O_6$, contains six carbon atoms, twelve hydrogen atoms, and six oxygen atoms.

Atomic Structure

While atoms are the smallest units that possess all the properties of the particular elements, each atom is composed of still smaller particles.

Protons, Neutrons, and Electrons

There are several kinds of subatomic particles, but only three are crucial to the properties of elements we will consider here: protons, neutrons, and electrons. *Protons* and *neutrons* are located in a central area of the atom called the *nucleus*. *Electrons* are located at various distances from the nucleus. Protons have a positive electric charge, and electrons have a negative charge; thus, protons attract electrons. Neutrons have no electric charge.

While protons and neutrons have nearly the same mass, their mass is much greater than the mass of electrons (more than 1,800 times as great). Chemists customarily use the *Atomic Mass Unit* (AMU), or *dalton*, to describe the mass of subatomic particles. Both the proton and the neutron have masses of approximately one dalton.

Each atom of an element has a characteristic number of protons in its nucleus, and this number is known as the element's *atomic number*. For example, because a carbon atom has six protons in its nucleus, its atomic number is six. You can see the atomic numbers of some biologically important elements in table 2.1.

Table 2.1. Some elements important in biology.

Element	*Symbol*	*Atomic number*	*Atomic weight*
Hydrogen	H	1	1.01
Carbon	C	6	12.01
Nitrogen	N	7	14.01
Oxygen	O	8	16.00
Phosphorus	P	15	30.97
Sulfur	S	16	32.06
Sodium	Na	11	23.00
Magnesium	Mg	12	24.31
Chlorine	Cl	17	35.45
Potassium	K	19	39.10
Calcium	Ca	20	40.08
Iron	Fe	26	55.85
Copper	Cu	29	63.54
Zinc	Zn	30	65.37

Normally, the number of electrons outside the nucleus of the atom is equal to the number of protons in the nucleus. Because the positive charges of the protons equal the negative charges of the electrons, the charges cancel one another and the atom is neutral.

The *mass number* of an element indicates how many protons and neutrons are in the nucleus of one of its atoms. For example, most carbon atoms contain six protons and six neutrons. Carbon's mass number, 12, is symbolized with a superscript numeral preceding the element's symbol: ^{12C} (read "carbon-12"). The *atomic weight* is the *actual measured weight* of an element and is usually very close to, but not exactly the same as, the mass number.

Isotopes

Use of the word isotope has become common in our society. We hear almost daily about the use of isotopes, particularly radioactive isotopes (or simply "radioisotopes") in medical diagnosis and cancer treatment. We also hear these terms in discussions of nuclear power plants and nuclear waste disposal. What is meant by the term "isotope," and what are radioactive isotopes?

All atoms of a given element have the same number of protons, but different atoms of that element can have different numbers of neutrons. Recall that the mass number of an element is equal to the number of protons plus the number of neutrons in each of its atomic nuclei. The atoms of different forms, or *isotopes*, of a given element have different numbers of neutrons and different mass numbers.

As we saw above, the most common form of carbon atom, with six protons and six neutrons in its nucleus, has a mass number of 12 (^{12}C). A carbon isotope with six protons and seven neutrons is ^{13}C, while a carbon isotope with six protons and eight neutrons is ^{14}C.

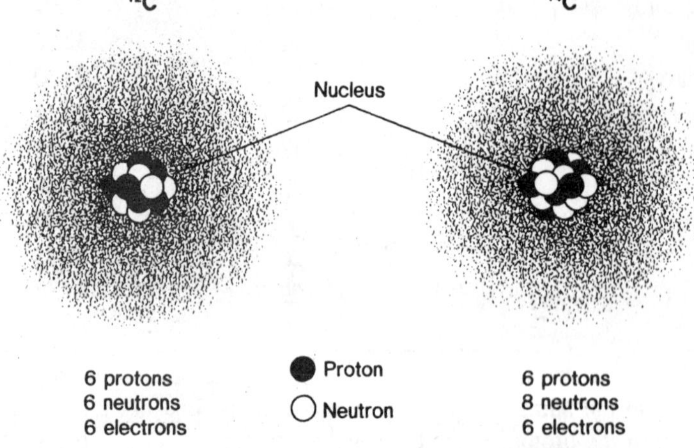

Fig. 2.1. Nuclei of carbon isotopes differ only in the number of neutrons present.

Some isotopes are stable, while others are unstable and tend to decay (break down) by spontaneously emitting small particles. These unstable isotopes are *radioactive*. ^{12}C and ^{13}C are stable isotopes, while ^{14}C is radioactive.

Radioactive isotopes of some elements emit particles with a great deal of energy. Some of the radioisotopes produced in nuclear reactors emit very high energy radiation and are extremely dangerous to handle and difficult to store safely.

Certain isotopes that emit lower-energy radiation are very useful in biological research and medical diagnosis. Radioactive isotopes can be used as chemical "labels" to mark certain compounds in cells or tissues. Because the different isotopes of an element react chemically in the same way, radioactive isotopes may be substituted for stable isotopes to make chemical substances under observation easier to detect.

Electrons and Orbitals

Although electrons have a much smaller mass than protons and neutrons, they are by no means less important in determining the chemical properties of atoms. The location of electrons in relation to the nucleus has a great influence on the way atoms and molecules react with each other.

Because electrons are continually in motion around the nucleus in every atom, we cannot say exactly where a particular electron is located at any one moment. We can say, however, where electrons are most likely to be located most of the time. We call the space in which an electron is located 90 percent of the time the electron's *orbital*.

Each orbital can contain a maximum of two electrons, and orbitals are grouped into *shells*. The electrons located in the various shells have different energy levels. The shell closest to the nucleus has the lowest energy level, and each shell beyond has a progressively higher energy level as the distance between nucleus and shell increases.

The first shell, the one closest to the nucleus, contains only one orbital (a spherical one), and thus can contain only two electrons. The second shell contains four orbitals, one spherical and the other three shaped like dumbbells. This second shell, then, can contain up to eight

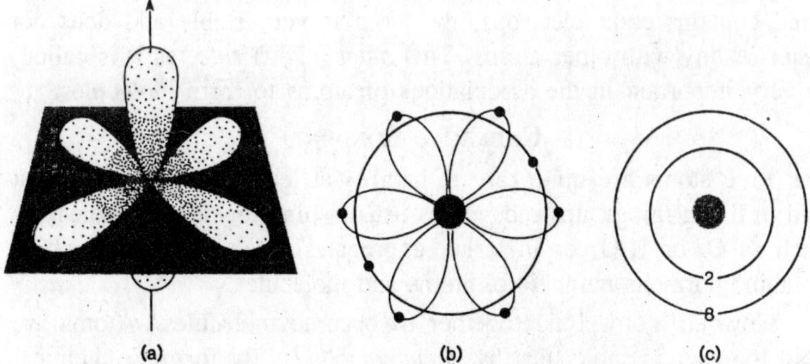

Fig. 2.2. Electron orbitals. (a) The three dumbbell-shaped orbitals of the second shell are at right angles to each other. (b) All four orbitals of the second shell on a flat plane. (c) A simpler way of representing the orbital shells with the number of electrons contained in each

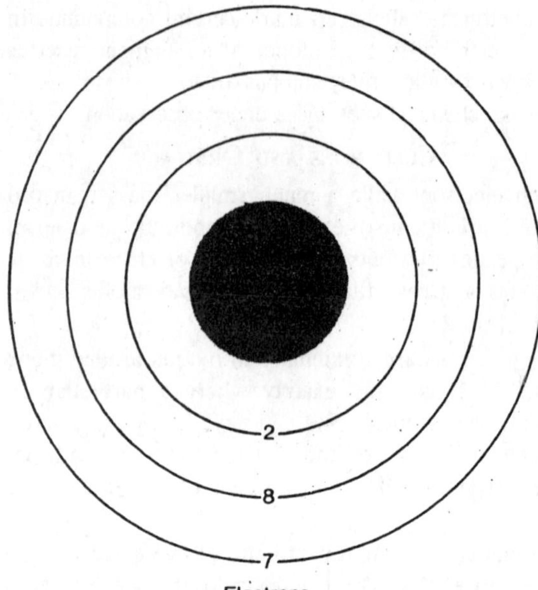

Fig. 2.3. Electron orbital shells of a chlorine atom.

electrons. Although the third shell and those beyond can hold more than eight electrons, these shells are in a particularly stable condition when they do hold exactly eight electrons.

Under normal circumstances, electrons will be in the innermost shells in which orbital space is available. In this way, each lower-energy shell is filled before the next shell is. When an atom's outermost shell contains eight electrons, the atom is very stable and does not react readily with other atoms. This *octet* (eight) *rule*, as it is called, is very important in the associations of atoms to form molecules.

Chemical Bonding

Free atoms are quite rare in nature—at least in the earth's crust and in living things. Instead, atoms usually bind together in molecules such as O_2 or H_2O, or in larger aggregates, such as crystals which contain enormous numbers of atoms and molecules.

How do atoms bind together to become molecules? Atoms are held together in molecules by *chemical bonds*. By forming chemical bonds, atoms tend to achieve a more stable electron configuration than they have by themselves. A stable molecule is formed when the uniting atoms can reshuffle their electrons so that each atom has a complete, or at least more stable, outer shell.

Atoms combining into molecules rearrange their electrons through different means. In *ionic bonding*, electrons are *transferred* from one atom to another. In *covalent bonding*, pairs of electrons are *shared* between atoms.

Ionic Bonding

To illustrate ionic bonding, let us examine the reaction between sodium (Na) and chlorine (Cl) in which sodium chloride (NaCl), common table salt, is formed. Sodium's atomic number is eleven—a sodium atom has eleven protons in its nucleus and eleven electrons orbiting around the nucleus. The inner shell, containing two electrons, is complete, as is the second shell with eight electrons. But the third shell of a sodium atom contains only one electron. An atom with only one electron in its outer shell tends to be an *electron donor*—that is, it tends to get rid of its lone electron.

Chlorine has an atomic number of seventeen, which means that a chlorine atom has two electrons in its first shell, eight in its second shell, and seven in its outer shell. Because it needs only one electron to complete its outer shell (remember the octet rule), a chlorine atom tends to be an *electron acceptor*. When a sodium atom and a chlorine atom come together, an electron is transferred from the sodium atom

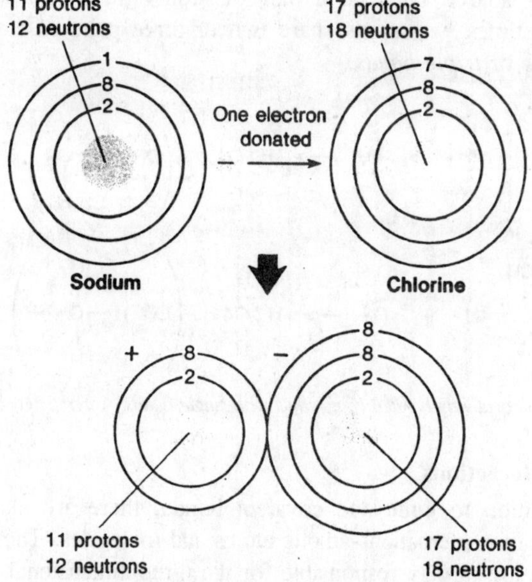

Fig. 2.4. Ionic bonding

to the chlorine atom. But this electron transfer changes the balance between the protons and electrons within each of the two atoms. The sodium atom ends up with one more proton than it has electrons, and the chlorine atom with one more electron than it has protons. The sodium atom is left with a net charge of +1 (symbolized Na^+), while the chlorine atom's net charge is –1 (symbolized Cl^-). Charged atoms are called *ions* and are attracted to one another by their opposite charges. Thus, an ionic compound such as sodium chloride (NaCl) is held together by the attraction between its positive and negative ions; that is, by an *ionic bond.*

The ionic bonds of sodium chloride are quite strong when the salt exists as a dry solid, but when NaCl is dissolved in water, it separates into Na^+ and Cl^- ions. Ionic compounds are most commonly found in this *dissociated* (*ionized*) form in the watery environment inside living things.

Covalent Bonding

In *covalent bonds* electrons are not donated from one atom to another, but are shared between atoms. Thus, a covalent bond consists of a pair of electrons that are shared between two atoms and that occupy two stable orbitals, one of each atom. If a single pair of electrons is shared between a pair of atoms, it is called a *single bond.* Sometimes two atoms share two or three pairs of electrons to form *double* or *triple bonds*.

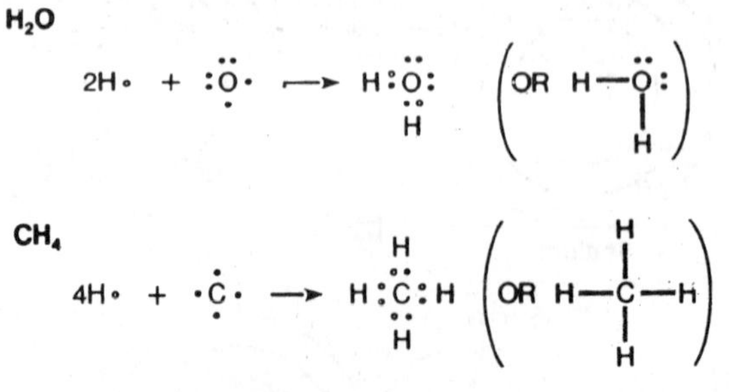

H_2O

2H• + :Ö• ⟶ H:Ö: (OR H—Ö: with H below)

CH_4

4H• + •Ċ• ⟶ H:C:H with H above and below (OR H—C—H with H above and below)

Fig. 2.5. Two covalently bonded compounds with single bonds, water (H_2O) and methane (CH_4).

Weaker Interactions

In addition to ionic and covalent bonds, there are also several types of weaker interactions among atoms and molecules. These weaker interactions are largely responsible for the three-dimensional shapes of many large biological molecules, such as proteins, and for some of

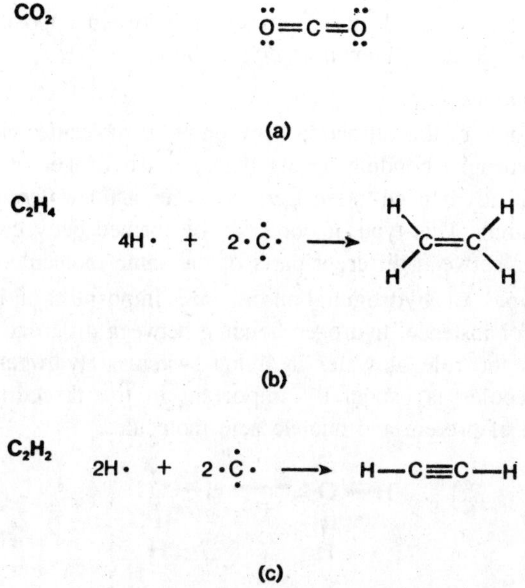

CO_2 $\quad \ddot{\underset{..}{O}}=C=\ddot{\underset{..}{O}}$

(a)

C_2H_4 $\quad 4H\cdot + 2\cdot\dot{\underset{\cdot}{C}}\cdot \longrightarrow H_2C=CH_2$

(b)

C_2H_2 $\quad 2H\cdot + 2\cdot\dot{\underset{\cdot}{C}}\cdot \longrightarrow H{-}C{\equiv}C{-}H$

(c)

Fig. 2.6. Some compounds with multiple covalent bonds.

their functional properties. In some molecules formed by covalent bonds, the electrons tend to be attracted more toward one atom than the other. This unequal amount of attraction among the electrons is called *polarity*. Polarity results in molecules that have a small positive charge at one end and a small negative charge at the other. Because of these slight charges at their ends or poles, polar molecules are attracted to

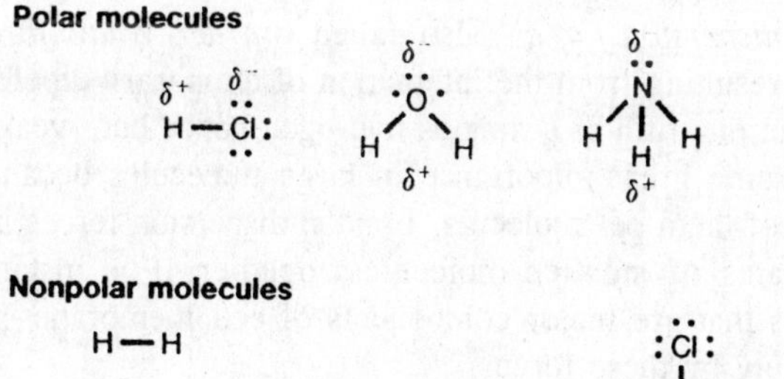

Polar molecules

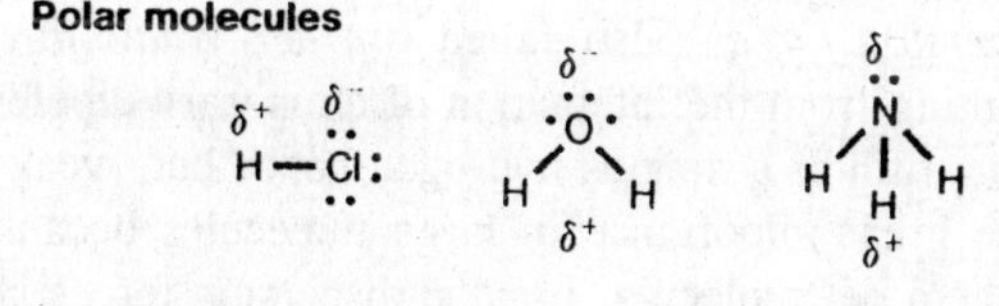

Nonpolar molecules

H—H

:Cl—Cl:

Nonpolar bonds

O=C=O

:Cl—C(Cl)(Cl)—Cl:

Polar bonds, but nonpolar molecule; bond dipoles cancel by symmetry

Fig. 2.7. Examples of polar and nonpolar molecules.

each other. The weak electrical attractions between opposite charges on polar molecules are known as *dipole forces*.

Hydrogen bonding

An example of this attraction between polar molecules is *hydrogen bonding*. Hydrogen bonding occurs between a hydrogen atom that is already covalently bonded to nitrogen or oxygen and a *different* nitrogen or oxygen atom. This type of bond can be formed between different molecules or between different parts of the same molecule.

Both types of hydrogen bonding are important in biological processes. For instance, hydrogen bonding between different molecules is crucial to the role of water in living systems. Hydrogen bonding within molecules is especially important in the threedimensional organization of protein and nucleic acid molecules.

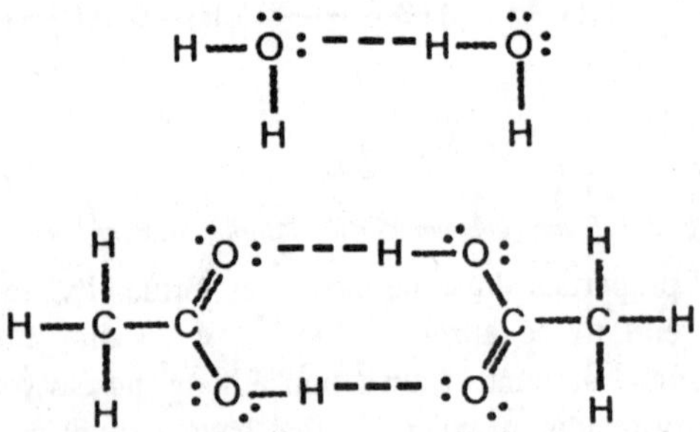

Fig. 2.8. Hydrogen bonding.

London dispersion forces

London dispersion forces (also called *van der Waals forces*) are weak forces resulting from the interaction of temporary dipoles. They are only about one-fifth as strong as hydrogen bonds but, weak as they are, they assume great importance in large molecules because there are so many of them per molecule. London dispersion forces also help hold aggregates of smaller molecules together. For instance, the phospholipids that are major components of cell membranes are held together mainly by these forces.

WATER IN LIVING SYSTEMS

Water is the most abundant material in living things, which are 60 to 95 percent water by weight. Understanding the properties of water is essential to understanding virtually all processes that occur in living systems.

Some Properties of Water

The structure of the water molecule is nonlinear (bent) with a bond angle of 104.5° between the HOH (hydrogen-oxygen hydrogen) atoms. Water is a polar molecule, with the oxygen atom forming the

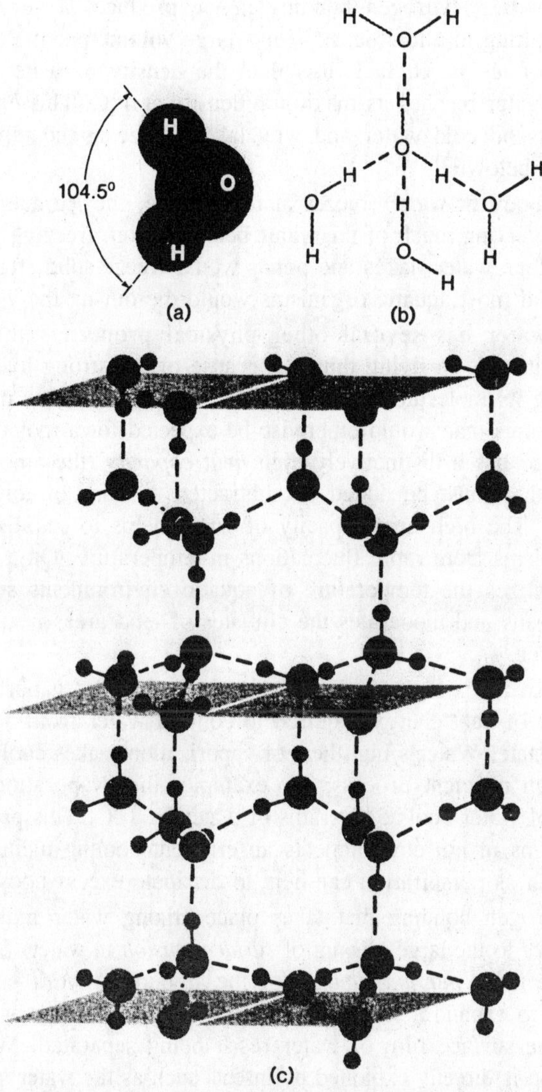

Fig. 2.9. (a) Structure of a water molecule. (b) Tetrahedral hydrogen bonding of water molecules in ice. (c) The open structure of ice.

negative end and the hydrogen atoms the positive end. Therefore, hydrogen bonding occurs between water molecules. We can see the importance of hydrogen/ bonding in water most strikingly by examining the structure of ice. In ice, each oxygen atom is surrounded tetrahedrally by four hydrogen atoms, two close (covalent bonds) and two distant (hydrogen bonds). Hydrogen bonding in ice produces a very open structure, resulting in a low density (or a large volume per molecule). The density of ice is, in fact, less than the density of water in its liquid state. Water reaches its maximum density at 4°C. This explains why ice floats on cold water and why lakes freeze at the top first, leaving water below.

When a body of water freezes at the surface, the ice acts as an insulator, preventing much of the water below it from freezing. If ice were denser than water, lakes and ponds would freeze solid, from the bottom up, and most aquatic organisms would die during the winter.

Liquid water has several other physical properties that are important to its role in living things. Because of the strong hydrogen bonds holding its molecules together, water has much higher melting and boiling points than would otherwise be expected for a molecule of its size. It also has a distinctively high *heat capacity* (the amount of heat energy that must be added or subtracted to raise or lower its temperature). The high heat capacity of water helps to stabilize and protect organisms from rapid fluctuations in temperature. On a larger scale, it stabilizes the temperature of aquatic environments such as lakes and oceans and moderates the climates of land areas near large bodies of water.

Water also has a high *heat of vaporization*. Heat of vaporization is the amount of heat energy required to convert water from a liquid to a gaseous state. Water's high heat of vaporization makes cooling by evaporation an efficient process; for example, the evaporation of a single gram of water cools 540 grams of water by 1°C. This property gives organisms in hot environments an efficient cooling method, as processes such as perspiration can help to dissipate excess body heat.

The hydrogen bonding that takes place among water molecules also contributes to the large amount of *surface tension* in water. Surface tension is the *force per unit length* (or the amount of work per unit area) needed to expand a surface. Put simply, the molecules of water making up the surface film of water resist being separated. Water's surface tension is directly exploited by insects such as the water strider, which can "skate" on a water surface, its small weight supported by surface tension. The mutual attraction of water molecules is also important in many other biological processes, including the movement of fluids in plants.

Water as a Solvent

Major parts of all living things consist of watery solutions. In a *solution*, there is a uniform mixture of the molecules of two or more substances. In living things, water is by far the most important *solvent*, the dissolving substance that is present in the greatest amount. Thus, molecules of various *solutes* (dissolved substances) are dispersed among molecules of water in solutions in living things.

How does water function as a solvent, and what determines which substances will or will not dissolve in water? Water's properties as a solvent relate to the polarity of water molecules. Recall that one end (pole) of each water molecule has a slight positive charge, and the other a slight negative charge. Many molecules that are important in living things are polar or have polar regions. The positive regions of such molecules are attracted to the negative ends of water molecules and vice versa, and water molecules are similarly attracted to them. The attractive forces involved are called *hydrophilic* ("water-loving") interactions. Because of these hydrophilic interactions with water molecules, such polar molecules become interspersed among water molecules. Sugars, salts, some alcohols, and ammonia are examples of water-soluble substances.

Other molecules, however, are nonpolar or have large nonpolar parts, and those molecules do not dissolve in water. This is because the forces attracting water molecules to one another are stronger than those acting between water and nonpolar molecules. Thus, the nonpolar molecules are "squeezed out" from between water molecules. Because nonpolar molecules do not interact with water molecules, they cannot become dispersed among water molecules to form a solution. Fats and oils are good examples of such substances. No matter how vigorously you stir oil in water, the oil remains in droplets and does not go into solution. The interactions between water molecules and nonpolar molecules or nonpolar parts of molecules are referrerd to as *hydrophobic* ("water-hating") interactions.

When an ionic compound, such as sodium chloride (NaCl), is dissolved in water, its ions separate because the charged poles of the water molecules attract the ions more than the ions attract each other. Not only are the ions pulled apart in the process, but they become surrounded by layers of water molecules called *hydration spheres*. The

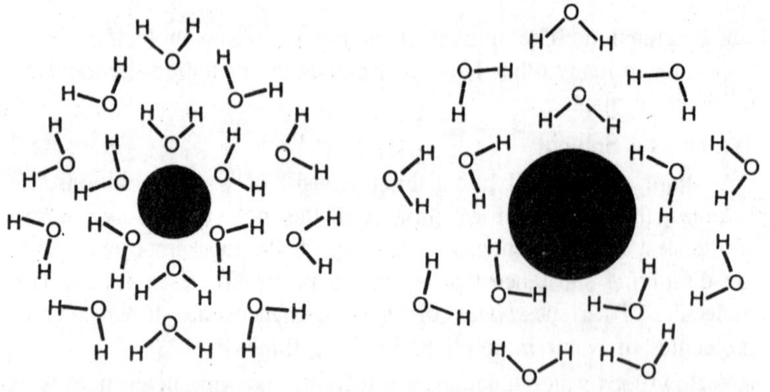

Fig. 2.10. Hydration of ions, using Na^+ and Cl^- ions as examples.

water molecules around the ions are oriented so that the negative (O) end of each water molecule is adjacent to the positive ion (Na^+). Similarly, the positive (H) end of each water molecule is attracted to the negative ion (Cl^-). Hydration spheres effectively increase the size of ions and affect the movement of ions in biological systems, including the passage of ions through cell membranes. The hydration process also affects the structure of water, because the water molecules in each sphere are oriented in one direction—with either their positive or negative ends attracted toward the ion. Because of their fixed orientation, the water molecules cannot, for example, take part so readily in ice formation. This explains why the higher the salt content of water, the more resistant it is to freezing—solutes depress the freezing point of water.

What happens when a molecule with both a small, polar (hydrophilic) portion and a large, nonpolar (hydrophobic) portion is added to water? When molecules with polar "heads" and long, nonpolar "tails" are mixed with water, they form structures known as *micelles*. In a micelle the nonpolar tails turn toward each other and away from the H_2O molecules. However, hydrophilic attraction does occur between the polar heads and adjacent H_2O molecules. Detergent action is a familiar, everyday example of micelle formation. A detergent disperses nonpolar grease by incorporating the grease molecules into the interior of micelles.

The phenomenon of micelle formation reveals much about the structure of biological membranes because biological membranes are largely made up of molecules that have both hydrophilic and hydrophobic ends.

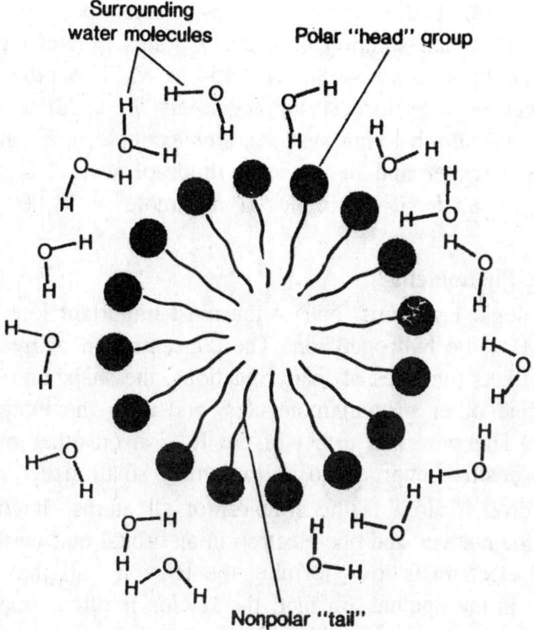

Fig. 2.11. The formation of a micelle by molecules that have a polar "head" and a nonpolar "tail."

Concentrations of Solutions

The properties of a solution depend not only on the natures of the solvent and solute, but also on their concentrations. We can express the concentrations of solutions in several ways. Stating the concentration as a *percentage* (normally, weight of solute to volume of solvent) is one simple approach. Physiological saline solution, for example, is 0.9 percent (weight-to-volume) NaCl in water.

In describing chemical reactions, however, the *mole* is a more useful unit of measurement for the amounts of different substances. A mole is the mass of substance, expressed in grams, that is numerically equal to the *molecular* (or *formula*) *weight*. The formula weight of any molecule is equal to the sum of the atomic weights of the atoms in the molecule. Thus, the formula weight of NaCl is 58.5, and one mole of NaCl is 58.5 grams:

23.0 (atomic weight of Na) + 35.5 (atomic weight of Cl) = 58.5

The mole is useful for expressing quantities of substances. One mole of any substance contains Avogadro's number (6.02×10^{23}) of molecules of the substance.

Molarity (*M*) is the number of moles of solute per liter of solution. Physiological saline solution contains 9 grams of NaCl per liter of solution, and thus, it is a 9/58.5 or 0.154 M NaCl solution. Moles and molarity can be used to describe accurately the small concentrations of many materials in living systems. For example, the concentrations of ions in seawater and in the body fluids of several organisms are compared in terms of millimoles (1 millimole = 1/1,000 mole) per liter.

Acid-Base Phenomena

In biological systems, one of the most important ions is also the simplest: H^+, the hydrogen ion. The concentration of hydrogen ions strongly affects the rates of many reactions, the shape and function of enzymes and other protein molecules, and even the integrity of the cell itself. The powerful effect of the H^+ ion on other molecules is due to its positive charge and its extremely small size.

A hydrogen atom is the simplest of all atoms. It contains one proton in the nucleus and one electron in an orbital outside the nucleus. When the electron is lost (forming the H^+ ion), all that remains is the proton in the nucleus. In fact, the H^+ ion is often simply called a proton.

In the internal environment of organisms, the most abundant molecule is water, which carries two pairs of electrons that are not involved in bonding:

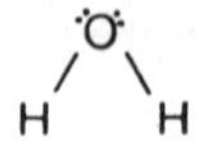

$$\begin{array}{c} \ddot{\text{O}} \\ / \quad \backslash \\ \text{H} \qquad \text{H} \end{array}$$

A proton (H^+) is strongly attracted to one of these pairs of electrons in a water molecule, thus forming a *hydronium ion* (OH_3^+).

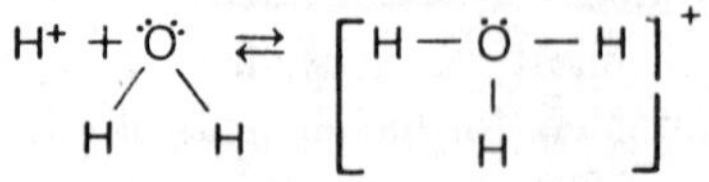

$$H^+ + \begin{array}{c} \ddot{\text{O}} \\ / \quad \backslash \\ \text{H} \qquad \text{H} \end{array} \rightleftarrows \left[\begin{array}{c} \text{H} - \ddot{\text{O}} - \text{H} \\ | \\ \text{H} \end{array} \right]^+$$

In fact, this proton attraction to H_2O is so strong that *no free* H^+ ions exist in aqueous solution: the H^+ ion is always attached to a water molecule. For convenience, however, we will continue to represent the hydrogen ion as H^+.

H^+ concentration and pH

At any given time, a small proportion of the molecules in pure water are ionized to produce hydrogen ions (H^+) and hydroxide (OH^-) ions:

$$H_2O \rightleftarrows H^+ + OH^-$$

This reaction is reversible, as indicated by the arrows pointing in both directions. We say that this reaction is at *equilibrium* because in pure water, the rate of ionization of water molecules

$$H_2O \rightarrow H^+ + OH^-$$

is equal to the rate of reaction among the ions.

$$H_2O \leftarrow H^+ + OH^-$$

Thus, at equilibrium there is no further *net* change in concentration, even though both reactions occur continually.

In pure water, this ionization occurs to such an extent that

$$[H^+] = 10^{-7} \text{ M and } [OH^-] = 10^{-7} \text{ M}$$

(Square brackets are conventionally used to denote concentrations.) Of course, the concentration of H^+ ions in pure water must equal the concentration of OH^- ions because neither can be formed without the other.

A convenient way of designating the H^+ concentration in a solution is provided by the *pH scale*, which is expressed as the negative logarithms of hydrogen ion concentrations. A *logarithm* is the exponent that indicates the power to which one number must be raised to obtain another number. (For example, the logarithm of 100 to the base 10 is 2.) Thus,

$$pH = \log_{10} \frac{1}{[H^+]}$$

$$(\text{or, } pH = -\log_{10} [H^+])$$

As an example, if $[H^+] = 10^{-3}$ M (.001 M) then pH = 3.

A solution is *neutral* if its H^+ and OH^- concentrations are equal. For water, $[H^+] = [OH^-]$ when both exist in concentrations of 10^{-7} M; that is, when the pH is 7. A solution is *acidic* if it has a higher H^+ concentration than OH^- concentration, and *basic* if its H^+ concentration is lower than its OH^- concentration. Therefore, acidic solutions have pH values below 7, and basic solutions have pH values above 7.

Acids and bases

Substances that change the concentrations of H^+ or OH^- ions when they dissolve in water are known as *acids* and *bases*. An *acid* is a *proton* (*hydrogen ion*) *donor*; it causes an increase in H^+ ion concentration when it ionizes in solution. For example, hydrochloric acid (HCl) is ionized quite completely by the reaction:

$$HCl \rightarrow H^+ + Cl^-$$

Adding hydrochloric acid to a solution increases the relative number of H^+ ions and, therefore, lowers the pH of the solution. The solution of gastric juices in the human stomach, for example, contains a great deal of hydrochloric acid and sometimes may have a pH as low as 1.0. This is such an acidic solution that stomach contents coming in contact with skin could cause a serious acid burn.

A *base* is a *proton acceptor* and has the opposite effect on a solution. For example, sodium hydroxide (NaOH) ionizes almost completely in solution and strongly increases the relative concentration of OH^- ions. Hydroxide ions (OH^-) combine with H+ ions to form water, and thereby decrease the H^+ ion concentration (thus raising the pH of the solution).

Buffers

Even small pH changes may have great effects on biological processes. Because many processes are sensitive to pH changes, it is important that the pH of body fluids remains stable. Cell interiors have an average pH of 7.0 to 7.3; and in humans, for example, blood and tissue fluids have a pH of 7.4 to 7.5. These values remain very stable despite the fact that many biochemical reactions either release or incorporate H^+ ions.

How does pH remain so constant in living things? Such pH stability is possible because organisms have built-in mechanisms to prevent pH change. The most important of these mechanisms are *buffers*. A buffer resists pH change by removing H^+ ions when the H^+ ion concentration rises, and releasing H^+ ions when the H^+ ion concentration falls.

An example of a buffer is the carbonic acid-bicarbonate ion buffer system, which is involved in the buffering of human blood:

$$\underset{\text{Carbonic acid}}{H_2CO_3} \rightleftarrows \underset{\text{Bicarbonate ion}}{H^+ + HCO_3^-}$$

In this example, if H^+ ions are added to the system, they combine with HCO_3^- to form H_2CO_3. This reaction removes extra H^+ ions and keeps the pH from changing. If H^+ ions are removed from the system (for example, if OH^- ions are added and H_2O is formed), more H_2CO_3 will ionize and replace the H^+ ions that were used. Again, pH stability is maintained. This and other buffer systems prevent potentially harmful pH changes in living things.

ORGANIC MOLECULES

Most of the thousands of kinds of chemical compounds in living organisms are *organic compounds*, compounds that contain carbon. The

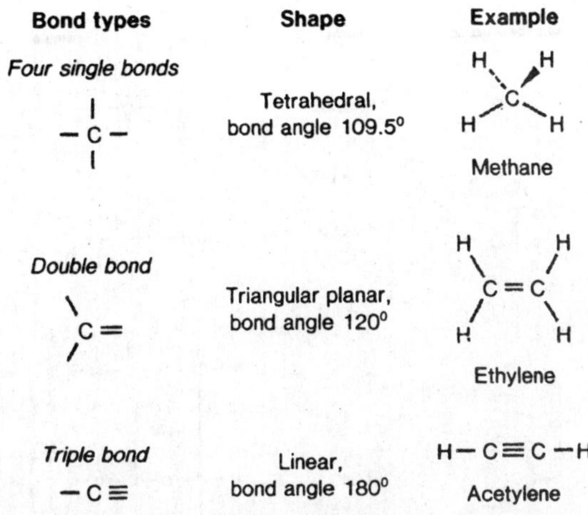

Fig. 2.12. Possible arrangements of bonds around the carbon atom.

most important characteristics of organic compounds depend on properties of this key element, carbon.

The carbon atom has an atomic number of six, with six protons and six electrons. Since such an atom has two electrons in the first shell and four electrons in the second shell, it can form covalent bonds with as many as four other atoms. A carbon atom can bond to a variety of elements, but it most commonly bonds to hydrogen, oxygen, nitrogen, and carbon.

The carbon-to-carbon bonding ability of carbon atoms makes possible carbon chains (sometimes called carbon skeletons) of various lengths and shapes. The carbon skeleton establishes the overall framework of an organic molecule. Carbon atoms can form single, double, or triple covalent bonds with one another, and such bonding patterns determine the characteristics of organic compounds.

Many organic compounds consist of just carbon and hydrogen, but many others contain *functional groups* as well. Functional groups are characteristic patterns of atoms other than those involved in carbon-carbon and carbon-hydrogen bonds. These functional groups are the parts of organic molecules that are most often involved in chemical reactions.

One reason there are so many organic compounds is that carbon forms chains of atoms better than any other element. An unusually long noncarbon chain might contain ten atoms. On the other hand, organic compounds with chains of more than fifty carbon-carbon bonded atoms can be found in living systems, and chains containing carbon

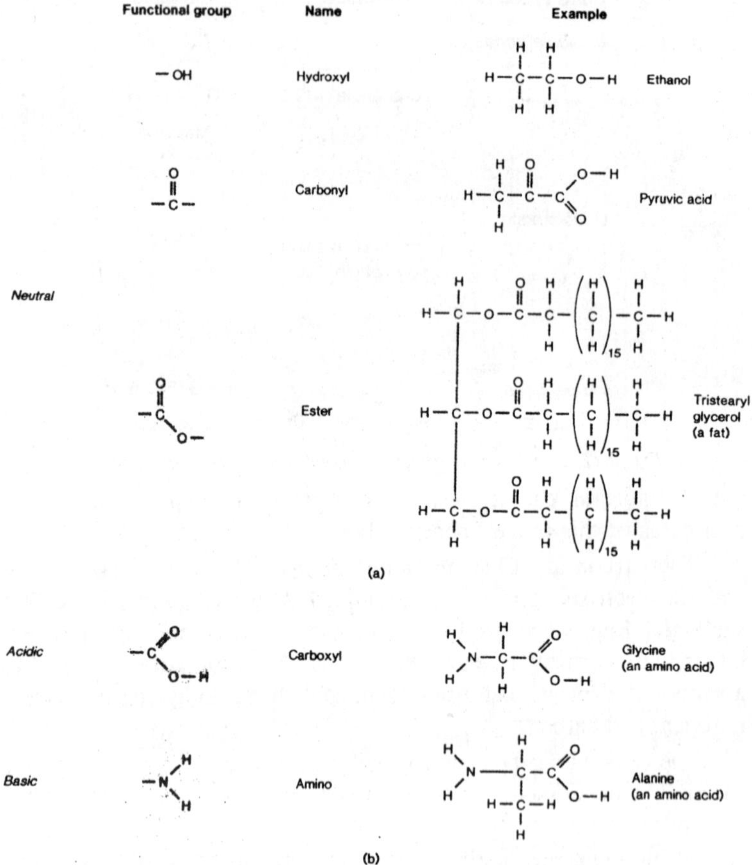

Fig. 2.13. Some examples of functional groups found in organic molecules. (a) Neutral functional groups. (b) Acidic and basic functional groups.

along with other elements such as nitrogen and oxygen are sometimes thousands of atoms long.

Another factor contributing to the diversity of organic compounds is the presence of numerous *isomers*. Isomers are compounds having the same molecular formulas but different atom arrangements. *Structural isomers* are completely different chemical compounds; they share the same molecular formula but have different pairs of atoms connected by bonds. *Geometric isomers* differ from each other only in the orientations of bonds between particular pairs of atoms. These differences are possible in the presence of certain rings or C=C double bonds. Although rotation around C—C single bonds occurs readily,

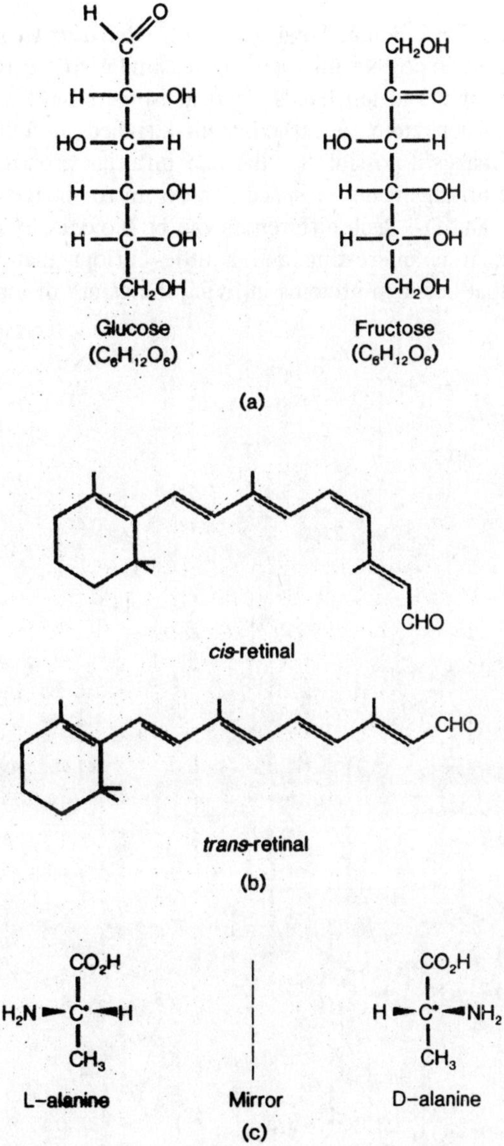

H
O
C
H—C—OH
HO—C—H
H—C—OH
H—C—OH
CH₂OH
Glucose
(C6H12O6)
CH2OH
C=O
HO—C—H
H—C—OH
H—C—OH
CH2OH
Fructose
(C6H12O6)
(a)
CHO
cis-retinal
CHO
trans-retinal
(b)
CO2H
H2N—C*—H
CH3
L-alanine
Mirror
CO2H
H—C*—NH2
CH3
D-alanine
(c)

Fig. 2.14. Some examples of isomers. (a) Structural isomers are molecules with identical molecular formula, but different pairs of atoms connected by bonds. (b) Geometric isomers are molecules that have different orientations of bonds in space, but that are not minor images of each other. (c) Optical isomers are pairs of molecules that are nonsuperimposable mirror images of each other.

rotation around C=C double bonds does not. *Optical isomers* occur as pairs of nonsuperimposible mirror images, similar to the relationship between your right and left hands. Optical isomers have at least one asymmetric carbon atom, a carbon atom attached to four different groups. This makes it possible for the four different groups to occupy two different arrangements in space. These mirror-image forms are designated L- and D-. Such differences can be biologically important. For example, it is interesting and a little curious that all of the amino acids that occur in proteins in living things are of the L- form.

3

MICROMOLECULES

Water spews from the Earth as geysers. It circles the globe as clouds and falls from the sky as the gentle rain or in thundering torrents. Frozen, it covers parts of Antarctica to a depth of 3,000 meters (10,000 feet) or more. Vast oceans of it submerge much of the planet. Water is one of the key ingredients that makes life on Earth possible. It makes up as much as 95 percent of the weight of some living things.

Water is a simple substance containing only three atoms-two of hydrogen and one of oxygen. What does the composition of water tell us about its characteristics, and why is water so important to living systems? We can't answer these questions without more information about atoms in general, and about hydrogen and oxygen atoms in particular. But it is not only the constituents of matter that are important. To understand the Behaviour of something as apparently simple as water, we need to know how its constituent atoms are linked together. The same is true of the other small molecules that are essential to living systems.

The first part of this chapter will address the constituents of matter: atoms—their variety, properties, and capacity to combine with other atoms. Then we'll consider how matter changes. In addition to changes in state (solid to liquid to gas), substances undergo changes that transform both their composition and their characteristic properties. When cells use. oxygen to "burn" glucose, the products are water, carbon dioxide, and energy to power life activities. This transformation is similar to what happens when the fuel propane is burned in a stove, a combustion reaction. By studying simple systems such as the combustion of propane, we can understand better what happens in systems as complicated as living cells. The discussion of general chemical principles and their application to small molecules aids our understanding of the large molecules that form the basis for the life of an organism.

Later in this chapter, we return to a consideration of the structure and properties of water and its relationship to acids and bases. We close with a bridge to the next chapter—a consideration of characteristic groups of atoms that contribute specific properties to larger molecules of which they are part.

Atoms: Constituents of Matter

More than a million million (1×10^{12}) atoms could fit in a single layer over the period at the end of this sentence. Each atom consists of a dense, positively charged nucleus, around which one or more negatively charged electrons move. The nucleus contains one or more protons and may contain one or more neutrons. Atoms and their component particles have mass. Mass is a property of all matter. Measuring mass measures the quantity of matter present. The greater the mass, the greater the quantity of matter.

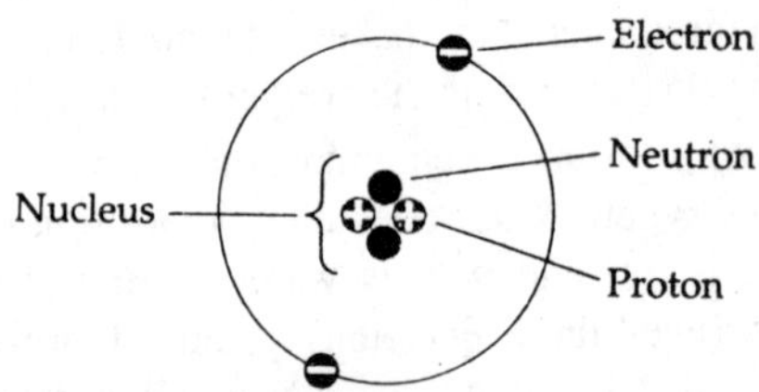

The mass of a proton serves as a standard unit: the atomic mass unit (amu), or dalton (named after the English chemist John Dalton). A single proton or neutron has the mass of 1 dalton, which is 1.7×10^{-24} grams (0.0000000000000000000000017 g). The mass of an electron is 9×10^{-28} g (0.0005 dalton). Because the mass of an electron is so much less than the mass of a proton or a neutron, the contribution of electrons to the mass of an atom can usually be ignored.

The positive electric charge on a proton is defined as a unit of charge. An electron has a charge equal and opposite to that of a proton. Thus the charge of a proton is +1 unit, that of an electron is -1 unit. Unlike charges attract each other; like charges repel. The neutron, as its name suggests, is electrically neutral, so its charge is 0 unit. Because the number of protons in an atom equals the number of electrons, the atom itself is electrically neutral.

An Element is Made Up of Only One Kind of Atom

The element hydrogen consists only of hydrogen atoms; the element iron consists only of iron atoms. An element is a pure substance that

The six elements highlighted in yellow make up 98% of the mass of any living organism.

Chemical symbol
Atomic number
Atomic mass

Elements shown in orange are present in tiny amounts in many organisms.

1 H 1.0079																	2 He 4.003
3 Li 6.941	4 Be 9.012											5 B 10.81	6 C 12.011	7 N 14.007	8 O 15.999	9 F 18.998	10 Ne 20.179
11 Na 22.990	12 Mg 24.305											13 Al 26.982	14 Si 28.086	15 P 30.974	16 S 32.06	17 Cl 35.453	18 Ar 39.948
19 K 39.098	20 Ca 40.08	21 Sc 44.956	22 Ti 47.88	23 V 50.942	24 Cr 51.996	25 Mn 54.938	26 Fe 55.847	27 Co 58.933	28 Ni 58.69	29 Cu 63.546	30 Zn 65.38	31 Ga 69.72	32 Ge 72.59	33 As 74.922	34 Se 78.96	35 Br 79.909	36 Kr 83.80
37 Rb 85.4778	38 Sr 87.62	39 Y 88.906	40 Zr 91.22	41 Nb 92.906	42 Mo 95.94	43 Tc (99)	44 Ru 101.07	45 Rh 102.906	46 Pd 106.4	47 Ag 107.870	48 Cd 112.41	49 In 114.82	50 Sn 118.69	51 Sb 121.75	52 Te 127.60	53 I 126.904	54 Xe 131.30
55 Cs 132.905	56 Ba 137.34	57–71 La–Lu	72 Hf 178.49	73 Ta 180.948	74 W 183.85	75 Re 186.207	76 Os 190.2	77 Ir 192.2	78 Pt 195.08	79 Au 196.967	80 Hg 200.59	81 Tl 204.37	82 Pb 207.19	83 Bi 208.980	84 Po (209)	85 At (210)	86 Rn (222)
87 Fr (223)	88 Ra 226.025	89–103 Ac–Lr	104	105	106	107	108	109									

Vertical columns have elements with similar properties.

Lanthanide series	57 La 138.906	58 Ce 140.12	59 Pr 140.9077	60 Nd 144.24	61 Pm (145)	62 Sm 150.36	63 Eu 151.96	64 Gd 157.25	65 Tb 158.924	66 Dy 162.50	67 Ho 164.930	68 Er 167.26	69 Tm 168.934	70 Yb 173.04	71 Lu 174.97
Actinide series	89 Ac 227.028	90 Th 232.038	91 Pa 231.0359	92 U 238.02	93 Np 237.0482	94 Pu (244)	95 Am (243)	96 Cm (247)	97 Bk (247)	98 Cf (251)	99 Es (252)	100 Fm (257)	101 Md (258)	102 No (259)	103 Lr (260)

Fig. 3.1. The periodic table.

contains only one type of atom. The atoms of each element have certain characteristics or properties that distinguish them from the atoms of other elements. The more than 100 elements found in the universe are arranged in the periodic table. The periodic table arranges elements with similar properties in vertical columns in order of their increasing size. Although there are more than 100 elements in the world, about 98 percent of the mass of a living organism (bacterium, turnip, or human) is composed of just six elements—carbon, hydrogen, nitrogen, oxygen, phosphorus, and sulfur. The chemistry of these elements will be our primary concern.

A substance (such as oxygen gas) that contains only one kind of atom is an elemental substance. A substance that contains more than one kind of atom is a *compound*. Most substances of biological interest are compounds.

Number of Protons Identifies the Element

An atom is distinguished from other atoms by the number of its protons, which does not change. This number is called the atomic number. An atom of hydrogen contains 1 proton, a helium atom has 2 protons, carbon has 6 protons, and plutonium has 94. The atomic numbers of these elements are thus 1, 2, 6, and 94, respectively.

Every atom except hydrogen has one or more neutrons in its nucleus. The *mass number* of an atom equals the total number of protons and neutrons in its nucleus. Because the mass of an electron is infinitesimal compared with that of a neutron or proton, electrons are ignored in calculating the mass number. The nucleus of a helium atom contains 2 protons and 2 neutrons; oxygen has 8 protons and 8 neutrons. Helium, therefore, has a mass number of 4 and oxygen a mass number of 16. The mass number may be thought of as the weight of the atom, in daltons.

Each element has its own one- or two-letter symbol. For example, H stands for hydrogen, He for helium, and O for oxygen. Some symbols come from other languages: Fe (from Latin *ferrum*) stands for iron, Na (Latin *natrium*) for sodium, and W (German *Wolfram*) for tungsten. The periodic table gives the symbols for all of the 92 natural elements, as well as those for 14 elements that do not occur naturally.

Isotopes Differ in Number of Neutrons

We have been speaking of hydrogen and oxygen as if each had only one atomic form. But this is not true. Not all atoms of an element have the same mass number. The different atomic forms of a single element are called *isotopes* of the element. Isotopes of an element differ in the number of neutrons in the atomic nucleus. The common form of hydrogen is ^{1}H, but about one out of every 6,500 hydrogen atoms on Earth has a neutron as well as a proton in its nucleus and is thus ^{2}H, called deuterium. Furthermore, it is possible to create ^{3}H, tritium, which has *two* neutrons and a proton in its nucleus. Because all three types of hydrogen atoms have only one proton, they all have the atomic number 1. Deuterium, tritium, and common hydrogen have virtually identical chemical properties, although ^{2}H is twice and ^{3}H three times as heavy as ^{1}H.

In nature, many elements exist as several isotopes. For example, the natural isotopes of carbon are ^{12}C, ^{13}C, and ^{14}C. Unlike the hydrogen isotopes, the isotopes of most other elements do not have distinct names. Rather they are written in the form shown here and are referred to as carbon-12, carbon-13, and carbon-14, respectively. Most carbon atoms are ^{12}C, about 1.1 percent are ^{13}C, and a tiny fraction are ^{14}C. An element's atomic *mass* (*atomic* weight) is the average of the mass numbers of a representative sample of atoms of the element, with all isotopes in their normal proportions. For example, the atomic mass of carbon is 12.011. In biology, one encounters the terms "weight" and "atomic weight" more frequently than "mass" and "atomic mass"; therefore, we will use "weight" for the remainder of this book. Thus we say that the atomic weight of carbon is 12.011.

Some isotopes, called *radioisotopes*, are unstable and spontaneously give off energy as α (alpha), β (beta), or γ (gamma) radiation from the atomic nucleus. Such radioactive decay transforms the original atom into another type, usually of another element. For example, uranium-238 loses an alpha particle to form thorium-234, and carbon-14 loses a beta particle to form nitrogen-14. Biologists can incorporate radioisotopes into molecules and use the emitted radiation as a tag to identify changes that the molecules undergo in the body or to identify the locations of molecules within the cell. Some radioisotopes commonly used in biological experiments are ^{3}H (tritium), ^{14}C (carbon-14), and ^{32}P (phosphorus-32).

Although radioisotopes are useful for experiments and medicine, even low doses of radiation from radioisotopes have the potential to damage molecules and cells. Gamma radiation from cobalt-60 (^{60}Co) is used medically to damage or kill rapidly dividing cancer cells. In addition to these applications, radioisotopes can be used to date fossils.

Electron Behaviour Determines Chemical Bonding

In atoms, biologists are concerned primarily with electrons. To understand organisms, biologists study chemical changes that occur in living cells. These changes, called *chemical reactions* or just *reactions*, are changes in the atomic composition of substances. They occur because of the way in which electrons behave. The characteristic number of electrons in each atom of an element determines how the atom reacts with other atoms. All chemical reactions involve changes in the relationships of electrons with each other.

The location of a given electron in an atom at any given time is impossible to determine. We can only describe a volume of space within the atom where the electron is likely to be. The region of space within which the electron is found at least 90 percent of the time is the electron's orbital. In an atom, a given orbital can be occupied by at most two electrons. Thus any atom larger than helium (atomic number 2) must have electrons in two or more orbitals. The different orbitals have characteristic forms and orientations in space.

The orbitals constitute a series of *electron shells*, or energy levels, around the nucleus. The innermost electron shell, called the *K* shell, consists of only one orbital, called an *s* orbital. The *s* orbital fills first, and its electrons have the lowest energy. Hydrogen ($_1$H) has one *K* shell electron; helium ($_2$He) has two. All other atoms have two *K* shell electrons, as well as electrons in other shells. The *L* shell is made up of four orbitals (an *s* orbital and three *p* orbitals) and hence can hold up to eight electrons. The M, N, O, P, and Q shells have different numbers of orbitals, but the outermost orbitals can usually hold only eight electrons.

In any atom, the outermost shell determines how the atom combines with other atoms; that is, these outermost electrons determine how an atom behaves chemically. When an outermost shell consisting of four orbitals contains eight electrons, the atom is stable and will

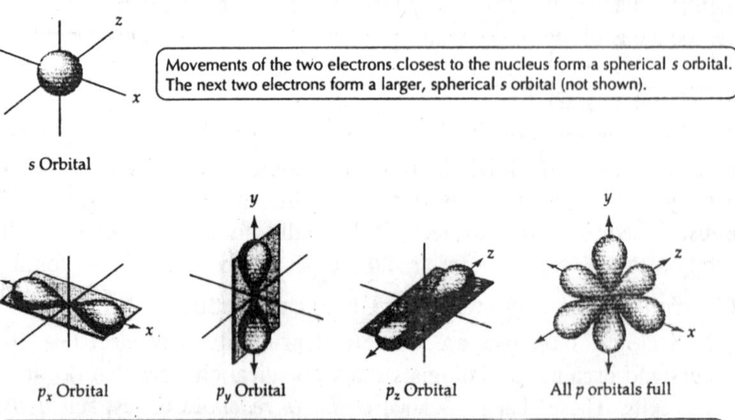

Fig. 3.2. Electron orbitals.

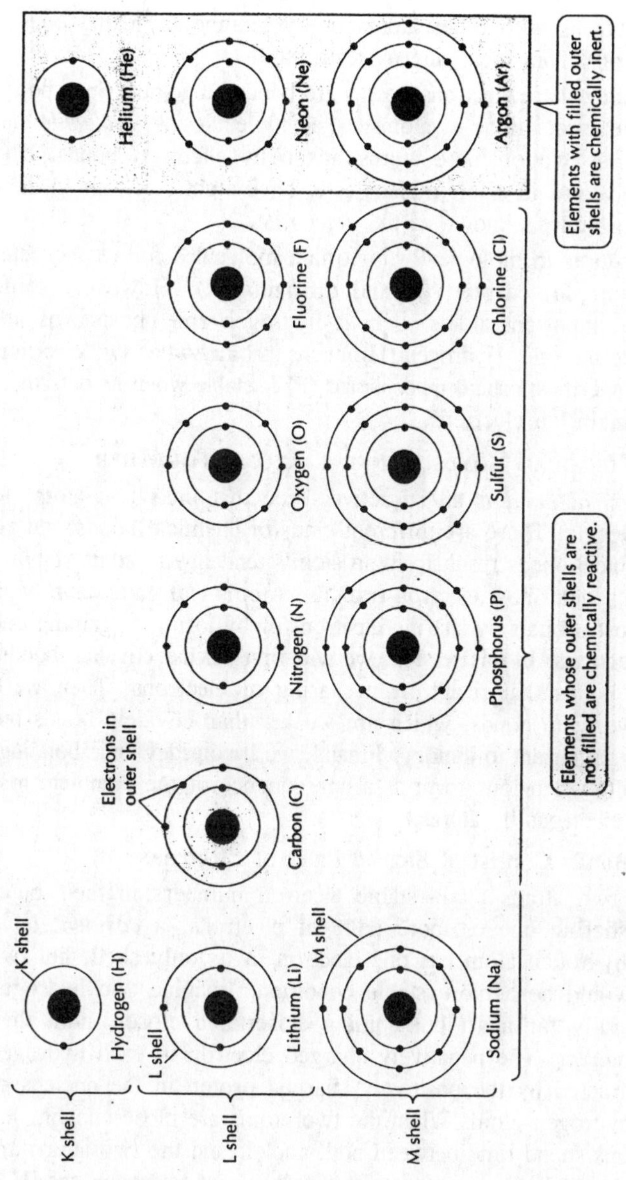

Fig. 3.3. Electron shells determine the reactivity of atoms.

not react with other atoms. Examples of some chemically inert elements are neon and argon, in each of which the outermost shell contains eight electrons. The atoms of other elements seek to attain the stable condition of having filled outer orbitals. They attain this stability by sharing electrons with other atoms or by gaining or losing one or more electrons from their outermost orbitals.

When they share electrons, atoms are bonded together. Such bonds create stable associations of atoms called molecules. A *molecule* can be defined as two or more atoms linked by chemical bonds. The tendency of atoms in stable molecules to have eight electrons in their outermost orbitals is known as the *octet rule*.

Many atoms in biologically important molecules follow the octet rule-for example, carbon (C) and nitrogen (N). However, some biologically important atoms such as hydrogen and phosphorus are exceptions to the rule. Hydrogen (H) attains stability when two electrons occupy its outermost orbital; phosphorus (P) is stable when its outermost orbitals contain ten electrons.

Chemical Bonds: Linking Atoms Together

A *chemical bond* is an attractive force that links two atoms to form a molecule. There are different kinds of chemical bonds, but all strong chemical bonds result from an atom's tendency to attain stability by filling its outermost electron orbitals. Atoms can gain stability in the outermost orbitals by sharing electrons or by losing or gaining one or more electrons. In this section, we will first discuss covalent bonds, the strong bonds that result from sharing of electrons. Then we'll examine hydrogen bonds, which are weaker than covalent bonds but enormously important to biology. Finally, we'll consider ionic bonding, which results when ions form as a consequence of the complete loss or gain of electrons by atoms.

Covalent Bonds Consist of Shared Pairs of Electrons

When two atoms attain stable electron numbers in their outer shells by sharing one or more pairs of electrons, a *covalent bond* forms. A hydrogen atom has one electron in its only shell, but two electrons would be a more stable condition. Imagine two hydrogen atoms, initially far apart but coming closer and closer, until they begin to interact. The negatively charged electron in each hydrogen atom is attracted by the positively charged proton in the nucleus of the other hydrogen atom. When the two atoms are close enough, the two electrons spend time between both nuclei, and the two atoms are covalently bonded together, forming a molecule of hydrogen gas (H_2). The two hydrogen nuclei share the two electrons equally and completely.

The two atoms do not come *too* close together, because their positively charged nuclei strongly repel each other. A certain distance between the coupled atoms gives the most stable arrangement. Pulling the atoms slightly farther apart would require an input of energy because of the "gluing" effect of the shared electrons. Pushing the atoms closer together would require energy because of the mutual repulsion of the protons. So the most stable arrangement of the covalently bonded hydrogens can also be described as an arrangement that has a minimum amount of energy and that is less reactive than are the individual atoms alone, each of which has an incompletely filled orbital in the K shell.

A carbon atom has a total of six electrons; two electrons fill its inner shell and four are in its outer L shell. Because the L shell can hold up to eight electrons, this atom can share electrons with up to four other atoms. Thus it can form four covalent bonds. When an atom of carbon reacts with four hydrogen atoms, a substance called methane (CH_4) forms, resulting from the overlapping of electron orbitals. Thanks to electron sharing, the outer shell of methane's carbon atom is filled with eight electrons, and the outer shell of each hydrogen atom is also filled. Thus four covalent bonds—each bond consisting of a shared pair of electrons—hold methane together.

Orientation of Bonds in Space

Not only the number, but also the spatial orientation, of bonds is important. The four filled orbitals around the carbon nucleus of methane distribute themselves in space so that the bonded hydrogens are directed to the corners of a regular tetrahedron with carbon in the center. Although the orientation of orbitals and shapes of molecules differ depending on the kinds of atoms and how they are linked together, it is essential to remember that all molecules occupy space and have three-dimensional shapes. The shapes of molecules contribute to their biological functions.

Multiple Covalent Bonds

A covalent bond is represented by a line between the chemical symbols for the atoms. Bonds in which a single pair of electrons is shared are called *single bonds* (for example, H-H, C-H). When four electrons (two pairs) are shared, the link is a *double bond* (C=C). In the gas ethylene ($H_2C=CH_2$), two carbon atoms share two pairs of electrons. *Triple bonds* (six shared electrons) are rare, but there is one in nitrogen gas (N-N), the chief component of the air we breathe. In the covalent bonds in these five examples, the electrons are shared more or less equally between the nuclei; consequently all regions of

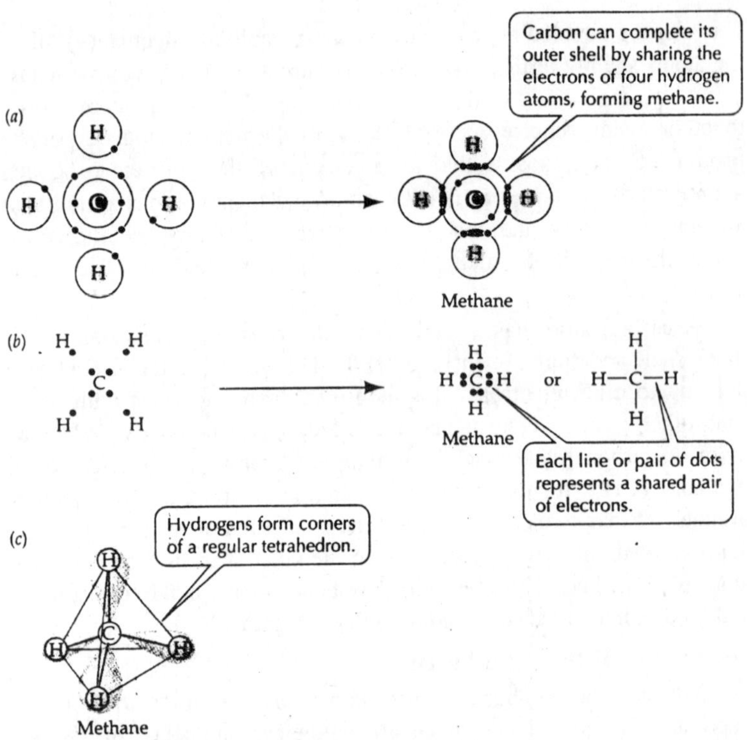

Fig. 3.4. Electrons are shared in covalent bonds.

the bonds are identical. However, when electrons are shared unequally in a covalent bond, regions of partial electric charge exist.

Uniqual Sharing of Electrons

So far we have discussed the covalent bonds that result from the equal sharing of electrons between two nuclei. Now we want to consider the kind of covalent bond that results from unequal sharing of the electrons.

Some atoms hold electrons to themselves more firmly than other atoms do. This characteristic is called *electronegativity*. Highly electronegative atoms that form covalent bonds include oxygen and nitrogen. When these atoms are covalently bonded to atoms with weaker electronegativity, such as carbon and hydrogen, the bonding pair of electrons is unequally shared between the two atoms, and the result is a *polar covalent bond*. For example, when oxygen is bonded to hydrogen, the bonding electrons spend much more time near the oxygen nucleus than near the hydrogen nucleus. Consequently, the oxygen end of the bond is slightly negative (symbolized δ- and spoken as "delta negative," meaning a partial unit charge), and the hydrogen end is slightly positive (S+). The bond is polar because these opposite charges are separated at the two ends of the bond. The partial charges that result from polar covalent bonds produce polar *molecules* or polar regions of large molecules. Polar bonds greatly influence the interactions between molecules that contain them, as we see in the interaction of water molecules in the liquid state.

Table 3.1. Typical bonding capabilities of some biologically important elements

Elements	*Number of covalent bonds*
Hydrogen (H)	1
Oxygen (O)	2
Sulfur (S)	2
Nitrogen (N)	3
Carbon (C)	4
Phosphorus (P)	5

Hydrogen Bonds may Form between Molecules

In liquid water, the negatively charged oxygen (δ-) atom of one water molecule is attracted to the positively charged hydrogen (δ+) of another water molecule. (Remember, negative charges attract positive charges.) The bond resulting from this attraction is called a *hydrogen bond* and is usually symbolized by a series of dots. Hydrogen bonds are not restricted to water molecules. They may form between any covalently bonded hydrogen and an electronegative atom, usually oxygen or nitrogen: -H···O- or -H···N-. Hydrogen bonds form between small molecules or between different parts of large molecules. Covalent bonds and polar covalent bonds, on the other hand, are always found *within* molecules.

A hydrogen bond is a weak bond; it has about onetwentieth (5 percent) of the strength of a covalent bond between a hydrogen atom and an oxygen atom. However, where many hydrogen bonds form, they have considerable strength and greatly influence the structure and properties of substances. Later in this chapter we'll discuss further how hydrogen bonding in water contributes to many of the properties of water that are significant for living systems. Hydrogen bonds also play important roles in determining and maintaining the three-dimensional shapes of giant molecules such as DNA and protein.

Ions Form Bonds by Electrical Attraction

When one interacting atom is much more electronegative than the other, a complete transfer of one or more electrons may take place. For example, a sodium atom has only one electron in its outermost shell; this condition is unstable. A chlorine atom has seven electrons in its outer shell, another unstable condition. The reaction between sodium and chlorine makes both atoms more stable. When the two atoms meet, the highly *electronegative chlorine atom takes the single unstable* electron from the sodium atom. The result is two electrically charged particles, called *ions*. Ions are electrically charged particles that form when atoms gain or lose one or more electrons.

The sodium ion (Na^+) has a +1 unit charge because it has one less electron than it has protons. The outermost electron shell of the sodium ion is full, with eight electrons, so the ion is stable. The chloride ion (Cl^-) has a –1 unit charge because it has one more electron than it has protons. This additional electron gives Clan outer shell with a stable load of eight electrons. Negatively charged ions are called *anions*; positively charged ions are called *cations*.

Some elements form ions with multiple charges by losing or gaining more than one electron to achieve a stable electron configuration in their outer shell. Examples are Ca^{2+} (the calcium ion, created from a calcium atom that has lost two electrons), Mg^+ (magnesium ion), and Al^{3+} (aluminum ion). Two biologically important elements each yield more than one stable ion: Iron yields Fee^+ (ferrous ion) and Fe^{3+} (ferric ion), and copper yields Cu^+ (cuprous ion) and Cue^+ (cupric ion). Groups of covalently bonded atoms that carry an electric charge are called *complex ions*; examples include NH_4^+ (ammonium ion), SO (sulfate ion), and PO_4^{2-} (phosphate ion).

Once they form, ions 'are usually stable, and no more electrons are lost or gained. As stable entities, ions can enter into stable associations through ionic bonding. Thus stable solids such as sodium chloride (NaCl) and potassium phosphate (K_3PO_4) are formed: Although a very complex solid, bone has as one of its major components the simple ionic compound $Ca_3(PO_4)_2$.

Ionic Bonds: Electrical Attraction

Ionic bonds are the bonds formed by electrical attractions between ions bearing opposite charges. In solids such as table salt (NaCl), the cations and anions are held together by ionic bonds. In solids, the ionic bonds are strong because the ions are close together. However, when ions are dispersed in water, the distance between them can be

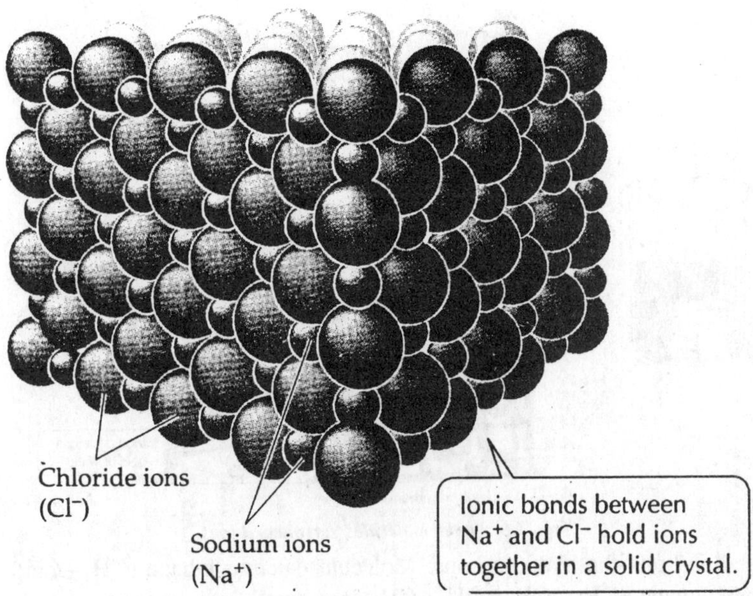

Fig. 3.5. Ionic binding in a solid.

large; the strength of their attraction is thus greatly reduced. Under the conditions that exist in the cell, an ionic bond is less than one-tenth as strong as a covalent bond that shares electrons equally, so an ionic bond can be broken much more readily than a covalent bond.

Not surprisingly, ions with one or more unit charges can interact with polar substances as well as with other ions. Such interaction results when table salt or any other ionic solid dissolves in water. The hydrogen bond that we described earlier is a weak type of ionic bond, because it is formed by electrical attractions. However, it is weaker than most ionic bonds because the hydrogen bond is formed by partial charges (δ- and δ+) rather than by whole unit charges (+1 unit, -1 unit).

Nonpolar Substances have no Attraction for Polar Substances

We have been discussing the bonds that result from electrical attractions between positive and negative charges (ionic bonds and hydrogen bonds). Now let's return to a brief consideration of substances that have "pure" covalent bonds.

These bonds form between atoms that have equal or nearly equal electronegativities such as carbon and hydrogenwhich share the bonding electrons equally. Such bonds are abundant in the compounds of hydrogen

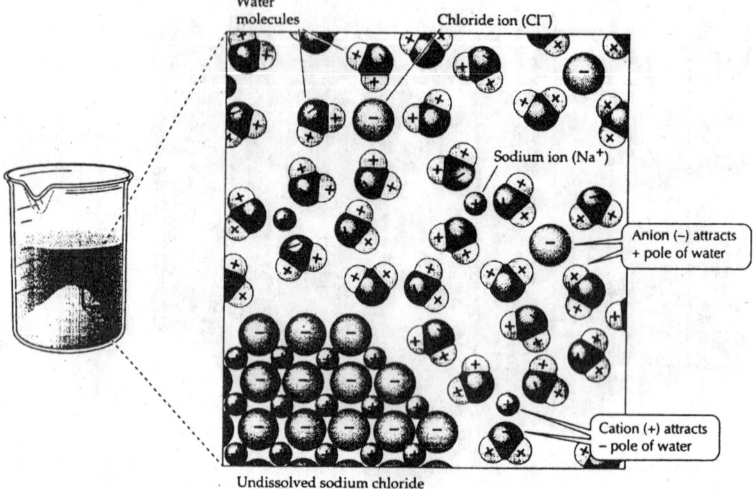

Fig. 3.6. Water molecules surrounded ions.

and carbon—the hydrocarbons. Molecules such as ethane (CH_3—CH_3) and butane (CH_3—CH_2—CH_2—CH_3) are small hydrocarbons, but in living systems, molecules exist with hydrocarbon chains consisting of 16 or more carbon atoms.

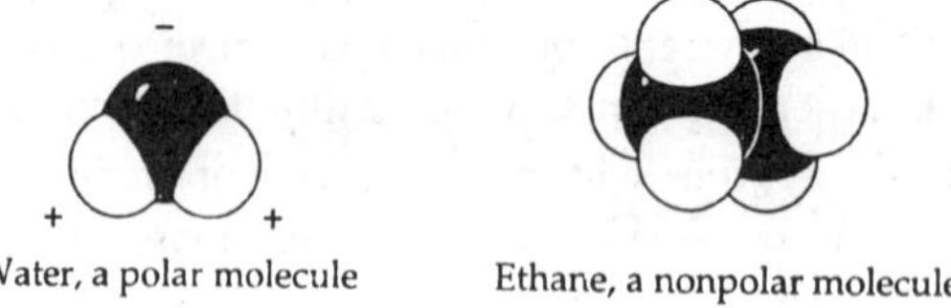

Fig. 3.7. Nonpolar molecules.

Attraction between Nonpolar Molecules

Nonpolar substances such as oils and fats show *van der Waals attractions* between molecules. These attractive forces operate only when nonpolar substances come very close to each other. The random variations in the electron distribution in one molecule create an opposite charge distribution in the adjacent molecule, and the result is a brief, weak attraction. Although each such interaction is brief and weak at any one site, the summation of many such interactions over the entire span of a nonpolar molecule can produce substantial attraction. Thus van der Waals interactions are important in holding together the long hydrocarbon chains that make up the inner portion of biological membranes. They also stabilize portions of the DNA double helix and the intricate folded structure of proteins.

Polar and Nonpolar Interactions

When electrons are shared equally, the resultant covalent bonds are nonpolar and they do not interact with the charges of polar covalent bonds. Substances with only nonpolar bonds, such as butane and oils, will not interact with substances that have polar bonds or ionic bonds. This explains why oils will not dissolve in water. Oils are nonpolar hydrocarbons, while water is a highly polar substance. Substances with nonpolar covalent bonds are said to be hydrophobic (literally "water-fearing"), which refers to the fact that there is no attraction between nonpolar substances and water or other electrically charged substances. Hydrophobic substances are also called *nonpolar substances*.

When hydrocarbons are dispersed in water, they slowly come together—dispersed molecules form droplets that form larger droplets. The forces that bring about this combining of molecules are sometimes called *hydrophobic interactions*, but this term is somewhat misleading. The interactions that combine nonpolar substances have less to do with forces between the nonpolar molecules than with the hydrogen bonding of the water that surrounds the molecules.

When nonpolar substances such as hydrocarbons are introduced into water, they cause a disruption in the usual hydrogen bonding between water molecules. In the vicinity of the hydrocarbon, the water molecules form a hydrogen-bonded "cage" that surrounds the nonpolar hydrocarbons and pushes them together. These water cages can bring together dispersed nonpolar molecules into larger groups.

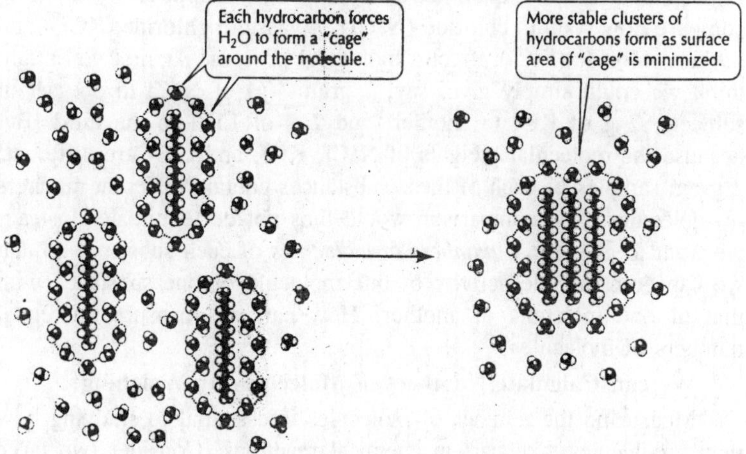

Fig. 3.8 Interaction of water and nonpolar substances.

Eggs by the Dozen, Molecules by the Mole

In modern biology, as in chemistry, the question "How much?" is as important as "What kind?" In this section, we will examine briefly how chemists and biologists deal quantitatively with atoms and molecules.

The molecular formula uses chemical symbols to identify different atoms, and subscript numbers to show how many atoms are present. For example, the molecular formula for methane is CH_4 (each molecule contains one carbon atom and four hydrogen atoms), that for oxygen gas is O_2, and that for sucrose (table sugar) is $C_{12}H_{22}O_{11}$. The hormone insulin is represented by the molecular formula $C_{2}5_{4}H_{377}N_{65}O_{76}S_{6}$! Although molecular formulas tell us what kinds of atoms and how many of each kind are present in the molecule, they tell us nothing about which atoms are linked to which. *Structural formulas* give us this information.

Each compound has a *molecular weight* (molecular mass): the sum of the atomic weights of the atoms in the molecule. The atomic weights of hydrogen, carbon, and oxygen are 1.008, 12.011, and 16.000, respectively. Thus the molecular weight of water (H_2O) is $(2 \times 1.008) + 16.000 = 18.016$, or about 18. What is the molecular weight of sucrose ($C_{12}H_{22}O_{11}$)? Your calculations should tell you that the answer is approximately 342. If you remember the molecular weights of a few representative biological compounds, you will be able to picture the relative sizes of molecules that interact with one another. Experiments require quantitative information, Suppose we want to compare how sodium chloride (NaCJ), potassium chloride (KCI), and lithium chloride (LiCI) affect a biological process. At first you might think we could simply give, say, 2 grams (g) of NaCl to one set of subjects, 2 g of KCI to another, and 2 g of LiCI to the third. But because the molecular weights of NaCl, KCI, and LiCI are different, 2-gram samples of each of these substances contain different numbers of molecules. The comparison would thus not be legitimate. Instead, we want to give *equal numbers of molecules* of each substance so that we can compare the activity of one molecule of one substance with that of one molecule of another. How can we measure out equal numbers of molecules?

We can Calculate Numbers of Molecules by Weighing

Measuring the number of molecules is essential to studying how many molecules take part in chemical reactions. Consider two large barrels, one filled with bolts and the other with pennies. How can we determine the number of units in each barrel? We could count them, but that would be tedious (and it is impossible to count individual molecules directly). However, if we know the weight of one bolt and one penny, we can calculate the number of units by dividing the total weight by the weight of each unit. The same principles apply to determining the number of molecules in a weighed sample.

To determine the number of molecules in a sample of a pure substance, we first determine the weight of the substance in grams, then we divide the grams by the relative weight of one molecule (the molecular weight defined earlier). Therefore, to measure out quantities of substances containing equal numbers of molecules, we weigh out in grams a quantity of each substance equivalent to its molecular weight. The molecular weight of NaCl is 58.45. Therefore 58.45 g of NaCl will have the same number of molecules as 74.55 g of KCI, whose molecular weight is 74.55.

If we divide the 74.55 g by the molecular weight of 74.55, the answer is 1. But one what? Surely not one molecule! The calculated numbers here are not the exact number, because we did not divide by the actual weight in grams of a molecule of potassium chloride (which would be about 1×10^{-22} g). Instead we divided by the relative weight. Therefore, instead of knowing the exact number of molecules, we know the relative number, which is called a *mole*. *One mole of a substance is an amount whose weight in grams is numerically equal to the molecular weight of the substance.* Potassium chloride (KCI) has a molecular weight of 74.55, so 1 mole of KCI weighs 74.55 g. Likewise, 1 mole of NaCl weighs 58.45 g, and 1 mole of LiCl, 42.40 g.

A mole of one substance contains the same number of molecules as does a mole of any other substance. This number, known as *Avogadro's number*, is 6.023×10^{23} molecules per mole. The concept of the mole is important for biology because it enables us to work easily with known numbers of molecules. Since we can neither weigh nor count individual molecules, we work with moles. The mole concept is analogous to the concept of a dozen or a gross. We buy some things, such as eggs, by the dozen (or another similar unit, such as the gross) because the individual items are inconvenient to count. Just as a dozen of anything contains twelve of that thing, a mole of any substance contains Avogadro's number of molecules of that substance.

Reactions Take Place in Solutions

In cells and in the laboratory, reactions take place in solutions formed when substances (solutes) are dissolved in water (the solvent).

The amount of substance dissolved in a given amount of water is the concentration of the solution. Knowing the concentrations of solutions is essential in performing experiments and interpreting their results.

A solution that contains 1 mole of solute per liter of solution has a one-molar concentration, abbreviated 1 *M*. A solution containing a half mole per liter is referred to as 0.5 *M*, or half-molar. How would you make 100 milliliters (ml) of a 0.5 *M* sucrose solution? The molecular weight of sucrose is 342, so 1 liter (1,000 ml) of a *1 M* sucrose solution contains 1 mole, or 342 g, of sucrose. You were asked to make just 100 ml of 0.5 *M* solution. Since 34.2 g of sucrose would make 100 ml of *1 M* sucrose, to make 100 ml of a 0.5 *M* sucrose solution, you would use 0.5×34.2 g = 17.1 g of sucrose.

Chemical Reactions: Atoms Change Partners

When atoms combine or change bonding partners, a chemical reaction is occurring. Consider the combustion reaction that takes place in the flame of a propane stove. When propane (C_3H_8) reacts with oxygen gas (O_2), the carbon atoms become bonded to oxygen atoms instead of to hydrogen atoms, and the hydrogen atoms become bonded to oxygen instead of to carbon. As the covalently bonded atoms change bonding partners, the composition of the matter changes, and propane and oxygen gas become carbon dioxide and water. This chemical reaction can be represented by the balanced equation

$$C_3H_8 + 5O_2 \rightarrow 3CO_2 + 4H_2O$$

In this equation, the propane and oxygen are the *reactants*, and the carbon dioxide and water are the *products*. The arrow symbolizes the chemical reaction. The numbers preceding the molecular formulas balance the equation and indicate how many molecules react or are produced. In this and all other chemical reactions, matter is neither created or destroyed. The total number of carbons on the left equals the total number on the right. However, there is another product of this reaction: energy.

The heat of the stove's flame and its blue light reveal that the reaction of propane and oxygen releases a great deal of energy. *Energy* is defined as the capacity to do work, but on a more intuitive level, it can be thought of as the capacity to change. Chemical reactions do not create or destroy energy, but *changes* in energy usually accompany chemical reactions. The energy released as heat and light was present in the reactants in another form, called *potential chemical energy*. In some chemical reactions, energy must be supplied from the environment (for example, some substances will react only after being heated), and some of this supplied energy becomes stored as potential chemical energy in the bonds formed in the reactants.

We can measure the energy associated with chemical reactions using the unit called a calorie (cal). A calorie is the amount of heat energy needed to raise the temperature of 1 g of pure water from 14.5°C to 15.5°C. (The nutritionist's Calorie, with a capital C, is what biologists call a kilocalorie (kcal) and is equal to 1,000 heat-energy calories.) Another unit of energy that is increasingly used is the joule (J). When you compare data on energy, always compare joules to joules and calories to calories. The two units can be interconverted: 1 J = 0.239 cal, and 1 cal = 4.184 J. Thus, for example, 486 cal = 2033 J, or 2.033 kJ. Although defined in terms of heat, the calorie and the joule are measures of any form of energy-mechanical, electric, or chemical.

Within living cells, chemical reactions called oxidation-reduction reactions take place that have much in common with this combustion of propane. The fuel for these biological reactions is different (the sugar glucose, rather than propane), and the reactions proceed by many intermediate steps that permit the energy released from the glucose to be harvested and put to use by the cell. But the products are the same: carbon dioxide and water. We will present and discuss oxidation-reduction reactions and several other types of chemical reactions that are prevalent in living systems in the chapters that follow.

WATER: STRUCTURE AND PROPERTIES

Water, like all other matter, can exist in three states: solid (ice), liquid, and gas (vapour). Liquid water is the medium in which life originated on Earth more than 3.5 billion years ago, and it is in water that life evolved for about a billion years. Today water covers three-fourths of Earth's surface, and all active organisms contain between 45 and 95 percent water. No organism can remain biologically active without water, which is important in both the outer and inner environments. Within cells, water participates directly in many chemical reactions, and it is the medium (or solvent) in which most reactions take place. In this section we will consider the structure and interactions of water molecules, exploring how these generate properties essential to life.

Water has a Unique Structure and Special Properties

The shape of a water molecule is determined by the distribution in space of the four pairs of electrons in the outer shell of the oxygen atom. Each pair of electrons is confined to an orbital. Two of these pairs are bonded to hydrogens, but the other two pairs (nonbonding pairs) also influence shape. Because of their negatively charged electrons, the four orbitals, each containing a pair of electrons, repel one another. In seeking to be as far apart as possible, they give the water molecule a nearly tetrahedral shape.

The shape of the water molecule, its polar nature, and the formation of hydrogen bonds between water molecules or with other substances give water its unusual properties. For example, ice floats, and compared to other liquids, water is an excellent solvent, making it an ideal medium for biochemical reactions. Water is both cohesive (sticking to itself) and adhesive (sticking to other things). And the energy changes that

Ice Floats

In its solid state (ice), water is held by its hydrogen bonds in a rigid, crystalline structure in which each water molecule is hydrogen-bonded to four other molecules. Although these molecules are held firmly in place, they are not as tightly packed as they are in liquid water. In other words, *solid water is less dense than liquid water,* which is why ice floats in water. If ice sank in water, as almost all other solids do in their corresponding liquids, ponds and lakes would freeze from the bottom up, becoming solid blocks of ice in winter and killing most of the organisms living in them. Once the whole pond had frozen, its temperature could drop well below the freezing point of water. However, because ice floats, it forms a protective insulating layer on the top of the pond, reducing heat flow to the cold air above. Thus fish, plants, and other organisms in the pond can survive the winter at temperatures no lower than 0°C, the freezing point of pure water.

Melting and Freezing

Compared to other nonmetallic substances of the same size, ice requires a great deal of heat energy to melt. Melting 1 mole of water molecules requires the addition of 5.9 kJ of energy. This value is high because more than a mole of hydrogen bonds must be broken for 1 mole of water to change from solid to liquid. In the opposite process, freezing, a great deal of energy must be lost for water to transform from liquid to solid. These properties help make water a moderator of temperature changes.

Another property of water that moderates temperature is the high heat capacity of liquid water. *Heat capacity* is the amount of heat energy that is required to raise the temperature of a substance 1°C.

Raising the temperature of liquid water takes a relatively large amount of heat. The temperature of a given quantity of water is raised only 1°C by an amount of heat that would increase the temperature of the same quantity of ethyl alcohol by 2°C, or of chloroform by 4°C. This phenomenon contributes to the surprising constancy of the temperature of the oceans and other large bodies of water through the seasons of the year. The temperature changes in coastal land masses are also moderated by large bodies of water. Indeed, water helps minimize variations in atmospheric temperature throughout the planet.

Cohesion and Surface Tension

In liquid water, the molecules are free to move about. The hydrogen bonds between the water molecules continually form and break. In other words, liquid water has a dynamic structure. On the average, every water molecule forms 3.4 hydrogen bonds with other water molecules. This number represents fewer bonds than exist in ice, but it is still a high number.

These hydrogen bonds explain the cohesive strength of water. The *cohesive strength* of water is what permits narrow columns of water to stretch from the roots to the leaves of trees more than 100 meters high. When water evaporates from leaves, the entire column moves upward in response to the pull of the molecules at the top.

Water also has a high *surface tension*, which means that the surface of water exposed to the air is difficult to puncture. The water molecules in this surface layer are hydrogen-bonded to other water molecules below. The surface tension of water permits a container to be filled slightly above its rim without overflowing, and it permits small animals to walk on the surface of water.

Evaporation and Cooling

Water has a high *heat of vaporization*; which means a lot of heat is required to change water from its liquid state to its gaseous state (the process of evaporation). This heat is absorbed from the environment in contact with the water. Evaporation thus has a cooling effect on the environment-whether a leaf, a forest, or an entire land mass. This effect explains why sweating cools the human body: As the sweat evaporates off the skin, it takes with it some of the adjacent body heat.

Water Molecules Sometimes Form Ions

The water molecule has a slight but significant tendency to come apart into a hydroxide ion (OH^-) and a hydrogen ion (a proton, H^+). Actually, *two* water molecules participate in ionization.

The hydronium ion is in effect a. hydrogen ion bound to a water molecule. For simplicity, biochemists tend to use a modified representation of the ionization of water:

$$H_2O \rightarrow H^+ + OH^-$$

Even though only about one water molecule in 500 million is ionized at any given time, the transformation is significant because H^+ and OH^- ions participate in many important biochemical reactions.

Acids, Bases, and the pH Scale

The ionization of water is very important for all living creatures. This fact may seem surprising, since the ionization is so slightonly one ionization out of 5×10^8 water molecules. But we are less surprised if we focus on the abundance of water in living systems and the reactive nature of H^+ produced by that ionization. Remember that H^+ is a proton, a tiny bit of charged matter-smaller than any atom or molecule in the cell.

Usually it attaches to another water molecule, but in the cell it can attach to other molecules and substantially change their properties. Because acids and bases donate and accept H^+, they can profoundly alter the water environment in which the chemical reactions and processes of life take place.

In this section we'll examine acids, bases, and their measurement using the pH scale. We'll close by looking at how buffers limit the changes in pH.

Acids Donate H^+, Bases Accept H^+

In pure water, the concentration of hydrogen ions exactly equals that of hydroxide ions (OH^-), and this "solution" is said to be *neutral*. Now suppose we add some HCl (hydrochloric acid). As it dissolves, the HCl ionizes, releasing H^+ and Cl^- ions:

$$HCl \rightarrow H^+ + Cl^-$$

Now there are more H^+ than OH^- ions. Such a solution is *acidic*. A *basic*, or alkaline, solution is one in which there are more OH^- than H^+ ions. A basic solution can be made from water by adding, for example, sodium hydroxide (NaOH), which ionizes to yield OH^- and Na^+ ions, thus making the concentration of OH^- ions greater than that of H^+ ions.

$$NaOH \rightarrow Na^+ + OH$$

An acid is any compound that can *release* H^+ ions in solution. HCl is an acid, as is H_2SO_4 (sulfuric acid). One molecule of sulfuric acid may ionize to yield two H+ ions and one SO_4^- ion. Biological compounds such as acetic acid and pyruvic acid, which contain -COOH (the carboxyl group) are also acids, because —COOH → -COO$^-$ + H$^+$.

Bases are compounds that can *accept* H$^+$ ions. These include the bicarbonate ion (HCO_3^-), which can accept a H$^+$ ion and become carbonic acid (H_2CO_3); ammonia (NH_3), which can accept a H+ ion and become an ammonium ion (NH_4^+); and many others.

Note that although —COOH is an acid, —COO$^-$ is a base, because -COO$^-$ + H$^+$ → —COOH. Acids and bases exist as pairs, such as —COOH and —COO$^-$, because any acid becomes a base when it releases a proton, and any base becomes an acid when it gains a proton.

You may have noticed that the two reactions just discussed are the opposites of each other. The reaction that yields —COO$^-$ and H$^+$ is reversible and may be expressed as

$$\text{—COOH} \rightleftharpoons \text{—COO}^- + \text{H}^+$$

A *reversible reaction* is one that can proceed in either direction-left to right or right to left-depending on the relative starting concentrations of reacting substances and products. In principle, all chemical reactions are reversible.

pH is the Measure of Hydrogen Ion Concentration

The terms "acidic" and "basic" refer only to *solutions*. How acidic or basic a solution is depends on the relative concentrations of H$^+$ and OH$^-$ ions in it. "Acid" and "base" refer to *compounds* and *ions*. A compound or ion that is an acid can donate H$^+$ ions; one that is a base can accept H$^+$ ions.

How do we specify how acidic or basic a solution is? First, let's look at the H$^+$ ion concentrations of a few contrasting solutions. In pure water, the H+ concentration is 10^{-7} *M*. In 1 M hydrochloric acid, the H$^+$ concentration is 1 *M;* and in 1 *M* sodium hydroxide, the H$^+$ concentration is 10^{-14} *M*. Because its values range so widely-from more than 1.0 *M* to less than 10^{-14} M the H$^+$ concentration itself is an inconvenient quantity. It is easier to work with the logarithm of the concentration, because logarithms compress this range. We indicate how acidic or basic a solution is by its pH (a term derived from "potential of Hydrogen"). The pH value is defined as the negative logarithm of the hydrogen ion concentration in moles per liter (molar concentration). In chemical notation, molar concentration is often indicated by putting brackets around the symbol for a substance; thus [H$^+$] stands for the molar concentration of H$^+$. The equation for pH is

$$\text{pH} = -\log_{10}[\text{H}^+]$$

Since the H^+ concentration of pure water is 10^{-1} *M,* its pH is -log(10^{-7}) = -(-7), or 7. A smaller negative logarithm means a larger number. In practical terms, a lower pH means a higher H^+ concentration, or greater acidity. In 1 *M* HCl, the H^+ concentration is 1 *M,* so the pH is the negative logarithm of 1 (–log 10^0), or 0. The pH of 1 *M* NaOH is the negative logarithm of 10^{-14}, or 14. A solution with a pH of less than 7 is acidic: It contains more H^+ ions than OH^- ions. A solution with a pH of 7 is neutral. And a solution with a pH value greater than 7 is basic. Because the pH scale is logarithmic, the values are exponential: A solution with a pH of 5 is 10 times more acidic than one with a pH of 6 (it has ten times as great a concentration of H^+); a solution with a pH of 4 is 100 times more acidic than one with a pH of 6.

Buffers Minimize pH Changes

An organism must control the chemistry of its cellsin particular, the pH of the separate compartments within cells. Animals must also control the pH of their blood. The normal pH of human blood is 7.4, and deviations of even a few tenths of a pH unit can be fatal. The control of pH is made possible in part by *buffers*, systems that maintain a relatively constant pH even when substantial amounts of acid or base are added. A buffer is a mixture of an acid that does not ionize completely in water and its corresponding base-for example, carbonic acid (H_2CO_3) and bicarbonate ions (HCO_3^-). If acid is added to this buffer, not all the H+ ions from the acid stay in solution. Instead, many of the added H^+ ions combine with bicarbonate ions to produce more carbonic acid, thus using up some of the H+ ions in the solution and decreasing the acidifying effect of the added acid:

$$HCO_3 + H^+ \rightleftharpoons H_2CO_3$$

If base is added, the reaction reverses. Some of the carbonic acid ionizes to produce bicarbonate ions and more H^+, which counteracts some of the added base.

In this way, the buffer minimizes the effects of added acid or base on pH. A given amount of acid or base causes a smaller change in pH in a buffered solution than in an unbuffered one. Buffers illustrate the reversibility of chemical reactions: Addition of acid drives the reaction in one direction; addition of base drives it in the other.

PROPERTIES OF MOLECULES

Molecules vary in size. Some are small, such as H_2 and CH_4. Others are larger, such as a molecule of table sugar (sucrose), which has 45 atoms. Still other molecules, such as proteins, are *gigantic*, sometimes containing tens of thousands of *atoms bonded* together in specific ways. Whether large, *medium*, or small, most of the molecules in living systems contain carbon atoms and are thus referred to as *organic molecules*. Most organic molecules include hydrogen and oxygen atoms, and many also include nitrogen and phosphorus.

All molecules have a specific three-dimensional shape. For example, the orientation of the bonding orbitals around the carbon atom gives the molecule of methane (CH_4) the shape of a regular tetrahedron, while in carbon dioxide (CO_2) the three atoms are in line with one another. Larger molecules have specific complex shapes that result from the number and kinds of atoms present and the ways in which these atoms are linked together. Some large molecules have compact ball-like shapes. Others are long, thin, ropelike structures. The shapes relate to the roles these molecules play in living cells.

In addition to size and shape, molecules have certain properties that characterize them and determine their biological roles. Chemists can use the characteristics of composition, structure (three-dimensional shape), reactivity, and solubility to distinguish a sample of one pure compound from another. For example, a compound such as sugar is soluble in water, which means that the solid disperses to form a uniform homogeneous mixture, but the same compound is insoluble in oil (the solid sugar remains a solid), and the resulting mixture is heterogeneous. Another substance, such as butane, is insoluble in water but highly soluble in oil. These solubility differences are due to the different atoms present and to their arrangement in these two kinds of molecules.

The presence of polar or charged sites on a molecule plays important roles in determining the molecule's solubility in water. Such sites can also determine the kinds of chemical reactions in which the molecule participates. The sizes, shapes, solubilities, and reactivities of molecules are significant to understanding the structures and operations of cells and organisms. Certain groups of atoms found together in a variety of molecules simplify our understanding of the reactions that molecules undergo.

Functional Groups give Specific Properties to Molecules

On the basis of atomic composition, structure, and reactivity, we can distinguish families of molecules from other families. Each member within a family of compounds has some characteristics in common with the other members of that family but differs from them in other characteristics. For example, organic acids are a family of carbon compounds that are all acidic but differ in other ways. They all contain a characteristic group of atoms called the *carboxyl group*,

$$-\overset{\displaystyle O}{\overset{||}{C}}-OH, \text{ or } COOH$$

that is the source of the H^+ that defines an acid:

$$-\overset{\displaystyle O}{\overset{||}{C}}-OH \rightarrow -\overset{\displaystyle O}{\overset{||}{C}}-O^- + H^+$$

The carboxyl group is a functional group. Functional groups are groups of atoms that are part of a larger molecule and have particular reactive characteristics. The same functional group may be part of very different molecules. In addition to the carboxyl group, you will encounter several other functional groups in your study of biology.

Several classes of biologically important compounds are defined by the functional groups they contain. When the functional group is a hydroxyl group (—OH), the product is an alcohol. Perhaps the most familiar alcohol is ethanol (also called ethyl alcohol, CH_3CH_2OH). Small alcohols like ethanol are soluble in water because of the hydrogen bonding possible between the polar hydroxyl group and water molecules, but larger alcohols are not soluble in water because of their long hydrocarbon chains.

Sugars contain both hydroxyl and carbonyl groups. The carbonyl *group* has a central carbon atom with a double bond to an oxygen atom (C=O). If one of the other two bonds of the carbon atom in a carbonyl group is attached to a hydrogen atom, the compound is an *aldehyde*.

Carboxyl groups are found in organic acids. Some organic bases, called amines, possess an *amino group* ($—NH_2$), which has a tendency to react with H^+ to produce the positively charged ammonium group ($—NH_3^+$). This H^+-accepting characteristic accounts for the classification of amines as bases.

Amino acids are important compounds that possess both a carboxyl group and an amino group attached to the same carbon atom, the a (alpha) carbon. Also attached to the a carbon atom are a hydrogen atom and a side chain designated by the letter R. Different side chains have different chemical compositions, structures, and properties. Each of the 20 amino acids found in proteins has a different side chain that gives it its distinctive chemical properties. Because they possess both carboxyl and amino groups, amino acids are simultaneously acids and bases. At the pH values commonly found in cells, both the carboxyl and the amino groups are ionized: The carboxyl group has lost α proton, and the amino group has gained one.

Two other functional groups deserve our attention: the sulfhydryl group and the phosphate group. The *sulfhydryl group* (—SH) is an important constituent of the side chains of two amino acids. When two of these sulfhydryl groups react, the two hydrogen atoms are lost and a covalent bond, called a disulfide bond, is formed between the sulfur atoms (—S—S—). As we'll see in the next chapter, disulfide bonds help maintain the three-dimensional structure of proteins that is required for normal protein functioning.

The *phosphate group* ($—OPO_3^-$) is found on many different kinds of molecules. It plays important structural roles in DNA and RNA and participates in reactions that transfer energy. The removal (hydrolysis) of phosphate groups from some molecules releases energy that can be used to fuel other energy-requiring reactions.

Isomers have Different Arrangements of the Same Atoms

Isomers are compounds that have the same chemical formula but different arrangements of the atoms. (The prefix "iso-" means "same and is encountered in many technical terms.) Of the different kinds of isomers, we will consider two: structural isomers and optical isomers.

Structural isomers are isomers that differ in how the atoms are joined together. Consider two simple molecules, each composed of four carbon and ten hydrogen atoms bonded covalently, with the formula C_4H_{10}. These atoms can be linked in two alternative ways in which carbon can form four bonds and hydrogen one bond. These two forms are called butane and isobutane. Their different bonding relationships are distinguished in structural formulas, and they have different chemical properties.

```
      H  H
      |  |
H3C—C—C—CH3
      |  |
      H  H
```

Butane

```
        CH3
         |
   H3C—C—CH3
         |
         H
```

Isobutane

Many molecules of biological importance, particularly the sugars and amino acids, have optical *isomers*. Optical isomers (also called enantiomers) are related to each other in the way an object is related to its mirror image. Optical isomers occur whenever a carbon atom has four *different* atoms or groups attached to it. These instances allow two different ways of making the attachments, each the mirror image of the other. Such a carbon atom is an asymmetric carbon, and the pair of compounds are optical isomers of each other. Your right and left hands are optical isomers. Just as a glove is specific for a particular hand, so some biochemical molecules can interact with a specific optical isomer of a compound but are unable to "fit" the other.

The a carbon in an amino acid is an asymmetric carbon because it is bonded to four different groups. Therefore, amino acids exist in two isomeric forms, called D-amino acids and 1.-amino acids. "D" and "L" are abbreviations for right and left, respectively. Only L-amino acids are commonly found in most proteins of living things.

The compounds discussed in this chapter include some of the more common ones found in organisms.

Between these small molecules and the world of the living stands another level, that of the giant macromolecules. These huge molecules—the proteins, lipids, carbohydrates, and nucleic acids—are the subject of the next chapter.

4

MACROMOLECULES

The lives of cells are dances and dramas with tens of thousands of different kinds of molecules. Their dramatic choreography is what scientists reveal as they investigate the molecules of living systems, their physical properties, and their chemical reactions. What molecules are present? What are their chemical structures and properties? And what biological functions do these molecules perform? In different ways, we will be concerned with these questions throughout much of this book.

In this chapter, we'll look at gigantic molecules called macromolecules (*macro-*, "large") and the subunits from which they are constructed. Macromolecules may contain hundreds or thousands of atoms and have very large molecular weights. For example, human hemoglobin has a molecular weight of 64,500 and contains more than 6,000 atoms. The molecular complexity of such a structure can be understood in terms of the structures and properties of its subunits, which are fewer in number and easier to understand. Macromolecules are chains of small, individual units called monomers (*mono-*, "one"; *-mer,* "unit") covalently bonded together to form a polymer (*poly-*, "many").

In living systems there are three types of macromolecules: polysaccharides, proteins, and nucleic acids. *Polysaccharides* are constructed from sugar monomers such as glucose; *proteins* are constructed from amino acids; and *nucleic acids* are composed of nucleotides. Another important group of molecules that can form large structures is the *lipids*. But because the individual components (the lipids) do not covalently bond together, the resulting large structures are not true macromolecules.

The monomers that form the basis of polysaccharides, proteins, and nucleic acids are identical in different species. For example, a molecule of glucose in human blood is identical to a molecule of glucose from a reptile or a cabbage plant. The same group of 20 different amino acids is used to construct proteins for all living things—from bacteria to birds, bats, and humans. However, this cannot always be said of the larger molecular forms built from these monomers.

Proteins and nucleic acids in particular are informational macromolecules that are different in different species. For example, human hemoglobin is different from the hemoglobin found in fish or birds. No animal acquires its macromolecules directly from its food. Instead, it uses the subcomponents of its food to construct new macromolecules suited to its unique needs. For example, we eat proteins constructed by other animals and plants, but we break these proteins down into their amino acids and then reassemble the amino acids into the chemically different proteins of our own bodies. This process is like picking up Lego toys that somebody else has made, taking them apart, and putting the parts together again to make the toys we want.

Before we turn to the main focus of this chapter—the macromolecules: carbohydrates, proteins, and nucleic acids—let's first take a look at lipids and the large macromolecule-like structures that they form.

Lipids: Water-Insoluble Molecules

Lipids are a chemically diverse group of hydrocarbons. The property they all share is an insolubility in water that is due to the presence of many nonpolar covalent bonds. Nonpolar molecules can associate together and form massive structures, but these structures are not considered macromolecules, because the bonds between the separate molecules are not covalent bonds. These nonpolar hydrocarbon molecules are literally pushed together by surrounding water molecules, which are not attracted to the nonpolar substance. When the nonpolar molecules are sufficiently close together, weak but additive van der Waals forces hold them together. Although insoluble in water, lipids are soluble in other lipids or in nonpolar solvents such as ether or benzene.

In addition to their diverse chemical structures, lipids have many different biological roles. Some of them store energy (the fats and oils). Others play important structural roles in cell membranes (phospholipids). The carotenoids help plants capture light energy, and

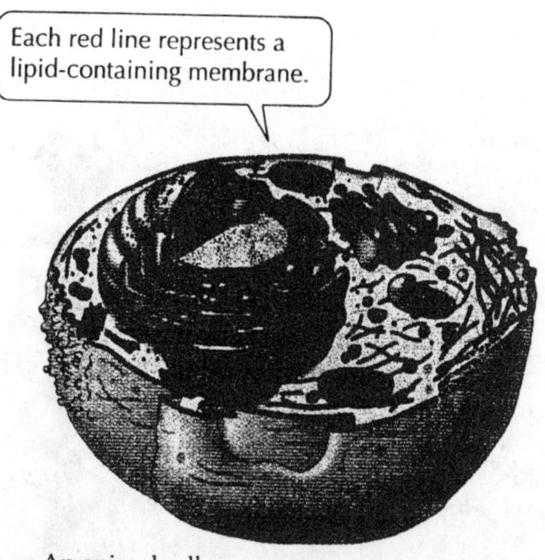

Fig. 4.1. Lipids from cellular membranes.

the steroids and some lipids play regulatory roles. We'll begin our discussion by considering the structures and properties of the fats and oils.

Fats and Oils Store Energy

Chemically, fats and oils are triglycerides, also known as *simple lipids*. Triglycerides that are solid at room temperature (20°C) are called fats; those that are liquid at room temperature are called *oils*.

Triglycerides are composed of two types of building blocks: fatty acids and glycerol. *Glycerol* is a small molecule with three hydroxyl (—OH) groups. Fatty *acids* have long nonpolar hydrocarbon tails and a polar carboxyl functional group (—COOH). Fatty acids can be of different lengths (usually 16 to 20 carbon atoms) and may be either Theturated or unsaturated.

In *saturated fatty acids*, all the bonds between the carbon atoms in the hydrocarbon chain are single bonds-there are no double bonds. That is, all the bonds are *saturated* with hydrogen atoms. Saturated fatty acids include palmitic acid, which has 16 carbon atoms, and stearic acid, which has 18. These molecules are relatively rigid and straight, and they pack together tightly, like pencils in a box.

In *unsaturated fatty acids*, the hydrocarbon chain contains one or more double bonds. Oleic acid is a monounsaturated fatty acid that has 18 carbon atoms, and one double bond near the middle of the hydrocarbon chain causes a kink in the molecule. Fatty acids, such as

(*a*) Palmitic acid

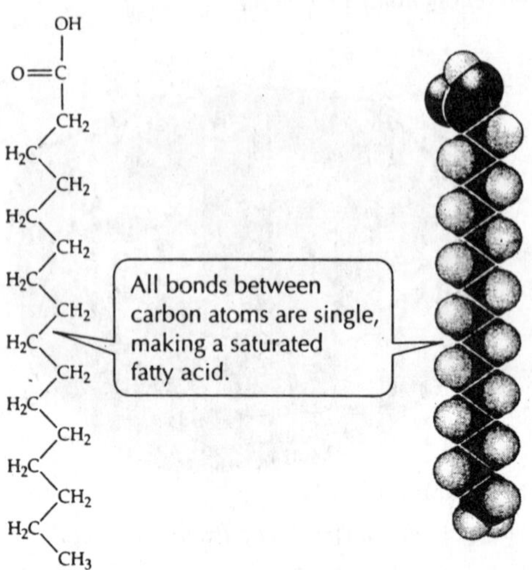

(*b*) Oleic acid

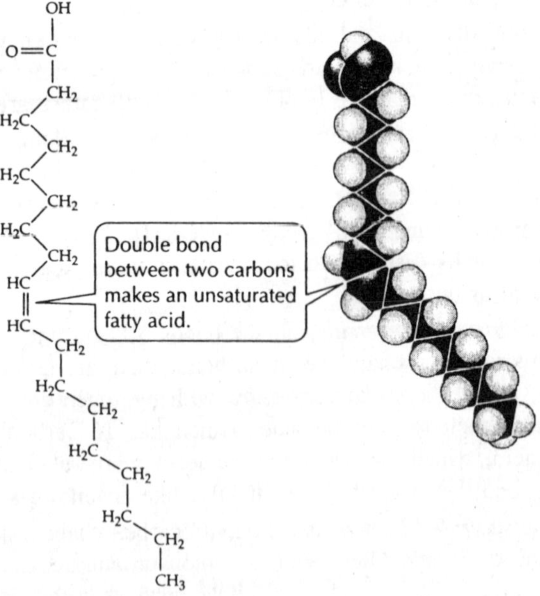

Fig. 4.2. Fatty acids.

linoleic acid, that have more than one double bond are polyunsaturated and have multiple kinks. These kinks prevent the molecules from packing together tightly. The two fatty acids that humans cannot synthesize and must obtain from their diet are both unsaturated fatty acids. Diets favouring unsaturated fatty acids over saturated ones tend to reduce the incidence of arteriosclerosis, which is a narrowing of the arterial wall that restricts blood flow to the heart and may lead to heart attacks. Diets rich in Theturated fatty acids have the opposite effect.

Three fatty acid molecules combine with a molecule of glycerol to form a molecule of a triglyceride. The carboxyl group of each fatty acid reacts with a hydroxyl group of the glycerol to form an *ester* (the reaction product of an acid and an alcohol) and water.

The ester linkage looks like this:

$$\overset{\displaystyle O}{\overset{\|}{—C}}—O—C—$$

The three fatty acids in one triglyceride molecule need not all have the same length, nor do they all have to be either saturated or unsaturated. The kinks associated with double bonds are important in determining the fluidity and melting point of a lipid. Animal fats such as lard and tallow are usually solids, and their triglycerides tend to have many long-chain saturated fatty acids. These fats are solids at room temperature because their fatty acids pack well together. The triglycerides of plants tend to have short or unsaturated fatty acids. Because of their kinks, these fatty acids pack poorly together and these triglycerides are usually liquid at room temperature. Natural peanut butter, for example, contains a great deal of oil. However, to market a more convenient and solid product, manufacturers often subject the oil to hydrogenation, which adds hydrogens to double bonds and converts unsaturated fatty acids to saturated fatty acids.

Lipids are marvelous storehouses for energy. By taking in excess food, many animal species deposit fat (lipid) droplets in their cells as a means for storing energy. Some plant species, such as olives, avocados, sesame, castor beans, and all nuts, have substantial amounts of lipids in their seeds or fruits that serve as energy reserves for the next generation.

Phospholipids Form the Core of Biological Membranes

Because lipids and water do not interact, a mixture of water and lipids forms two distinct phases. Many biologically important substances—such as ions, sugars, and free amino acids—that are soluble in water are insoluble in lipids. These two properties—water solubility and lipid insolubility—have proved essential to distinguishing the interior of cells from their external environment.

Suppose that you must design water-filled compartments, separated from each other and from their environment by barriers that limit the passage of materials. Given the properties of lipids, a seemingly effective way to accomplish this task is to use membranes that contain a special class of lipid called the phospholipid. This is the system that has evolved in nature. Molecular traffic within an organism or into and out of its compartments is constrained by the properties of the lipid portion of the surrounding membrane. Compounds that dissolve readily in lipids can move rapidly through biological membranes, but compounds that are insoluble in lipids are prevented from passing through the membrane or must be transported across the membrane by specific proteins.

Like triglycerides, phospholipids have fatty acids bound to glycerol by ester linkages. In phospholipids, however, any one of several phosphate-containing compounds may replace one of the fatty acids. Many phospholipids are important constituents of biological membranes. If you think carefully about the structure and properties of phospholipids, you will find it easy to understand how they are oriented in membranes. The phosphate functional group has negative electric charges, so this portion is hydrophilic, attracting polar water molecules. But the two fatty acids are hydrophobic, so they are pushed together by water.

In a biological membrane, phospholipids line up in such a way that the nonpolar, hydrophobic "tails" pack tightly together to form the interior of the membrane, and the phosphate-containing "heads" face outward (some to one side of the membrane and some to the other), where they interact with water, which is excluded from the interior of the membrane. The phospholipids thus form a bilayer, a sheet two molecules thick. Because membranes are so important.

Because the word "lipid" defines compounds in terms of their solubility rather than their structural similarity, a great variety of different chemical structures are included as lipids. The next two lipid classes we'll discuss—the carotenoids and the steroids—have chemical structures very different from the structures of triglycerides and phospholipids and from the structures of each other.

Carotenoids Trap Light Energy

The *carotenoids* are a family of light-absorbing pigments found in plants and animals. Betacarotene (β-carotene) is one of the pigments that traps light energy in leaves during photosynthesis. It is the β-carotene in plants that senses light and causes their parts to grow toward or away from the light. In humans, a molecule of β-carotene can be broken down into two vitamin A molecules, from which we make the pigment rhodopsin, which is required for vision. Carotenoids are responsible for the colour of carrots, tomatoes, pumpkins, egg yolks, and butter.

Steroids are Signal Molecules

The *steroids* are a family of organic compounds whose multiple rings share carbons. Some steroids are important constituents of membranes. Others are hormones, chemical signals that carry messages from one part of the body to another. *Testosterone* and the *estrogens* are steroid hormones that regulate sexual development in vertebrates. *Cortisol* and related hormones play many regulatory roles in the digestion of carbohydrates and proteins, in the maintenance of salt balance and water balance, and in sexual development.

Cholesterol is synthesized in the liver and contributes to the structure of some cellular membranes. It is the starting material for making testosterone and other steroid hormones, as well as the bile salts that help break down dietary fats so that they can be digested.

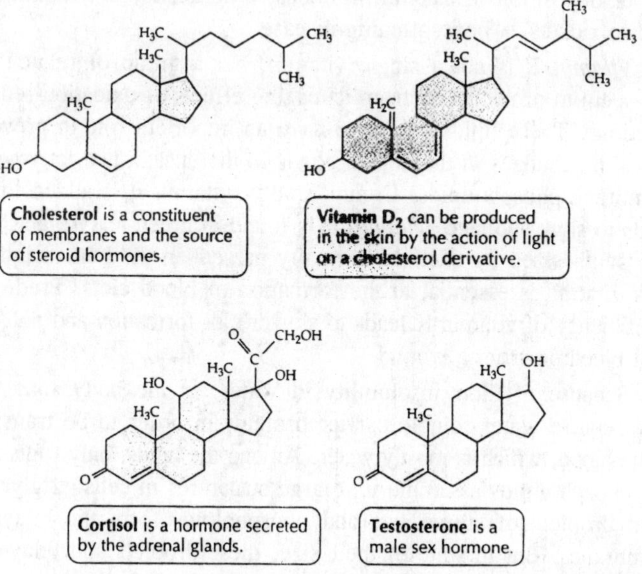

Fig. 4.3. All steroids have the same ring structures.

We absorb cholesterol from foods such as milk, butter, and animal fats. When we have too much cholesterol in our blood, it is deposited in our arteries (along with other substances), a condition that may lead to arteriosclerosis and heart attack.

Some Lipids are Vitamins

A large group of fat-soluble substances, including the carotenoids and steroids, are synthesized by covalent linking and chemical modification of isoprene to form a series of isoprene units:

$$CH_2 = \overset{\overset{\displaystyle CH_3}{|}}{C}—CH = CH_2$$

The fat-soluble vitamins (A, D, E, and K) are formed in this manner by plants and bacteria. These substances are not synthesized by humans and must be acquired from dietary sources. *Vitamin A* is formed from β-carotene found in green and yellow vegetables. Among other roles, vitamin A is directly involved in the reception of light by our eyes. Deficiency in vitamin A leads to dry skin, eyes, and internal body surfaces; retarded growth and development; and night blindness, which is a diagnostic symptom for the deficiency. *Vitamin D* regulates the absorption of calcium from the intestines. It is necessary for the proper deposition of calcium in bones; a deficiency of vitamin D can lead to rickets, a bonesoftening disease.

Vitamin E is not a single vitamin, but a group of related lipids that seem to protect cells from damaging effects of oxidation-reduction reactions. These lipids appear to have an important role in preventing unhealthy changes in the double bonds in the unsaturated fatty acids of membrane phospholipids. Commercially, vitamin E is added to some foods to slow spoilage. *Vitamin K* is found in green leafy plants and is also synthesized by bacteria normally present in the human intestine. This vitamin is essential to the formation of blood clots. Predictably, a deficiency of vitamin K leads to slower clot formation and potentially fatal bleeding from a wound.

Because of their insolubility in water, in the body some lipids may require water-soluble carrier proteins in order to be transported in the blood, which is mostly water. Among the lipids, only triglycerides and phospholipids assemble into large structures in cells: triglycerides form droplets of stored fat, and phospholpids form the bilayers of membranes. But in spite of their size, these droplets and bilayers are not usually considered macromolecules.

Now that we have seen the macromolecule-like structures that lipids form, let's turn to the large structures that constitute true macromolecules.

MACROMOLECULES: GIANT POLYMERS

As noted earlier, *macromolecules* are giant *polymers* constructed by the covalent linking of smaller molecules called *monomers*. These monomers may or may not be identical, but they always have similar chemical structures. Molecules with molecular weights exceeding 1,000 are usually considered macromolecules, and the polysaccharides, proteins, and nucleic acids of living systems certainly fall into this category.

In the cell, each type of macromolecule performs some combination of a diversity of functions: energy storage, structural support, protection, catalysis, transport, defense, regulation, movement, and heredity. These roles are not necessarily exclusive. For example, both carbohydrates and proteins can play structural roles-supporting and protecting cells and organisms. However, only nucleic acids specialize in information and function as hereditary material, carrying both species and individual traits from generation to generation.

The functions of macromolecules are directly related to their shapes and the chemical properties of their monomers. Some macromolecules, such as catalytic and defensive proteins, fold into compact spherical forms with surface features that make them watersoluble and capable of intimate interaction with other molecules. Other proteins and carbohydrates form long, fibrous systems that provide strength and rigidity to cells and organisms. Still other long, thin assemblies of proteins can contract and cause movement.

Because macromolecules are so large, they contain many different functional groups. For example, a large protein may contain hydrophobic, polar, and charged groups. These groups give specific properties to local sites on a macromolecule. As we will see, this diversity of properties determines the shapes of macromolecules and their interactions with both other macromolecules and smaller molecules.

Macromolecules Form by Condensation Reactions

The polymers of living things are constructed by a series of reactions called *condensation reactions*, or dehydration hydration reactions (both words refer to the loss of water). Condensation reactions covalently bond monomers. Water is lost in the reaction:

$$A\text{—}H + B\text{—}OH \rightarrow A\text{—}B + H_2O$$

A—H is a molecule with a reactive hydrogen; B—OH is a molecule with a hydroxyl group. The products of the reaction are A—B and H_2O. The atoms that make up the water molecule are derived from the reactants: one hydrogen atom from one reactant, and an oxygen atom and the other hydrogen atom from the other reactant. Condensation reactions are not limited to polymer formation; for example, we encountered them in the linkage (esterification) of fatty acids to glycerol.

The condensation reactions that produce the different kinds of macromolecules differ in detail, but in all cases polymers will form only if energy is added to the system. In living systems, specific energy-rich molecules supply the energy, and there are additional steps to the reaction. The reverse of a condensation reaction is a *hydrolysis reaction* (*hydro-*, "water"; *-lysis,* "breakage"). These reactions digest polymers and produce monomers. Water reacts with the bonds that link the monomers together, and the products are free monomers. The elements of the reactant H_2O become part of the products. Hydrolysis reactions of all sorts are very important in cellular functioning and are not limited to the digestion of polymers.

Carbohydrates: Sugars and Sugar Polymers

Carbohydrates are a diverse group of compounds based on the general formula CH_2O. Some are relatively small, with molecular weights less than 100. Others are true macromolecules, with molecular weights of hundreds of thousands. There are four categories of biologically important carbohydrates. *Monosaccharides* (*mono-*, "one"; *saccharide,* "sugar") such as glucose or fructose—are the monomers out of which the larger forms are constructed. *Disaccharides* (*di-*, "two") consist of two monosaccharides. *Oligosaccharides* (*oligo-*, "several") have several monosaccharides (3 to 20). *Polysaccharides* (*poly-*, "many")-such as starch, glycogen, and cellulose—are composed of hundreds of thousands of glucose units.

The relative proportions of carbon, hydrogen, and oxygen indicated by the general formula for carbohydrates, CH_2O, is true for monosaccharides. However, for disaccharides, oligosaccharides, and polysaccharides, these proportions differ slightly from this general formula because two hydrogens and an oxygen are lost during the condensation reactions.

Monosaccharides are Simple Sugars

All living cells contain the monosaccharide glucose, whose formula is $C_6H_{12}O_6$. Green plants produce glucose by photosynthesis, and other

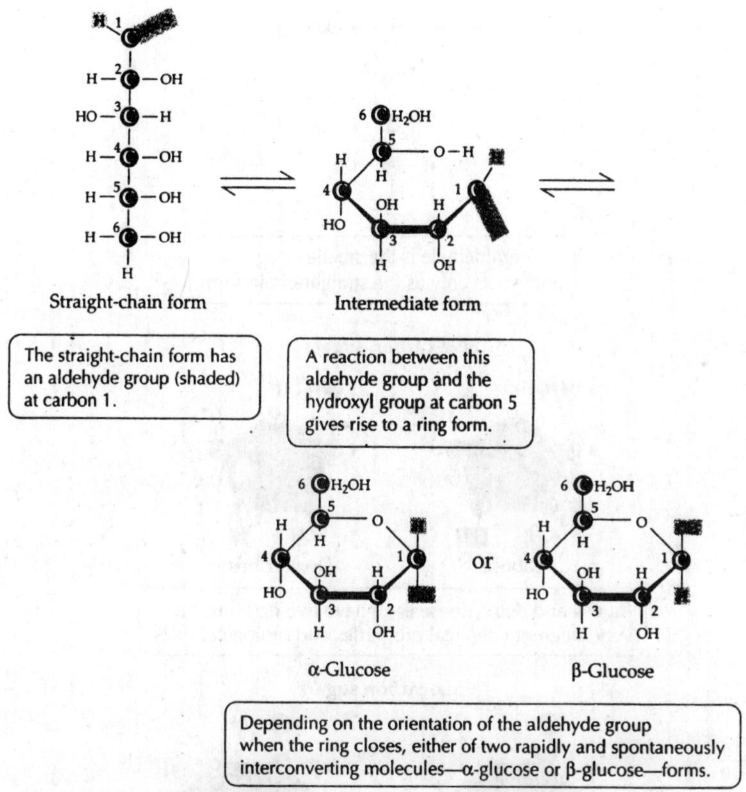

Fig. 4.4. Glucose: from one form to the other.

organisms acquire it directly or indirectly from plants. Cells use glucose as an energy source, changing it through a series of reactions that release stored energy and produce water and carbon dioxide.

Two forms of glucose, the straight chain and the ring, exist in equilibrium with each other when dissolved in water, but the ring form predominates (>99%). The two distinct ring forms (a and b-glucose) differ only in the placement of the —H and —OH attached to carbon 1. Most of the monosaccharides found in living systems belong to the D series of optical isomers. But there are structural isomers-composed of the same kinds and numbers of atoms, but with the atoms combined differently in each. All *hexoses* (*hex*-, "six"), a group of structural isomers, have the formula $C_6H_{12}0_6$. Included among the hexoses are fructose (so named because it was first found in fruits), mannose, and galactose.

Three-carbon sugar

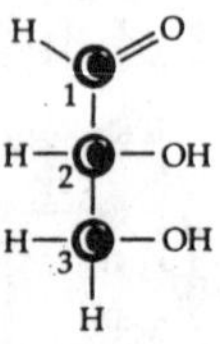

Glyceraldehyde is the smallest sugar and exists only as the straight-chain form.

Five-carbon sugars

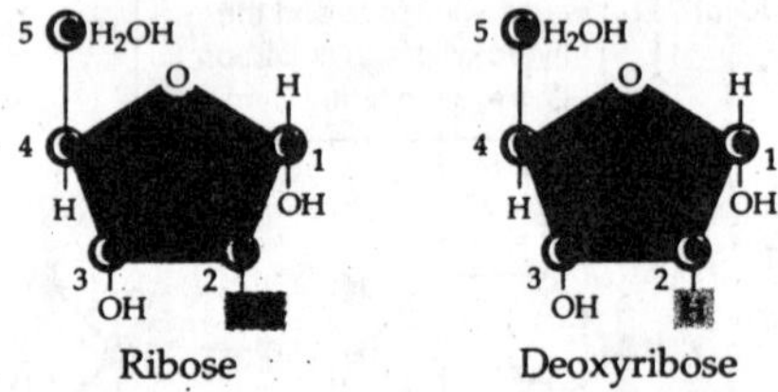

Ribose and deoxyribose each have five carbons, but very different chemical properties and biological roles.

Six-carbon sugars

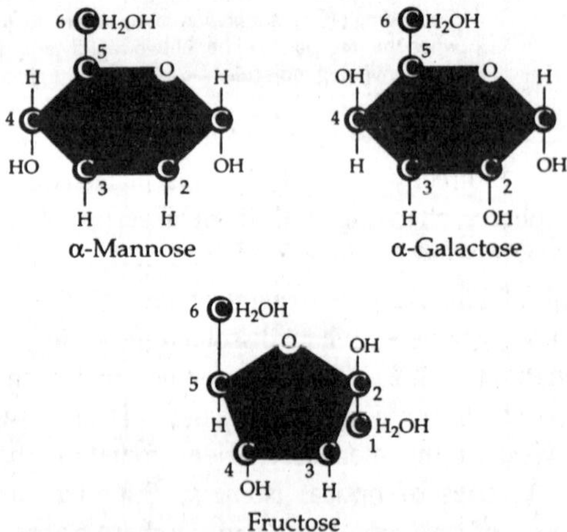

These hexoses are isomers. All have the formula $C_6H_{12}O_6$–but each has distinct chemical properties and biological roles.

Fig. 4.5. Monosaccharides are simple sugars.

Pentoses (*pent-*, "five") are five-carbon sugars. Some pentoses are found primarily in the cell walls of plants, as are several of the hexoses. Two pentoses are of particular importance: *Ribose* and *deoxyribose* form part of the backbones of RNA and of DNA, respectively. These two pentoses are not isomers; rather, one oxygen atom is missing from carbon 2 in deoxyribose (de, "absent"). The absence of this oxygen atom has enormous consequences for the functional distinction of RNA and DNA.

Glycosidic Linkages bond Monosaccharides Together

Monosaccharides are covalently bonded together by condensation reactions that form *glycosidic linkages*. Such a linkage between two simple sugars forms a *disaccharide*. For example, a molecule of *sucrose* (table sugar)—the sugar that is transported to different parts of plants—is formed from a glucose and a fructose molecule, while *lactose* (milk sugar) contains glucose and galactose. The disaccharide *maltose* contains two glucose molecules.

But maltose is not the only disaccharide that can be made from two glucoses. When glucose molecules form glycosidic linkages, the disaccharide product must be one of two types: α-linked or β-linked, depending on whether the molecule that bonds by its 1 carbon is α-glucose or β-glucose. An α linkage with carbon 4 of a second glucose molecule gives *maltose*, whereas a β linkage gives *cellobiose*. Maltose and cellobiose are disaccharide isomers, and both have the formula $C_{12}H_{22}O_{11}$. However, they are different compounds with different properties. They undergo different chemical reactions and are recognized by different catalytic proteins (enzymes).

Oligosaccharides contain several monosaccharides linked by glycosidic linkages at various sites. Many oligosaccharides gain additional functional groups, which give them special properties.

Oligosaccharides are often covalently bonded to proteins and lipids on the outer cell surface, where they serve as cell recognition signals.

Polysaccharides are Energy Stores or Structural Materials

Polysaccharides are giant chains of monosaccharides connected by glycosidic linkages. Starch is a polysaccharide of glucose with linkages in the a orientation. *Cellulose* is a similar giant polysaccharide made up solely of glucose, but its individual units are connected by β linkages instead of a linkages. Cellulose is the predominant component of plant cell walls and is by far the most abundant organic compound on this planet. Both starch and cellulose are composed of nothing but glucose, but their biological functions and chemical and physical properties are entirely different.

Starch can be more or less easily degraded by the actions of chemicals or catalytic proteins (enzymes). Cellulose, however, is very stable because of its β glycosidic linkages. Thus starch is a good storage form that can be easily degraded to supply glucose for energy producting reactions, while cellulose is an excellent structural component that can withstand harsh environmental conditions without changing.

Starch is not just one chemical substance, but large family of giant molecules of broa structure. All starches are large polymers of glucose with a linkages, but different starches can be distinguished by the amount of branching between carbons 1 and 6. Some starches are highly branched; others are not. Plant starches, called *amylose*, are not highly branched. The polysaccharide glycogen, which stores glucose in animal livers and muscles, is highly branched.

What do we mean when we say that starch and glycogen are storage compounds for energy? Very simply, these compounds can readily be hydrolyzed to yield glucose monomers. Glucose, in turn, can enter into a series of reactions that liberate its stored energy in forms that can be used for cellular activities. However, glucose can also serve the cell's need for raw materials—that is, carbon atoms. Glucose can undergo other chemical reactions that disassemble parts of the molecule and rearrange its carbon atoms to form the skeletons of other compounds needed by cells. Glycogen and starch are thus storage depots for carbon atoms as well as for energy.

Derivative Carbohydrates Contain other Elements

Sometimes carbohydrates are modified by chemical changes in their structure or by the addition of functional groups such as phosphate and amino groups, thus becoming *derivative carbohydrates*. For example, carbon 6 in glucose may be oxidized from —CH_2OH to a carboxyl group (—COOH), producing glucuronic acid. Or a phosphate group may be added to one or more of the —OH sites. Some of these *sugar phosphates*, such as fructose 1,6-bisphosphate, are important intermediates in cellular energy reactions. When an amino group is substituted for an —OH group, *amino sugars* such as glucosamine and galactosamine are produced. Galactosamine is a major component of cartilage, the material that forms caps on the ends of bones and stiffens the protruding parts of the ears and nose. A derivative of glucosamine produces the polymer *chitin*, which is the principal structural polysaccharide in the skeletons of insects, crabs, and lobsters, as well as in the cell walls of fungi. Fungi and insects (and their relatives) constitute more than 80 percent of the species ever described, and chitin is one of the most abundant substances on Earth.

Proteins: Amazing Polymers of Amino Acids

Proteins are an extraordinary group of macromolecules. Constructed of amino acids, proteins are fascinating because they have such intricate and diverse structures and because they perform so many different functions for cells. Among the cellular functions of macromolecules listed earlier, only energy storage and heredity are not functions of proteins. Proteins are involved in structural support, protection, catalysis, transport, defense, regulation, and movement. Of particular importance are the catalytic proteins, called *enzymes*, that increase the rates of chemical reactions in cells. In general, each chemical reaction requires a different enzyme, because proteins show great specificity for the smaller molecules with which they will interact.

Proteins range in size from the small RNA-digesting enzyme *ribonuclease A*, which has a molecular weight of 5,733 and 51 amino acid residues, to gigantic molecules such as the cholesterol transport protein *apolipoprotein B*, which has a molecular weight of 513,000 and 4,636 amino acid residues. (The word "residue" is used for a monomer when it is part of a polymer.) Each of these proteins consists of one chain of amino acids folded into a specific three-dimensional shape that is required for protein function. Some proteins have more than one polymer chain. For example, the oxygen-carrying protein *hemoglobin* has four chains that are folded separately and associate together to make the functional protein. The largest protein complex known has more than 40 separate chains of amino acids.

All of these different proteins have a characteristic amino acid composition, but every protein contains neither all 20 amino acids nor an equal number of different amino acids. Nor is there a simple regular sequence in which the amino acids are linked. The diversity in amino acid content and sequence is the source of the diversity in protein structures and functions. In some proteins, different kinds of chemical substances, called *prosthetic groups*, may be attached to the protein. These prosthetic groups include carbohydrates, lipids, phosphate groups, the iron-containing heme group, and metal ions such as copper and zinc. Whether they are small or large, and whether or not they have prosthetic groups, there is nothing "casual" about the structures of proteins. Proteins have specific three-dimensional shapes that are necessary for their specific functions.

To understand this stunning variety of functions, we must explore protein structure. First, we will examine the properties of the 20 amino acids and the characteristics of how they link to form proteins. Then we will systematically examine the four levels of protein structure and look at how a linear chain of amino acids is consistently folded into a compact three-dimensional shape.

Proteins are Composed of Amino Acids

The 20 different amino acids commonly found in proteins show a wide variety of properties. We considered the structure of amino acids and identified four different groups attached to a central carbon atom: a hydrogen atom, an amino group, a carboxyl group, and a side chain, or R group. Since amino acids are linked together by reactions between their amino and carboxyl groups, these groups are not exposed and do not give the protein its distinguishing properties and functional specificity-the side chains fill this role.

The *side chains* of amino acids are a protein's reactive groups and show a wide variety of chemical properties. Side chains control the function of a protein and contribute to its structure. The sequence of side chains determines how a protein folds into a three-dimensional shape (which we will discuss shortly). Although side chains are very important, they are often omitted from diagrams of proteins when the focus is on other aspects of structure. In these cases, the side chains are identified by the letter R (for "residue") and are sometimes called R groups.

One useful classification of amino acids is based on whether their side chains are electrically charged (+1, -1), polar (δ+, δ-), or nonpolar and hydrophobic. The five amino acids that have electrically charged side chains attract water and oppositely charged ions of all sorts. The four amino acids that have polar side chains tend to form weak hydrogen bonds with water and with other polar or charged substances. Eight amino acids have side chains that are nonpolar hydrocarbons or very slightly modified hydrocarbons. In the watery environment of the cell, the hydrophobic side chains may cluster together.

Three amino acids—cysteine, glycine, and proline—are special cases, although their side chains are generally hydrophobic. Two *cysteine* side chains, which have terminal —SH groups, can react to form a covalent bond in a *disulfide bridge* (—S—S—). Hydrogen bonds and disulfide bridges help determine how a protein chain folds. When cysteine is not part of a disulfide bridge, its side chain is very hydrophobic. The *glycine* side chain consists of a single hydrogen atom; thus glycines may fit into tight corners in the interior of a protein molecule, where a larger side chain could not fit. *Proline* differs from other amino acids because it possesses a modified amino group.

Peptide Linkages Covalently Bond Amino Acids Together

When amino acids polymerize, the carboxyl group of one amino acid reacts with the amino group of another, undergoing a condensation reaction that forms a peptide linkage. Gives a simplified description of the reaction. (In living cells, other molecules must activate the reactants in order for this reaction to proceed, and there are intermediate steps.) A linear polymer of amino acids connected by peptide linkages is a *polypeptide*. A *protein* is made up of one or more polypeptides.

At one end of the polypeptide is a free amino group. This end is the N terminus, named for the nitrogen atom in the amino group. At the other end of the polypeptide—the C terminus—is a free carboxyl group. The other amino and carboxyl groups are bound in peptide linkages. Thus a protein has direction. For example, the dipeptide glycine-alanine, in which glycine has the free amino group, differs from alanine-glycine, in which alanine has the free amino group. In cells, the synthesis of polypeptide chains begins with the N terminus.

In the peptide linkage, the C=O oxygen carries a slight negative charge (δ-), whereas the N—H hydrogen is slightly positive (δ+). This asymmetry of charge favours hydrogen bonding within the protein molecule itself and with other molecules, contributing to both the structure and the function of many proteins.

Primary Structure of a Protein is its Amino Acid Sequence

Protein structure is elegant and complex—so complex that it is described as consisting of four different levels: *primary*, *secondary*, *tertiary*, and *quaternary*. However, proteins are always linear chains; there are no branches. The precise sequence of amino acids in a polypeptide constitutes the primary *structure* of a protein. The *peptide backbone* of this primary structure consists of a repeating sequence of three atoms (—N—C—C—) from the amino group, the central carbon, and the carboxyl group of each amino acid.

In cells, the primary structure of a protein is dictated by the precise sequence of nucleotides in a linear segment of a DNA molecule. The elucidation of this relationship between DNA primary structure and protein primary structure was one of the triumphs of molecular biology.

The theoretical number of different proteins is enormous. Since there are 20 different amino acids, there are 20 × 20 = 400 distinct dipeptides, and 20 × 20 × 20 = 8,000 different tripeptides. Imagine this process of multiplying by 20 extended to a protein made up of 100 amino acids (which is considered a small protein): There could be 20^{100} of these small proteins, each with its,own distinctive primary structure. How large is the number 20^{100}? In the entire universe there aren't that many electrons!

At the higher levels of protein structure, local coiling and folding give the final functional shape of the molecule, but all of these levels derive from the primary structure. The different properties associated with a precise sequence of amino acids determine how the protein can twist and fold. By twisting and folding, each protein adopts a specific stable structure that distinguishes it from every other protein.

Secondary Structure of a Protein Requires Hydrogen Bonding

Although the primary structure of each protein is unique, the secondary structure of many different proteins may be the same. A protein's *secondary structure* consists of regular, repeated patterns in different regions of a polypeptide chain. One type of secondary structure, the *α helix* (alpha helix), is a right-handed coil that is "threaded" in the same direction as a standard wood screw. The amino acid side chains extend outward from the peptide backbone of the helix.

This helical structure of a polypeptide chain results from hydrogen bonds between elements of the peptide bonds that are distributed along the back bone of the chain. Hydrogen bonds form between the slightly positive hydrogen of the N—H of one peptide bond and the slightly negative oxygen of the C=O of another peptide bond. When this pattern of hydrogen bonding is established repeatedly over a segment of the protein, it stabilizes the twisted form, resulting in an a helix. However, the ability of a protein to form an a helix depends on its primary structure. Amino acids with larger side chains that distort the coil or otherwise prevent the formation of the necessary hydrogen bonds will keep the a helix from forming.

Alpha helical secondary structure is particularly evident in the fibrous structural proteins called *keratins*. Keratins constitute many of the protective materials found in mammals and birds, such as fingernails and claws, skin, hair, wool, and feathers. Hair can be stretched because this stretching requires that only hydrogen bonds in an a helix, and not covalent bonds, be broken; when the tension on the hair is released, both the helix and the hydrogen bonds reform.

Another type of secondary structure, the β pleated sheet, is found in the protein silk. In silk, rather than being coiled, the protein chains are almost completely extended and lie next to one another, stabilized by hydrogen bonds between the elements of the peptide linkages. The R pleated sheet may be found between separate polypeptide chains, as in silk, or between different regions of the same polypeptide that is bent back on itself. Many enzymes contain regions of a helix and of β pleated sheet in the same polypeptide chain.

Tertiary Structure of a Protein is Formed by Bending and Folding

In most proteins, to establish the compact structure the polypeptide chain must be bent at specific sites and folded back and forth. The resulting overall shape of a protein is its tertiary structure. Although the a helices and β pleated sheets contribute to the tertiary structure, more frequently only limited portions of the molecule have these secondary structures, and large regions consist of structures unique to a particular protein.

A complete description of the tertiary structure specifies the location of every atom in the molecule in three-dimensional space, in relation to all the other atoms. Bear in mind that this tertiary structure and the secondary structure derive from the protein's primary structure. If lysozyme is heated carefully, causing only the tertiary structure to break down, the protein will return to its normal tertiary structure when it cools. The only information needed to specify the unique shape of the lysozyme molecule is the information contained in its primary structure.

Hemoglobin is the protein that transports oxygen (O_2) from the lungs to the tissues. In muscle tissue, hemoglobin delivers O_2 to myoglobin, a smaller but similar protein, for storage. The hemoglobin molecule consists of four similar polypeptide chains that have tertiary structures made up almost entirely of a helices but that lack any disulfide bridges. The helical segments bend and fold against each other, forming a pocket that encloses a *heme group*, an ironcontaining prosthetic group that binds O_2. The stabilizing interaction of hydrophobic side chains on the inner sides of the a helices helps ensure that the helices fold against one another correctly as each polypeptide assumes its tertiary structure.

Quaternary Structure of a Protein Consists of Subunits

As we mentioned earlier, some proteins have two or more polypeptide chains folded into their own unique tertiary structures.

Each of these folded polypeptides is considered a protein subunit. *Quaternary structure* results from the ways in which these protein subunits fit together and interact; it can be illustrated by hemoglobin. Hydrophobic interactions, hydrogen bonds, and ionic bonds all help hold the four subunits together to form the functional hemoglobin molecule. As the hemoglobin molecule takes up one O_2 molecule, the four subunits shift their relative positions slightly, changing the quaternary structure. Ionic bonds are broken, exposing buried side chains that enhance the binding of additional O_2 molecules.

Each subunit of hemoglobin is folded like a myoglobin molecule, suggesting that both hemoglobin and myoglobin are evolutionary descendants of the same oxygen-binding ancestral protein. But on the surfaces where its subunits come in contact with each otherregions that on myoglobin are exposed to aqueous surroundings and are hydrophilic-hemoglobin has hydrophobic side chains. Again, the chemical nature of side chains on individual amino acids determines how the molecule folds and packs in three dimensions.

Molecular Chaperones Help Shape Proteins

The primary structure of a protein constrains the secondary, tertiary, and quaternary structures (if subunits exist). By determining the primary structure, DNA also determines the higher levels of structure. However, other factors also affect the tertiary structure that is required for proper protein function.

Elevated temperatures, pH changes, or altered salt concentrations can cause a protein to adopt a different, biologically inactive tertiary structure. Biological function depends on a specific three-dimensional structure. Increased temperature causes more rapid molecular movement and thus can break weak hydrogen bonds and hydrophobic interactions. Altered pH can change the pattern of ionization of carboxyl and amino groups in the side chains of amino acids, thus disrupting the pattern of ionic attractions and repulsions that contribute to normal tertiary structure.

The loss of appropriate tertiary structure is called *denaturation*, and it is always accompanied by a loss of the normal biological function of the protein. Denaturation can be caused by heat or high concentrations of polar substances such as urea that disrupt the hydrogen bonding that is crucial to protein structure. Nonpolar solvents may also disrupt normal structure. Usually denaturation is irreversible, particularly if many denatured proteins interact in random nonspecific ways. However, in some cases denaturation is reversible, and upon return to normal environmental conditions, the protein may return to its active form, as does lysozyme. This return is called *renaturation*.

How can such a complicated molecule spontaneously fold into its normal tertiary structure? Biologists are just beginning to be able to predict how a protein will fold, given its primary structure. The study of protein folding has a long way to go. The situation in the living cell is complex-even more complex than we first imagined. The protein doesn't form all at once. In a human cell it may take minutes from the beginning of synthesis until the entire protein squeezes into the compartment where it folds-and things are very crowded in that compartment. There, other proteins present inappropriate potential partners for interaction (in the form of amino acid side chains).

A special group of proteins, called the *chaperone proteins*, at least in part help prevent newly formed proteins from reacting inappropriately. The chaperones attach to a new protein while it is forming and protect it from interactions with other proteins until the new molecule has its correct shape. Most chaperone proteins can act on many different forming proteins. Some act on only a few, and others may act on only a single, specific protein.

What if the chaperone proteins fail to do their job, or if for other reasons a protein folds abnormally? Evidence suggests that abnormal protein folding underlies certain infectious diseases, including mad cow disease and, in humans, Creutzfeldt-Jakob disease. Alzheimer's disease may also result from protein misfolding.

Nucleic Acids: Informational Macromolecules

The nucleic acids are linear polymers specialized for the storage, transmission, and use of information. There are two types of nucleic acids: DNA (deoxyribonucleic acid) and RNA (ribonucleic acid). DNA molecules are giant polymers that encode hereditary information and pass it from generation to generation (through reproduction). The information encoded in DNA is also used to make specific proteins through the intermediate RNA. RNA molecules of various types copy the information in segments of DNA to specify the sequence of amino acids in proteins. Information flows from DNA to DNA in reproduction. But for nonreproductive activities of the cell, information flows from DNA to RNA to proteins, which ultimately carry out these functions. What compositions, structures, and properties of nucleic acids permit them to play these fundamental roles in living systems?

Nucleic Acids have Characteristic Structures and Properties

Nucleic acids are composed of monomers called *nucleotides*, each of which consists of a pentose sugar, a phosphate group, and a nitrogen-containing base-either a pyrimidine or a purine. Molecules consisting of a pentose sugar and a nitrogenous base, but no phosphate group, are called nucleosides. In DNA, the pentose sugar is deoxyribose, which differs from the ribose found in RNA by only one oxygen atom.

In both RNA and DNA, the backbone of the molecule consists of alternating sugars and phosphates (—sugar—phosphate—sugar—phosphate—). Bases are attached to the sugars and project from the chain. The nucleotides are joined by covalent bonds in what are called *phosphodiester* linkages between the sugar of one nucleotide and the phosphate of the next ("-diester" refers to the two bonds formed by reacting —OH groups with acidic phosphate groups). The phosphate groups link carbon 3' in one ribose to carbon 5' in the adjacent ribose.

Most RNA molecules are *single-stranded*, consisting of only one polynucleotide chain. DNA, however, is usually *double-stranded*; it has two polynucleotide chains held together by hydrogen bonding between their complementary nitrogenous bases. The two strands of DNA run in opposite directions. You can see what this means by drawing an arrow through the phosphate group from carbon 3' to carbon 5' in the next ribose. If you do this for both strands, the arrows point in opposite directions. This antiparallel orientation is necessary for the strands to fit together.

Uniqueness of a Nucleic Acid Resides in its Base Sequence

Only four nitrogenous bases—and thus only four nucleotides—are found in DNA. The DNA bases and their abbreviations are: adenine (A), cytosine (C), guanine (G), and thymine (T). A key to understanding the structures and functions of nucleic acids is the principle of *complementary base pairing* through hydrogen bond formation. In double-stranded DNA, adenine and thymine always pair (AT), and cytosine and guanine always pair (CG).

Base pairing is complementary because of two factors: the corresponding sites for hydrogen bonding and the molecular sizes of the paired bases. Adenine and guanine are both purines consisting of two fused rings. Thymine and cytosine are both pyrimidines consisting of only one ring. The pairing of a large purine with a small pyrimidine ensures a stable and consistent dimension to the double-stranded molecule of DNA. We'll discuss in more detail how complementary strands of DNA separate and are faithfully copied.

Ribonucleic acids also have four different monomers, but the nucleotides differ from those of DNA. In RNA the nucleotides are termed ribonucleotides. They contain ribose rather than deoxyribose, and instead of the base thymine, RNA uses the base uracil. The other three bases are the same as in DNA.

Although RNA is generally single-stranded, complementary hydrogen bonding between ribonucleotides can take place. These bonds play important roles in determining the shapes of some RNA molecules and in the associations between RNA molecules during protein synthesis. During the DNA-directed synthesis of RNA, complementary base pairing also takes place between ribonucleotides and the bases of DNA. In RNA, guanine and cytosine pair as in DNA, but adenine pairs with uracil. Adenine in an RNA strand can pair either with uracil (in another RNA strand) or with thymine (in a DNA strand).

The three-dimensional appearance of DNA is strikingly regular. Through hydrogen bonding, the two complementary polynucleotide strands pair and twist to form a double helix. Compared to the complex and varied tertiary structures of different proteins, this formation seems amazingly regular. But this structural contrast makes sense in terms of the functions of these two classes of macromolecules.

DNA is a purely informational molecule. *The information in DNA is encoded in the sequence of bases carried in its chains.* This sequence is, in a sense, like the tape of a tape recorder. The message must be read easily and reliably, in a specific order. A uniform molecule like DNA can be interpreted by standard molecular machinery, and any cell's machinery can read any molecule of DNA-just as a tape player can play any tape of the right size.

Table 4.1. Distinguishing RNA from DNA

Nucleic acid	*Sugar*	*Bases*
RNA	Ribose	Adenine Cytosine Guanine Uracil
DNA	Deoxyribose	Adenine Cytosine Guanine Thymine

Proteins, on the other hand, have good reason to be so varied. In particular, different enzymes must recognize their own specific "target" molecules, They do this by having a unique three-dimensional form that can match at least a portion of the surface of their target molecules. In other words, structural diversity in the molecules with which enzymes react requires corresponding diversity in the structure of the enzymes themselves. *In DNA the information is in the sequence of bases; in proteins the information is in the shape.*

DNA is a Guide to Evolutionary Relationships

Because DNA carries hereditary information between generations, a series of DNA molecules with changes in base sequences stretches back- through time. Of course, we cannot study all of these DNA molecules, because many of their organisms have become extinct. However, we can study the DNA of living organisms, some of which are judged to belong to lineages that are more ancient than those of other living forms. Comparisons and contrasts of these DNAs add to the evolutionary record.

Closely related living species should have more similar base sequences than species judged by other criteria to be more distantly related. Indeed this is the case. The examination of base sequences confirms many of the evolutionary relationships that have been inferred from the study of microscopic or macroscopic structures or studies of biochemistry and physiology. For example, the closest living relative of humans (*Homo sapiens*) is the chimpanzee (genus *Pan*), which shares more than 98 percent of its DNA base sequence with human DNA.

Confirmation of well-established evolutionary relationships gives credibility to the use of DNA to elucidate relationships when studies of structure are not possible or are not conclusive. For example, DNA studies revealed a close evolutionary relationship between starlings and mocking-birds that was not expected on the basis of studies of anatomy or behaviour. DNA studies support the division of the prokaryotes into two kingdoms (Eubacteria and Archaebacteria). Each of these two groups of prokaryotes is as distinct from the other as either is from the other four kingdoms into which living things are classified. In addition, DNA comparisons support the hypothesis that certain subcellular compartments of eukaryotes (the organelles called *mitochondria* and *chloroplasts*) evolved from early bacteria that established a stable and mutually beneficial way of life inside larger cells.

Interactions of Macromolecules

We have been treating the classes of macromolecules as if each were completely separate from the others. In cells, however, certain macromolecules of different classes may be covalently bonded to one another. Proteins with attached oligosaccharides are called *glycoproteins* (*glyco*-, "sugar"). The specific oligosaccharide chain determines the placement of a glycoprotein within the cell.

Other carbohydrate chains bind to lipids, resulting in *glycolipids*, which reside in the cell surface membrane, with the carbohydrate chain extending out into the cell's environment. In humans, the substances that determine the ABO blood types are carbohydrates attached to either proteins or lipids. When a human body cell becomes cancerous, the glycolipids and glycoproteins on the cell surface are modified, and these changes serve as a recognition signal for the body defenses to destroy the abnormal cell.

Not all the associations among macromolecules or between a macromolecule and a smaller molecule are covalent. Weaker linkages such as ionic bonds, hydrogen bonds, and even van der Waals forces may also be involved. Enzymes bind to their reactants using a variety of weak bonds. Ionic bonds link proteins to DNA, forming *nucleoproteins* that regulate the activities of DNA. Still other proteins, in combination with cholesterol and other lipids, form *lipoproteins*. The lipoproteins make it possible to move very hydrophobic lipids through water-rich environments such as the blood and tissue fluid and deliver cholesterol to all the body's cells.

5

CHROMOSOME

We have previously discussed the control of gene expression in prokaryotes and bacteriophages. Compared to eukaryotes, bacteriophages and prokaryotes are relatively simple. Of fundamental importance is that, in these lower forms, the operon model of induction and repression of transcription is a unifying theme for control of gene expression. Despite nuances such as catabolite repression and attenuator control, the operon model provides a relatively clear picture of how genes are turned on and off in phages and prokaryotes. This model does not exist for eukaryotes. In attempting to elucidate models for control of gene expression in eukaryotes, we must take one very important factor into account: the complexity of the structure of the eukaryotic chromosome. In this chapter, we cover the current understanding of how these very large structures are organized.

EUKARYOTIC CELL

Eukaryotes and prokaryotes are the two superkingdoms of organisms. The following comparisons, using *E. coli* as a general model for prokaryotes, show how much more complex eukaryotes are:

1. An *E. coli* chromosome contains approximately 4.2×10^6 base pairs of DNA. The haploid human genome contains nearly one thousand times as much DNA.
2. An *E. coli* cell has very little internal structure. Eukaryotes have a number of internal organelles and an extensive lipid membrane system, including the nuclear envelope itself.
3. Eukaryotic DNA is in the form of nucleoprotein, a DNA-histone protein complex. Although a few histonelike proteins have been

found in *E. coli*, its chromosomal DNA is not complexed with protein to anywhere near the same extent.

4. An *E. coli* cell is small (0.5 to 5.0 μm in length for bacteria). Eukaryotic cells are generally larger than prokaryotes (10 to 50 μm in length for animal tissue cells).
5. The messenger RNA of *E. coli* is translated while it is being transcribed. Eukaryotic messenger RNA is modified within the nucleus before it is transported out for translation in the cytoplasm.
6. Almost no messenger RNA isolated from eukaryotic cells, including the messenger RNA of animal viruses, has been found to be polycistronic (containing many genes). Most prokaryotic messenger RNAs are polycistronic.
7. Most *E. coli* genes are parts of inducible or repressible operons; there are almost no operons in eukaryotes.
8. *E. coli* exists as a simple, single cell. Although some prokaryotes do aggregate, sporulate, and show a few other limited forms of differentiation, they are primarily one-celled organisms. And, although some eukaryotes are single-celled (e.g., yeast), the essence of eukaryotes is differentiation. In human beings, a zygote gives rise to every other cell type in the body in a relatively predictable manner.

Eukaryotic Chromosome

DNA Arrangement

Evidence that the eukaryotic chromosome is *uninemic*—that is, contains one double helix of DNA—comes from several sources.The best data are provided by radioactive-labeling studies, first done by J. Taylor and his colleagues in 1957. If a eukaryote is allowed to undergo one DNA replication in the presence of tritiated (^{3}H-) thymidine, each of the daughter chromatids would be expected to contain a double helix with one unlabeled DNA template strand and one labeled strand of newly synthesized bases. This configuration is expected on the basis of semiconservative replication, with each chromatid containing one double helix. A second round of DNA replication, in the absence of ^{3}H-thymidine, should produce chromosomes in which one chromatid would have unlabeled DNA and one would have labeled DNA. As expected, one chromatid of every pair is labeled and one is not.

In another kind of experiment, R. Kavenoff, L. Klotz, and B. Zimm demonstrated that *Drosophila* nuclei contained pieces of DNA of the size predicted from their DNA content, based on the premise

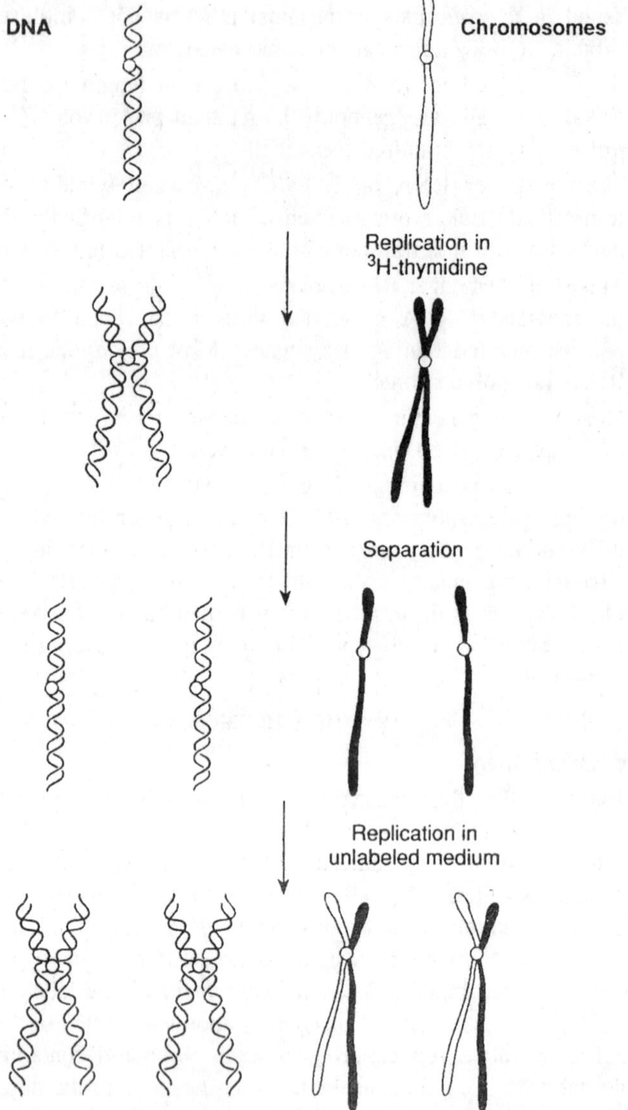

Fig. 5.1. Radioactive labeling of a uninemic eukaryotic chromosome following semiconservative replication.

that each chromosome contains one DNA molecule. They isolated the DNA and measured the size of the largest DNA molecules using the *viscoelastic* property of DNA, the rate at which stretched molecules relax. From other sources, primarily UV absorbance studies, it was estimated that the largest *Drosophila* chromosome had about 43 $\times$ 10^9 daltons of DNA. Results from the viscoelastic measurements indicated the presence of DNA molecules of between 38 and 44 $\times$ 10^9 daltons. Viscoelastic measurements of inversions, which changed the ratio of the arms but not the overall size of the chromosome, yielded similar results. However, a translocation that radically changed the size of the chromosome to 59 $\times$ 10^9 daltons resulted in an equivalent change in the viscoelastic estimates to between 52 and 64 $\times$ 10^9 daltons.

The conclusion from these studies is that the largest *Drosophila* chromosome, and by extension every eukaryotic chromosome, contains a single DNA molecule running from end to end, encompassing both arms. The viscoelastic values were corroborated by carefully isolating and measuring the lengths of long DNA molecules, an especially difficult task given DNA's propensity to break. The longest molecule that the investigators found was 1.2 cm long, equivalent to between 24 and 32 $\times$ 10^9 daltons, close to the predicted size. Thus, the evidence is in complete concordance with the simple uninemic model of eukaryotic chromosomal structure.

Nucleoprotein Composition

Nucleosome structure

Since each eukaryotic chromosome consists of a single, relatively long piece of duplex DNA, the average diploid cell contains many of these long pieces of DNA. For chromosomes to be properly distributed to each daughter cell during mitosis and meiosis, they must be condensed into structures that are more easily managed. Wrapping the DNA around "spools" of protein constitutes the first step in a series of coiling and folding processes that eventually result in the fully compacted chromosome we see at metaphase.

Interphase nuclei can be disrupted by placing them in a hypotonic liquid such as water. When this happens, chromatin material is released. When this material is observed under the electron microscope, small particles called *nucleosomes* can be seen. These are the spools that the DNA is wrapped around. They are made of *histone* proteins and associated DNA. The histones, a group of arginine- and lysine-rich basic proteins, have been well characterized. They are especially well suited to bind to the negatively charged DNA.

When chromatin is treated with micrococcal nuclease, individual nucleosomes can be isolated, indicating that the DNA between nucleosomes is accessible to digestion. The results of these studies

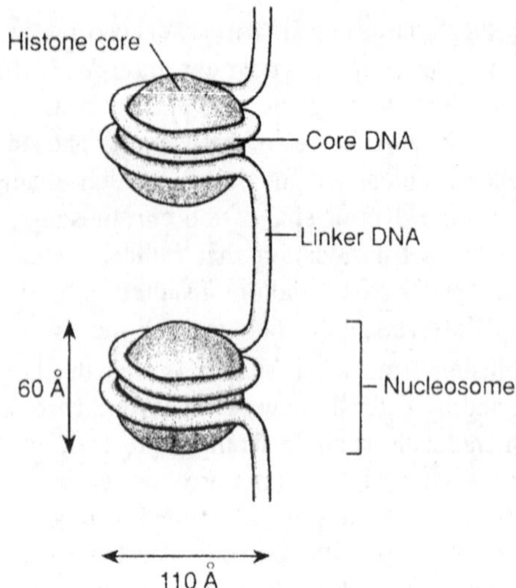

Fig. 5.2. The eukaryotic chromosome is associated with histone proteins to form nucleosomes.

indicate that a length of 168 base pairs (bp) of DNA, the core DNA, is intimately associated with the nucleosome, and another 50 to 75 base pairs, depending on species, connects the nucleosomes (linker DNA). When the quantities of the various histones were measured, there were two each of histones H2A, H2B, H3, and H4 per nucleosome and only one molecule of histone H1. Reconstitution and degradation studies have indicated that histone H1 is not a necessary component in the formation of nucleosomes. We believe that histone H1 is associated with the linker DNA as it enters and emerges from the nucleosome, although its exact position is not known with certainty. Histone H1 may be more off center and internally located than illustrated. The term *chromatosome* has been suggested for the core nucleosome plus the H1 protein, a unit that includes approximately 168 base pairs of DNA. Nucleosomes, then, are a first-order packaging of DNA; they reduce its length and undoubtedly make the coiling and contraction required during mitosis and meiosis more efficient.

When DNA is replicated, twice as many nucleosomes are needed since one double helix becomes two. Recent studies indicate that a parental nucleosome is partly disassembled during DNA replication and reassembled on one or the other daughter strand, apparently randomly. The other DNA strand has a new nucleosome constructed of histones from the cellular pool with the help of proteins called *chromatin assembly factors*; at least three of these factors are known.

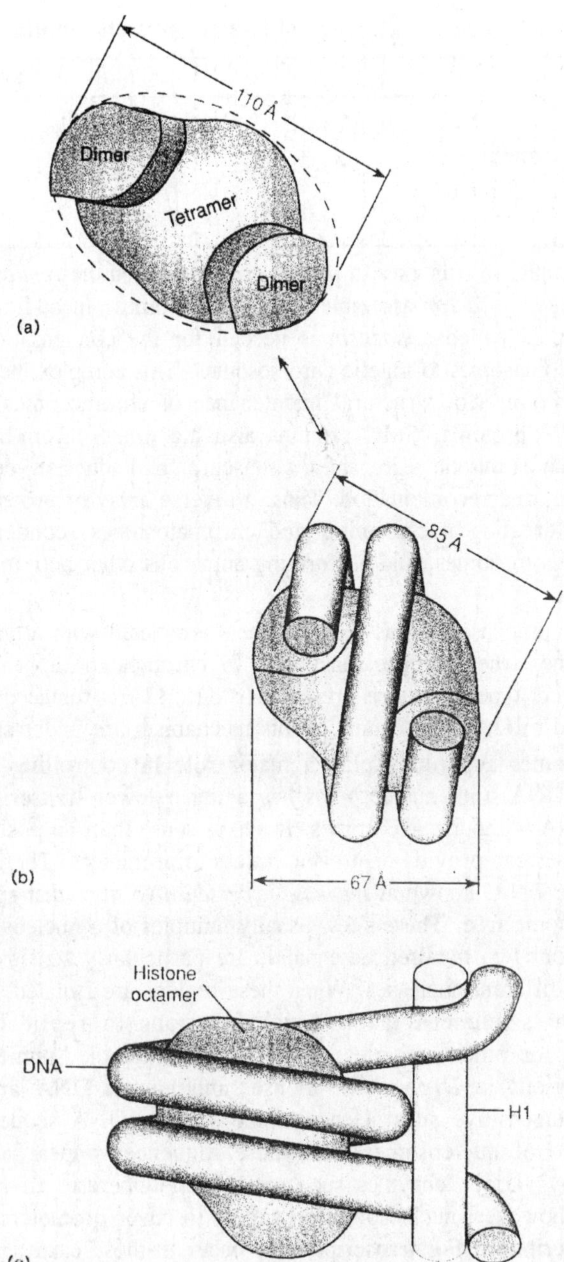

Fig. 5.3. Nucleosome structure.

Table 5.1. The constituency of calf thymus chromatin

Constituent	*Relative Weight*
DNA	100
Histone proteins	114
Nonhistone proteins	33
RNA	1

For example, in fruit flies, a protein complex called the *replication-coupling assembly factor* assembles new nucleosomes. In addition, a protein complex called *condensin* is needed for the condensation of interphasechromosomes to mitotic chromosomes. This complex includes two *SMC proteins* (for structural maintenance of chromosomes) and two non-SMC proteins. SMC proteins also aid other chromosomal activities, such as mitotic segregation, sister-chromatid adhesion, dosage compensation, and recombination. Thus, a diverse array of proteins is involved in creating nucleosomes and chromatosomes, condensing interphase chromosomes, and performing numerous other activities of chromosomes.

Dosage compensation has recently been associated with a change in nucleosome structure. The inactivated X chromosome appears to have a different type of histone present. Histone H2A is replaced by a variant called mH2A. The details of this mechanism are under study.

Nucleosomes apparently play a major role in controlling gene expression; DNA with nucleosomes has a much lower transcription rate than DNA without nucleosomes. It makes sense that the positions of nucleosomes can provide or prevent access to promoters. There are regions of the DNA, known as *nuclease-hypersensitive sites*, that appear to be nucleosome free. These sites, usually mutiples of a nucleosomal region of about two hundred base pairs, are particularly sensitive to digestion by different nucleases. When these regions are isolated, they usually have sequences that control functions in replication, transcription, or other activities of DNA. For example, numerous promoter regions in *Drosophila*, mouse, and human DNA are in nuclease-hypersensitive sites. Hence, some specific DNA sequences are kept free of nucleosomes, and these sequences appear to be recognized by various enzymes such as RNA polymerase. In many other cases, however, nucleosomes do appear to cover promoters and repress transcription. For transcription to occur in these cases, some form of *chromatin remodeling* must take place.

Two general classes of proteins are involved in chromatin remodeling. First are proteins that acetylate the N-terminal tails of the histones, a process that may cause the nucleosomes to bind the DNA less tightly and thus make it available for attachment of transcription factors. These enzymes are called *histone acetyl transferases* (HATs). Deacetylating enzymes have the reverse effect: They act to repress transcription. Second, a class of ATP-dependent proteins such as the SWI/SNF complex in yeast also affect chromatin remodeling. (Some workers called the proteins SWI because they were involved in mating type switching, and others called them SNF for sucrose nonfermenting.) The SWI/SNF complex is a group of eleven proteins involved in transcription activation in many genes, presumably allowing transcription factors to access promoters by remodeling chromatin. These proteins are able to reposition a nucleosome on DNA by sliding the nucleosome down the DNA.

We thus conclude that although nucleosomes serve as a general, first-order packing mechanism in eukaryotic DNA, they can be positioned precisely and can attenuate transcription. It is interesting to note that once transcription begins, RNA polymerase apparently moves along nucleosomed DNA by translocation of the histones by 75 to 80 base pairs without disrupting the nucleosome itself. This seems to be accomplished by the RNA polymerase moving the DNA and then re-forming the nucleosome in its wake.

Higher-order structure of chromatin

Since the nucleosome has a width of only 110 Å, and metaphase chromosomes appear to be constructed of a fiber having a diameter of about 2,400 Å, several additional levels of chromatin compaction lead to the metaphase chromosome. Various experiments, which change the ionic strength the chromatin is subjected to, indicate that the 110 Å DNA spontaneously forms a 300 Å, solenoidlike fiber with increased ionic strength. It seems that this fiber results from the coiling of the nucleosomal DNA. This 300 Å fiber is not, however, the final form of the DNA. We can account for the contraction of the 300 Å fiber to the 2,400 Å fiber found in metaphase chromosomes by the formation of a second solenoidlike structure from the winding of the 300 Å fiber.

If the histones are removed from a chromosome, the DNA billows out, leaving a proteinaceous structure termed a *scaffold*. This scaffold structure is formed from *nonhistone proteins*; two of them predominate, namely SC1 and SC2. SC1 has been identified as topoisomerase II. It

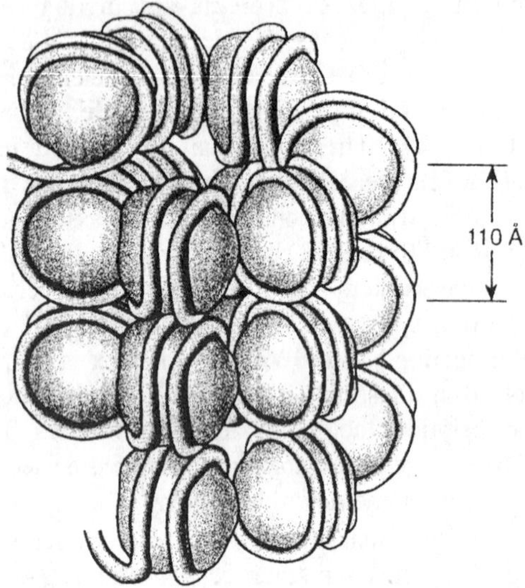

Fig. 5.4. Solenoid model for the formation of the 300-Å chromatin fiber.

would not be unreasonable to expect several hundred different proteins, many in minute quantities, to be associated with the chromosome and involved in replication, repair, and transcription.

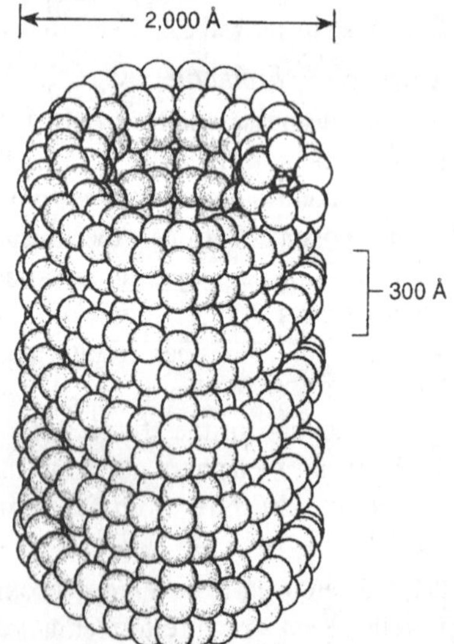

Fig. 5.5. The 2,000-Å fiber of the eukaryotic chromosome is a solenoidlike structure formed by the coiling of the 300-Å fiber, which itself is a solenoid.

Polyteny, puffs, and balbiani rings

Drosophila's salivary glands, as well as some other tissues of *Drosophila* and other diptera, contain giant banded chromosomes that result from the replication of the chromosomes and the synapsis of homologues without cell division (endomitosis). These chromosomes

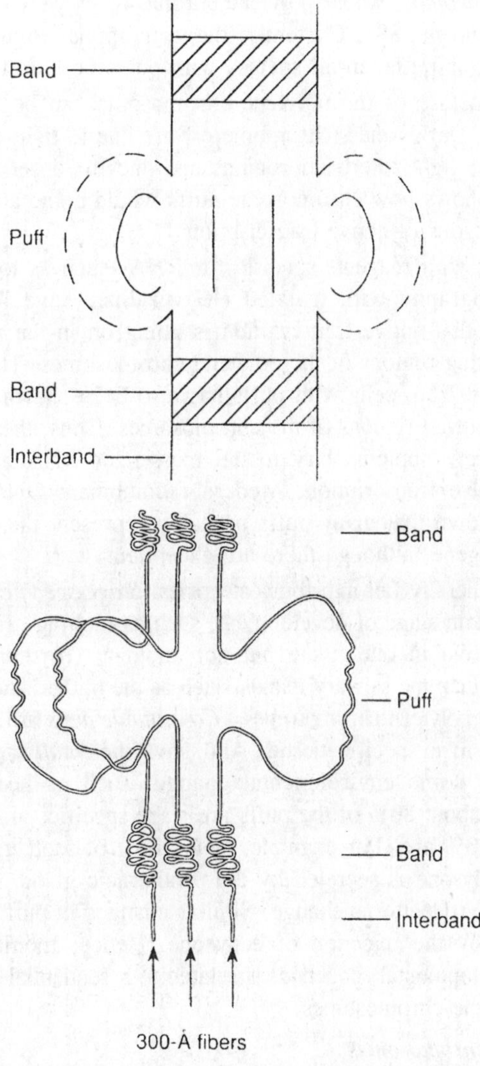

Fig. 5.6. Polytene chromosome with bands and a puff. Three of the approximately one thousand synapsed chromatids are shown diagrammatically on the below.

consist of more than one thousand copies of the same chromatid and appear as alternating dark bands and lighter interband regions. The dark bands are referred to as chromomeres. Also seen are diffuse areas called *chromosome puffs*. Chromosome puffs are also referred to as *Balbiani rings*. These rings were originally defined as puffs in the midge, *Chironomus*, whose polytene chromosomes were discovered by E. G. Balbiani in 1881. Currently, the term applies to all puffs, or at least the larger puffs, in all species with polytene chromosomes.

The structure of the polytene chromosome can be explained by the diagram. Dark bands (chromomeres) are due to tight coiling of the 300 Å fiber; light interband regions are due to looser coiling. The figure also shows how chromosome puffs would come about as fibers unfold in regions of active transcription.

Staining with reagents specific for RNA, such as toluidine blue, or autoradiography with tritiated (^{3}H) uridine, have been used to demonstrate that active transcription is going on in the puffs but not in neighbouring regions of the polytene chromosomes. The messenger RNA isolated from cells with puffs has also been shown to hybridize only to the puffed regions of the chromosomes. Thus, these regions of the DNA are complementary to the messenger RNA and represent areas of active transcription. Modern recombinant DNA techniques have also shown that many puffs probably represent the transcription of only one gene, although there are exceptions.

Puffs generally fall into four categories. *Stage-specific puffs* appear during a certain stage of development, such as molting. *Tissue-specific puffs* are active in one tissue but not another. (In dipteran larvae, tissues other than the salivary glands, such as the midgut and Malpighian tubules, have polytene chromosomes.) *Constitutive puffs* are active almost all the time in a specific tissue. And *environmentally induced puffs* appear after some environmental change, such as heat shock. In *Drosophila*, about 80% of the puffs are stage specific; in *Chironomus*, only about 20% are. For example, at the time of molt in insects, the hormone ecdysone is secreted by the prothoracic gland. At the same time, many puff patterns change. Similar changes in puff patterns can be induced by the injection of ecdysone. Hence, molting, a stage-specific developmental sequence, is related to a sequential transcription sequence in the chromosomes.

Lampbrush chromosomes

Lampbrush chromosomes, which occur in amphibian oocytes, are so named because their looped-out configuration has the appearance of a brush for cleaning lamps, now a relatively uncommon household item. The loops of the lampbrush chromosomes are covered by an RNA matrix and are the sites of active transcription. Presumably, the loops are unwindings of the single chromosome, similar to the unwindings in the polytene chromosome. Thus, under certain circumstances, such as in polytene chromosomal puffs and in lampbrush chromosomes, active transcription can be seen in the light microscope. Since only certain bands puff at any one moment in polytene chromosomes, and since the loops of lampbrush chromosomes are of various sizes (with some regions not looped at all), we have evidence of specific transcription. However, we have no indication, so far, of the nature of the control of that transcription.

Chromosomal Banding

Several chromosomal staining techniques reveal consistent banding patterns. By means of these patterns, all of the human chromosomes can be differentiated. Of possibly greater importance is the fact that these staining techniques have provided some insight into the structure of the chromosome. The techniques for staining the C, G, and R chromosomal bands will serve as an illustration.

G-bands are obtained with *Giemsa stain*, a complex of stains specific for the phosphate groups of DNA. Treatment of fixed chromatin with trypsin or hot salts brings out the G-bands. Giemsa stain enhances banding that is already visible in mitotic chromosomes. The banding pattern is caused by the arrangement of chromomeres. Under careful observation, the major G-bands prove to consist of many smaller chromomeres. This banding appearance has led D.

C-bands are Giemsa-stained bands after the chromosomes are treated with NaOH. The *C* is for "centromere," because these bands represent constitutive heterochromatin surrounding the centromeres. The DNA is also usually satellite rich. *Satellite DNA* differs in *buoyant density* from the major portion of cellular DNA. When eukaryotic DNA is isolated and centrifuged in CsCl, forming a density gradient, the majority of the DNA forms one band in the gradient at a single buoyant density. The buoyancy is determined by the G-C content of the DNA. However, smaller secondary bands are also usually present, indicating regions of DNA having sequences different from the majority of the cell's DNA. DNA isolated this way is referred to as *satellite DNA* because of the secondary, or satellite, bands formed in the density gradient. As we will see, this DNA is found primarily around centromeres and consists of numerous repetitions of a short sequence

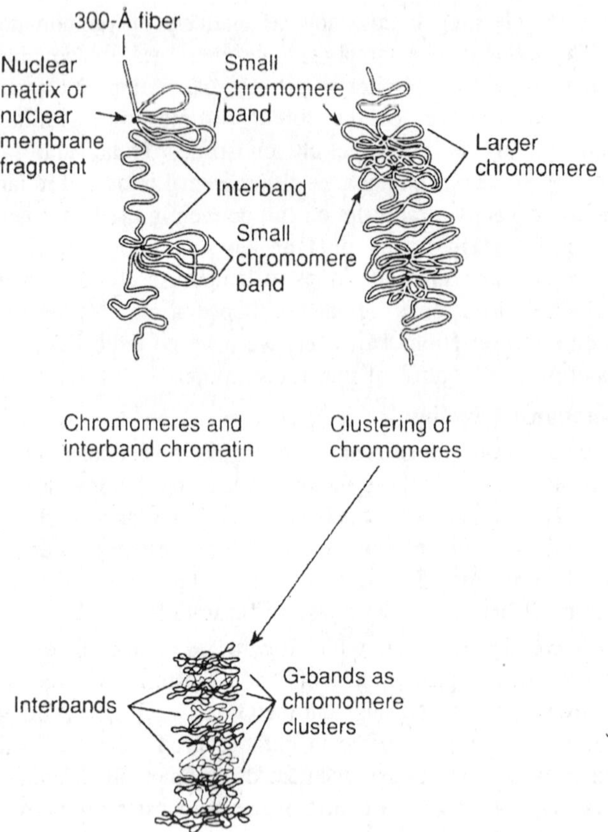

Fig. 5.7. Model of eukaryotic (mammalian) chromosomal banding.

R-bands are visible with a technique that stains the regions between G-bands. The chromosomes are fixed, stained with Giemsa, and then viewed with a phase contrast microscope. Since the dark-light pattern is the opposite of the G-band pattern, these bands are called *reverse bands*.

From the information gleaned from these staining techniques, D. Comings distinguished between three basic chromatin types: euchromatin, constitutive heterochromatin, and intercalary heterochromatin. Presumably, the only chromatin involved in transcription is *euchromatin*. *Constitutive heterochromatin* surrounds the centromere and is rich in satellite DNA. *Intercalary heterochromatin* is found throughout the chromosome. Thus, it becomes apparent that the eukaryotic chromosome is a relatively complex structure.

Centromeres and Telomeres

Centromeres

Two regions of the eukaryotic chromosome have specific functions—the centromere and the telomeres. The centromere is involved in chromosomal movement during mitosis and meiosis, whereas the telomeres terminate the chromosomes. As we pointed out, the terms *centromere* and *kinetochore*, while occasionally used interchangeably, are distinct. The kinetochore is the interface between the visible constriction in the chromosome (the centromere) and the microtubules of the spindle. The kinetochore of higher organisms (e.g., mammals) contains proteins and some RNA. Microscopically, it is a trilaminar structure, attached to chromatin at the inner layer and to microtubules at the outer layer.

Most of our knowledge of the genetics of centromeres has come from work in yeast (*Saccharomyces cerevisiae*). Cells did not maintain most artificially created yeast plasmids because they were lost during mitosis. However, plasmids were isolated that did replicate normally during cell division. Presumably, they contained centromeres, allowing them to replicate and move in synchrony with the host's chromosomes. Further genetic engineering made it possible to isolate smaller and smaller regions that could serve as centromeres. After sequencing the centromeres of fifteen of the sixteen yeast chromosomes, it was possible to conclude that the centromere from yeast is about 250 base pairs long with three consensus regions; we are defining a centromere as a sequence of DNA called the *CEN* locus or CEN region. Recent data indicate that this region may contain a single, modified nucleosome associated with region II. The 250 base-pair length of the CEN regions of yeast chromosomes is about 200 Å, the same as the diameter of a microtubule, indicating that only one micro-tubule attaches to each centromere during mitosis or meiosis in a yeast cell. This region is called a *point centromere*. Higher eukaryotes have larger centromeric regions that attach more microtubules. These regions are referred to as *regional centromeres*. Regional centromeres range from nineteen to one hundred kilobases (kb; 19,000–100,000 bases) with unique and satellite (repeated sequence) DNA that is heterochromatic and may include expressed genes. We know much less about regional centromeres than we do about point centromeres.

Telomeres

Since eukaryotic chromosomes are linear, each has two ends, referred to as *telomeres*, that not only mark the termination of the linear chromosome but also have several specific functions. Telomeres must prevent the chromosomal ends from acting in a “sticky” fashion, the way that broken chromosomal ends act. In other words, chromosomal ends must not elicit a DNA repair response. Telomeres must also prevent the ends of chromosomes from being degraded by exonucleases and must allow chromosomal ends to be properly replicated.

Most telomeres isolated so far are repetitions of sequences of five to eight bases. In human beings, the telomeric sequence is TTAGGG, repeated 300 to 5,000 times at the end of each chromosome. The human telomere was discovered by R. Moyzis and his colleagues when they probed the highly repetitive segment of human DNA. (Highly repetitive DNA, as its name implies, consists of numerous copies of a single sequence and usually comprises the satellite components of the cell’s DNA; see next section.) When a probe for this sequence was applied to human chromosomes, the sequence was found at the tip of each chromosome in roughly the same quantity.This is a highly conserved sequence, found in all vertebrates studied as well as in unicellular trypanosomes. Similar sequences are found in various other eukaryotes; the first sequence was isolated by E. Blackburn and J. Gall in 1978.

When a linear DNA molecule is replicated, the 3' → 5' strand can be replicated to the end. The 5' → 3' strand, however, is replicated with RNA primers that are then degraded, leaving a short gap on the progeny strand. It is always the G-rich strand of telomeric DNA that ends up single-stranded, forming a 3' overhang of twelve to sixteen nucleotides. Thus, the normal replication process of a linear DNA molecule leaves an incomplete terminus. Hence, scientists suspected that there would be a unique mechanism for the replication of telomeres.

Telomeric sequences appear to be added de novo without, DNA template assistance by an enzyme called *telomerase*, discovered by E. Blackburn and her colleagues. This was seen when telomeres from another species were engineered into yeast cells. After a cell cycle, the yeast telomeric sequence had been added on at the ends of the foreign chromosome, the result, presumably, of the telomerase enzyme.

When Blackburn and her colleagues isolated telomerase, they discovered that a segment of RNA, about 160 base pairs, is an integral part of the enzyme. That RNA has a region that is complementary to the G-rich repeat of the telomeric DNA sequence of the species. After careful experimentation, including modifying the gene for the

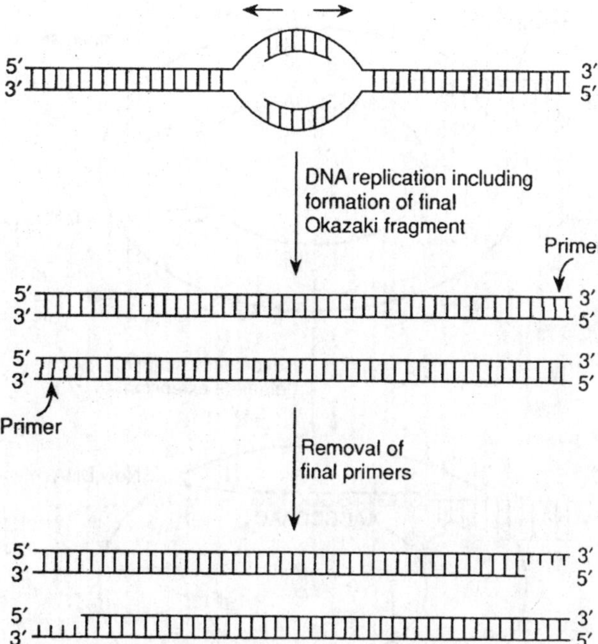

Fig. 5.8. Removal of final primers after the replication of linear DNA creates single-stranded ends.

telomerase RNA, Blackburn and her colleagues concluded that telomerase uses its RNA as a template for adding telomeric repeats to the ends of chromosomes. Telomerase is thus a reverse transcriptase, using RNA nucleotides as a template to polymerize DNA nucleotides.

Blackburn and her colleagues proposed that the first step in telomere extension is hybridization of the 3' end of the telomere with the RNA component of telomerase. Then, with the telomerase RNA as a template, the 3' end of the telomere is extended. Finally, a translocation step takes place that displaces the telomere in respect to the RNA, returning to the configuration at the beginning of the process. The single-stranded C-rich strand is then synthesized with DNA polymerase and DNA ligase.

Once telomeres have been added to the ends of eukaryotic chromosomes, different organisms use any of three different methods known to protect the ends of the chromosomes. First, the guanine-rich DNA can form complex structures. Biochemists have discovered that four guanines can form a planar *G-tetraplex*, with the four bases hydrogen bonded to each other. Several structures have been hypothesized

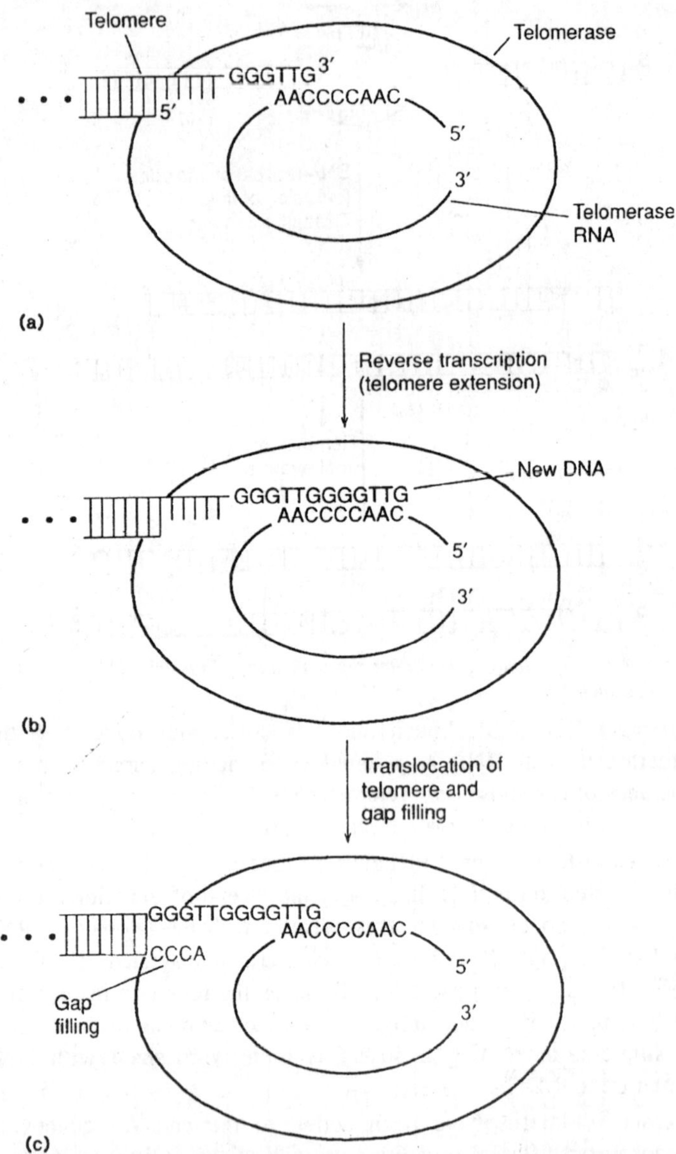

Fig. 5.9. Telomerase extends telomeres using telomerase RNA as a template Gap filling by DNA polymerase I and ligase complete the double helix.

to explain the novel ends of these chromosomes. Second, proteins have been discovered that bind to the 3' ends of telomeres. In the ciliate *Oxytricha nova*, a protein called the telomere end-binding protein (TEBP) attaches to the 3' ends of telomeres and protects them. Finally, a novel structure called the *t-loop* has been discovered at the end of mammalian telomeres. This loop forms at the ends of chromosomes under the direction of a protein called TRF2 (telomere repeat-binding factor), which causes the 3' end of the chromosome to loop around and interdigitate into the double helix, forming the loop.

How do cells keep track of the number of their telomeric repeats? Proteins have been isolated that bind to telomeres (Rap1 in *Saccharomyces cerevisiae*, TRF1 in human beings). By mutating these proteins or the telomeric sequences, scientists have changed the equilibrium number of telomeric repeats.This led to the current model that the cell counts the number of these proteins bound to the telomeres, not the number of telomeres directly, to know whether telomeres should be added. This is a very active area of research.

In yeast, protozoa, and other single-celled organisms, telomerase is active, keeping the ends of the chromo somes at the appropriate lengths. These cells can divide potentially forever. However, in most cells of higher organisms, telomerase is not active, and the ends of the chromosomes get shorter with each cell division. At a certain telomeric length, the cells no longer divide. However, if telomerase becomes active, and the ends of the chromosomes lengthen, a signal is conveyed to keep cells dividing, which can lead to cancerous growth. In fact, human telomerase was isolated from an immortal cell line (HeLa) derived from cervical cancer cells. Thus, attention is now turning to the possible clinical application of this knowledge: If telomerase can be deactivated in tumor cells, the cells may stop dividing or die, thereby eliminating the cancer. Further, studying normal telomere shortening, which appears to act as a biological clock, may help us understand the aging process and senescence.

C-Value Paradox

Why do eukaryotes have so much DNA, and why is there huge variation in the DNA content between species of comparable complexity? These questions define the *C-value paradox*, in which *C* refers to the quantity of DNA in a cell. For an example of the paradox, although human beings have 3.3 billion base pairs in the haploid genome, an amoeba has more than 200 billion base pairs. And although an average bony fish has over 300 billion base pairs of DNA in its haploid genome, the Japanese puffer fish has less than half a billion base pairs. If the basic bony fish pattern can be created with less than half a billion base pairs, why does the average bony fish have over 600 times that much DNA? What is this excess DNA doing? To explain the C-value paradox, researchers examined the repetitiveness of DNA, and more recently, probed and sequenced DNA to understand its properties.

DNA-DNA Hybridization

R. Britten and his colleagues, using the technique of *DNA-DNA hybridization*, first systematically analyzed the repetitiveness of the DNA within eukaryotes. When DNA is heated, it denatures or unwinds into single strands; when it cools, it renatures. The rate of renaturation depends on the DNA sequences. If the sample contains DNA with repeated sequences, it will hybridize faster than DNA that does not have repeated sequences. From these studies, Britten and his colleagues found that eukaryotic chromosomes contain regions of unique, moderately repetitive, and highly repetitive DNA. *Unique DNA* is, as its name implies, DNA with unrepeated sequences. *Repetitive DNA* is DNA whose sequences are repeated in the genome.

Satellite DNA, found around centromeres, is highly repetitive DNA with a unique repeat length of about two hundred base pairs. Given the quantity of satellite DNA per cell, there must be more than one million repetitions of this two-hundrednucleotide sequence in higher eukaryotes. At the other end of the spectrum is unique DNA, which makes up most of the transcribed genes of an organism. The rest of the DNA is repetitive DNA in a few to several hundred thousand copies. This repetitive DNA comprises at least three categories. One is "junk" DNA, DNA that is not useful to the organism, made up of untranscribed and parasitic sequences (selfish DNA). Another category is transcribed genes in many copies that have diverged from each other, such as antibody, collagen, and globin genes. We use the term *gene family* to refer to genes that have arisen by duplication, with or without divergence, from an ancestral gene. And finally, transcribed genes in many copies that are virtually identical, such as ribosomal RNA and histone genes, make up a third category of repetitive DNA.

Junk DNA

We saw that transposons in prokaryotes are generally viewed as selfish or parasitic: They serve no purpose to the cell. The transposons replicate on their own, increasing in number. Eukaryotic transposons are mostly *retrotransposons*, transposable elements that move by way of an RNA intermediate. That is, the retrotransposon is transcribed into RNA and then, by reverse transcription, converted to a cDNA that is then inserted into the genome. These elements can make up

50% of the eukaryotic genome, existing in hundreds of thousands of copies. They generally fall into two categories: LINES and SINES. *Long interspersed elements* (LINES), are up to seven thousand base pairs each and contain genes for reverse transcription, RNA binding, and endonuclease activity. They thus have the ability to jump by way of an RNA intermediate. Human DNA is believed to be composed of about 15% LINES.

Short interspersed elements (SINES) are generally derivatives of transfer RNA genes and do not have the ability to retrotranspose on their own. That is, in the past, their transcripts were modified, converted to cDNA by reverse transcription, and then reinserted into the host's genome. They rely on the reverse transcriptase provided by the genes of LINES or retroviruses. One group of SINEs not derived from transfer RNA is derived from the RNA of the signal recognition particle; members of this group occur in human beings in about five-hundred thousand copies of a three-hundred-basepair sequence. Because these sequences are cleaved by the restriction endonuclease *Alu*I, they are called the *Alu family*. The human genome is also permeated by remnants of at least a dozen distinct families of ancient retroviruses scattered throughout our chromosomes.

At this point, we can see some potential explanations for the C-value paradox. Much eukaryotic DNA is junk, apparently doing no harm. In some cases, 97% of the host genome is composed of junk DNA. Recent work seems to indicate that gross differences in DNA content between higher organisms may be due to the differing abilities of different species to rid themselves of this parasitic DNA. If it builds up without being removed, the DNA content of the species can soar. Thus, the wide differences in DNA content among higher eukaryotes mentioned at the beginning of this section have little to do with the complexity of the organism, but rather with the ability of the organism to remove junk DNA as it forms.

Expressed genes in many copies

Several types of genes create a product that is needed in such large quantity that one copy of the gene could not fulfill the cell's needs.We are familiar with the nucleolus, the site of the ribosomal RNA genes. Human beings have about two hundred copies of the major ribosomal RNA gene and about two thousand copies of the 5S ribosomal RNA gene. Fruit flies have about two hundred and one hundred copies, respectively, of the two genes. In some cases, the normal number of multiple copies of a gene is still not enough. The cell must then resort to *gene amplification*, a process whereby the cell increases the number of copies of the gene. For example, during oogenesis, ribosomal RNA genes (rDNA) are often amplified. In *Xenopus*, rDNA is amplified about one thousand times, which allows an oocyte to accumulate about 10^{12} ribosomes. The amplified DNA is in the form of small, circular, extrachromosomal molecules of DNA. Several models have been proposed as to how cells actually amplify their DNA. One model relies on unequal crossing over (as in *Bar* eye in *Drosophila*), whereas another model is based on unscheduled extra DNA replication in a region, followed by recombinational events that generate linear and circular forms of the excess DNA. It is not presently clear which model is correct.

In addition to ribosomal RNA genes, other genes are repeated, ensuring adequate gene products. The number and location of repeated genes are usually discovered by hybridization studies using probes, similar to the way that telomeric DNA was shown to be at the tips of the chromosomes. Repeated genes include the genes for transfer RNAs and histones. The average transfer RNA is repeated about a dozen times in *Drosophila*. Human beings have over thirteen hundred copies of transfer RNA genes in the haploid genome. In many species, the five histone genes form a repeated cluster, although each gene is transcribed independently, while prokaryotic operons are transcribed as a unit. The arrangement of histone genes may be more complex in higher forms. There are indications that in mammals, histone genes may lie in small groups or even as individual genes.

Several types of genes occur in similar but not identical forms—that is, an original gene was duplicated but, unlike histone or ribosomal RNA genes, the copies diverged in function. These gene families include globin genes, immunoglobulin genes, chorion protein (insect eggshell) genes, and *Drosophila* heat shock genes.

Globin gene family

Globins are oxygen-transporting and storage molecules found in animals, some plants, and microorganisms. In higher vertebrates, there are two types of globins: myoglobin, which stores oxygen in muscles, and hemoglobin, found in red blood cells. Myoglobins function as single molecules, whereas hemoglobins occur as tetramers, two each of two protein chains. Evolution in the globin gene family can be traced by comparative studies of globins in different species as well as molecular studies of globins within a species. Studying hemoglobins has provided a great deal of information on gene expression and evolution. We turn our attention to the globin gene family in human beings. During human development, four major hemoglobins appear: embryonic hemoglobin, Hb F, Hb A, and Hb A_2. Structurally, the ζ (Greek, zeta) subunit (a component of embryonic hemoglobin) is α-like, whereas the rest are β-like. Fetal hemoglobin has a higher affinity for oxygen than does adult hemoglobin, thus allowing fetuses to draw oxygen from their mother's blood. From a comparative study of the DNA sequences, the evolution of the various hemoglobin genes has been inferred.

The α genes are located in a cluster on chromosome 16; the β genes are located in a cluster on chromosome 11. These two clusters provide a clear case history of gene duplication, presumably by unequal crossing over, followed by divergence. Having a second or third copy of a gene allows one of the duplicates to diverge (and perhaps to become nonfunctional in the process), whereas the original still performs the required function.

Many diseases of genetic interest involve the hemoglobins. In fact, hemoglobinopathies, including sickle-cell anemia and the thalassemias, are the most common genetic disorders in the world population. The best-known mutation of a hemoglobin gene itself is the one that causes sickle-cell anemia, a mutation of the sixth amino acid of the β chain. In the homozygous state, the disease is usually fatal. However, heterozygotes show an increased resistance to malaria. One of the ramifications is that the sickle-cell allele is maintained at relatively high frequencies in malarial regions.

The *thalassemias* are a group of diseases that affect the regulation of the α and β hemoglobin genes. (*Thalassemia* comes from the Greek for "sea blood," because the disease is best known in individuals living around the Mediterranean Sea.) In α and β thalassemias, the α or β subunit, respectively, is present in very low quantities or entirely absent. Many of the genetic defects are deletions, possibly due to unequal crossing over within the globin gene complexes. T. Maniatis showed that β thalassemia is caused by a mutation in the β-globin gene that disrupts RNA splicing. The body compensates by forming γ_4 or β_4 hemoglobin in a thalassemias, or $\alpha_2\gamma_2$ or $\alpha_2\delta_2$ in β thalassemias. These are relatively unsuitable or inefficient responses; the diseases range from very mild to very severe and frequently fatal. More information is needed regarding the control of hemoglobin production in the thalassemias.

6

Genes and Chromosomes

Mendel's experiments made clear that the units of heredity are stable and particulate. However, at the time, the mechanics of the transmission of genes from one generation to the next were quite mysterious, and both the role of the nucleus in reproduction and the details of cell division were unknown. Once these phenomena became understood and chromosomes were seen by microscopy and recognized to be carriers of genes, new understanding came at a rapid pace. This chapter examines both the relation between chromosomes and genes and the mechanism of chromosome segregation during cell division.

Stability of Chromosome Complements

In the 1870s the importance of the nucleus and its contents was recognized by the observation that the nuclei of two gametes fuse in the process of fertilization. The next major advance was the discovery of *chromosomes*, which had been made visible by light microscopy, stained by basic dyes. Then, chromosomes were found to segregate into both gametes and daughter cells by an orderly process prior to cell division. Finally, three important regularities were observed about the chromosome complements (the complete set of chromosomes) of higher plants and animals:

1. The nucleus of each *somatic cell* (a cell of the body, in contrast with a *germ cell* or gamete) contains a fixed number of chromosomes typical of the particular species. However, the numbers vary tremendously among species and have little relationship to the complexity of the organism.
2. The chromosomes in the nuclei of somatic cells usually occur in pairs. Thus, the 46 chromosomes of humans consist of 7 pairs.

Furthermore, one chromosome of each pair comes from the maternal parent and the other from the paternal parent of the organism. Cells with nuclei of this sort, containing two similar sets of chromosomes are called *diploid*.

3. The germ cells of gametes that unite in fertilization to produce the diploid state of somatic cells have nuclei with only one set of chromosomes, consisting of one member of each pair—these nuclei are *haploid*.

The presence of a constant diploid chromosome number in cells of complex organisms that develop from single cells, and the formation of gametes with the haploid chromosome number indicate that there are two processes of nuclear division, one that maintains chromosome number—mitosis—and another that halves the number—meiosis. These two processes are examined in the following sections.

Mitosis

Mitosis is a precise process of nuclear division that ensures that each of two daughter cells receives a complement of chromosomes identical with the complement of the parent cell. The essential details of the process are the same in all organisms. Moreover, the basic process is remarkably simple: each chromosome, present as a doubled structure at the beginning of nuclear division, divides into identical halves that are separated from each other, and one of them goes into each of the two daughter nuclei that are formed.

In a cell not ready for mitosis chromosomes are not visible. This stage of the cell cycle is called *interphase*. In preparation for mitosis DNA synthesis occurs during a period of late interphase called *S*. DNA synthesis is accompanied by chromosome replication. Before and after S, there are periods, called G_1 and G_2, respectively, in which DNA synthesis does not occur. The *cell cycle*, or the life cycle of a cell, is commonly described in terms of these three interphase periods followed by mitosis, *M*—that is, $G_1 \rightarrow S \rightarrow G_2 \rightarrow M$. This is a somewhat arbitrary representation in which the final event in cell reproduction—The division of the cytoplasm into two approximately equal parts containing the daughter nuclei—is included in the M period.

The length of time required for a complete life cycle varies with cell type. In higher organisms the majority require 18-24 hr. The relative duration of the different periods in the cycle also varies considerably with cell type. Mitosis is usually the shortest period, requiring between 1/2 and 2 hr.

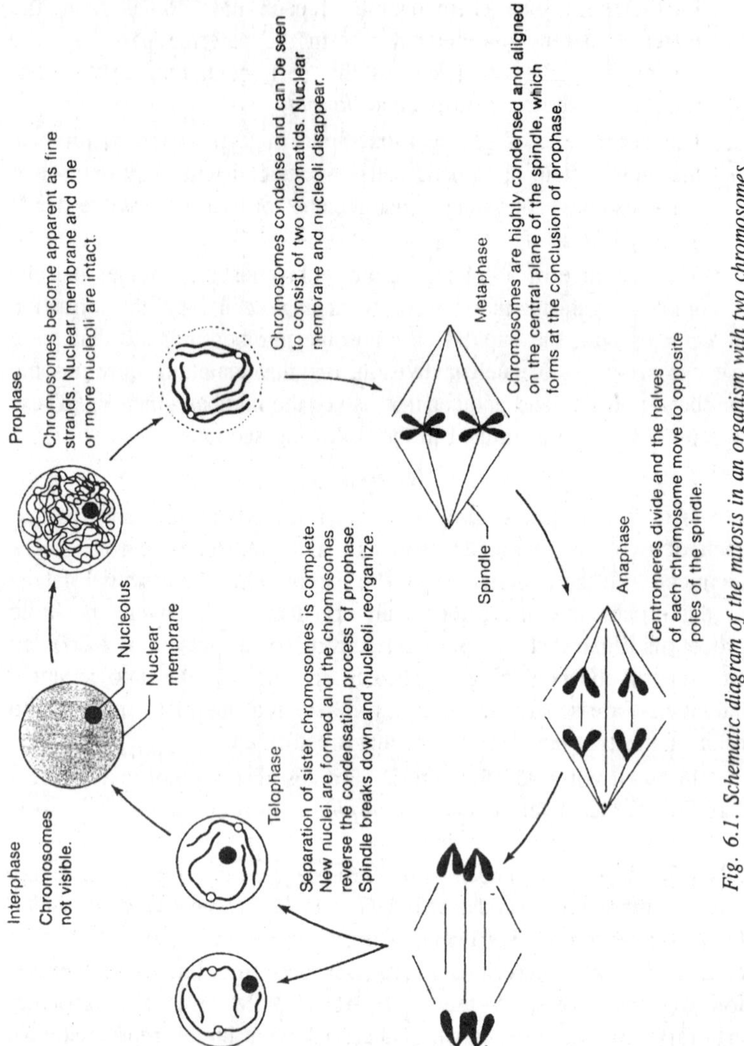

Fig. 6.1. Schematic diagram of the mitosis in an organism with two chromosomes.

The four stages—*prophase*, *metaphase*, *anaphase*, and *telophase*—into which mitosis is conventionally divided have the following characteristics:

Prophase

During interphase, the chromosomes have the form of extended filaments and cannot be seen as discrete bodies with a light microscope. Except for the presence of one or more conspicuous dark bodies (*nucleoli*), the nucleus has a diffuse, granular appearance. The beginning of prophase is marked by the condensation of chromosomes to form visibly distinct, thin threads within the nucleus. Occasionally, the chromosomes are seen rather early to be longitudinally double, consisting of two closely associated subunits called *chromatids*. Each pair of chromatids is the product of replication of one chromosome in the S period of interphase. The chromatids in a pair are held together at a specific region of the chromosome called the *centromere*. As prophase progresses, the chromosomes become shorter and thicker, as a result of intricate boiling. In later prophase, three phenomena occur: (1) the nucleoli disappear; (2) the membrane enclosing the nucleus disintegrates; and (3) the *mitotic spindle* forms. The spindle is a bipolar structure and consists of fiberlike bundles that extend between the poles. Each chromosome becomes attached to a spindle fiber at its centromere.

Metaphase

After the chromosomes are attached to spindle fibers they move toward the center of the cell, until all centromeres lie on a plane equidistant from the spindle poles. The period during which the chromosomes are located at the central plane of the spindle is called metaphase. At metaphase the chromosomes have reached their maximum contraction and are easiest to count and examine for difference in morphology.

Anaphase

In anaphase, the centromeres divide, and the two sister chromatids move toward opposite poles of the spindle. The movement results from a pulling to the poles by the spindle fibers that are attached to the centromeres. At the completion of anaphase, the newly separated chromosomes lie in two groups near the poles of the spindle. Each group contains the same number of chromosomes that was present in the original interphase nucleus.

Telophase

During telophase, a nuclear envelope forms around each compact group of chromosomes, the spindle disappears, nucleoli recognize, and the chromosomes undergo a reversal of the condensation process that occurred in prophase until they are no longer visible as discrete entities. Gradually, the two daughter nuclei assume a typical interphase appearance, and the cell divides.

Meiosis

Meiosis is a mode of cell division in which cells form that have *half* the number of chromosomes present in the premeiotic cell; typically cells with the haploid chromosome number are formed by division of a cell with the diploid number. Meiosis consists of two successive nuclear divisions. During the first division, the two members of each pair of chromosomes, which are said to be *homologous* to each other, become closely associated along their length. The homologous

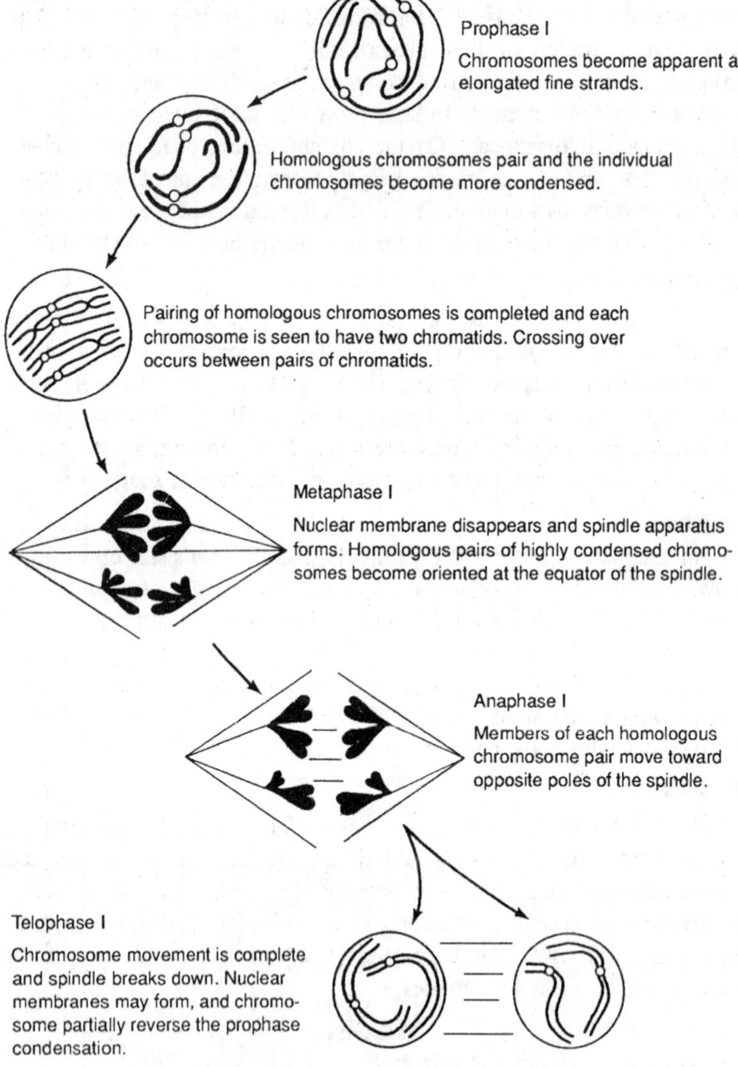

Fig. 6.2. First meiotic division.

chromosomes, each consisting of two sister chromatids that remain joined at the centromere, are then separated from one another into two nuclei, each containing a haploid set of duplex chromosomes. Without chromosome replication, a second division (resembling a mitotic division) occurs in which the chromatids of each chromosome are separated into different daughter nuclei. Consequently, the products of the two divisions are four daughter nuclei, each containing one complete haploid set of chromosomes that consists of a single chromatid from each pair of homologous chromosomes.

Although these details of meiosis are similar in all sexually reproducing organisms, only one of the four products develops into a

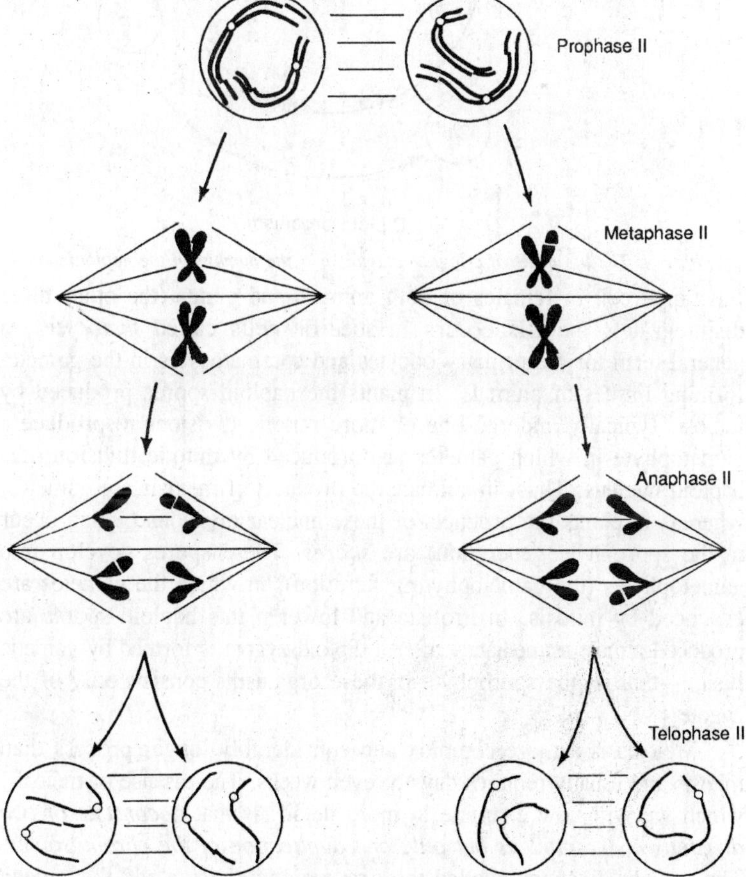

Fig. 6.3. Second meiotic division.

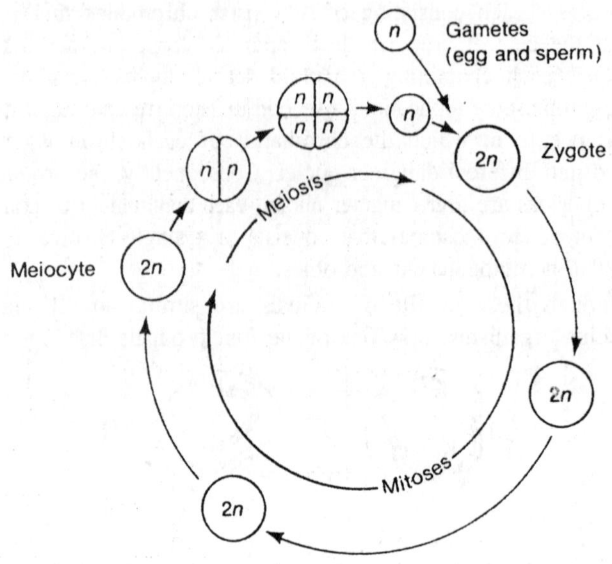

Fig. 6.4. The life cycle of a higher animal; n is the number in the haploid set.

functional cell in females of both animals and plants (the other three disintegrate); meiosis occurs in specific cells called *meiocytes*, a general term for the primary oocytes and spermatocytes in the gamete-forming tissues of animals. In plants the haploid spores produced by meiosis typically undergo one or more mitotic divisions to produce a gametophyte in which gametes are produced by mitotic division of a haploid nucleus. Thus, in animals the products of meiosis are gametes, whereas in plants the products of these nuclear divisions (which occur in the sporophyte generation) are *spores*. These spores develop into gametophytes (the gametophyte generation), in which the gametes are produced by mitosis. In Protista and lower plants haploid spores are produced almost immediately after a diploid zygote is formed by gametic fusion—that is, the sporophyte in these organisms consists only of the zygote.

Meiosis is a more complex and considerably longer process than mitosis and usually requires days or even weeks. The essence of meiosis, which we will now examine in more detail, is that *it consists of two divisions of the nucleus but only one duplication of the chromosomes*. The nuclear divisions—called the *first meiotic division* and the *second meiotic division*—can be separated into a sequence of stages similar

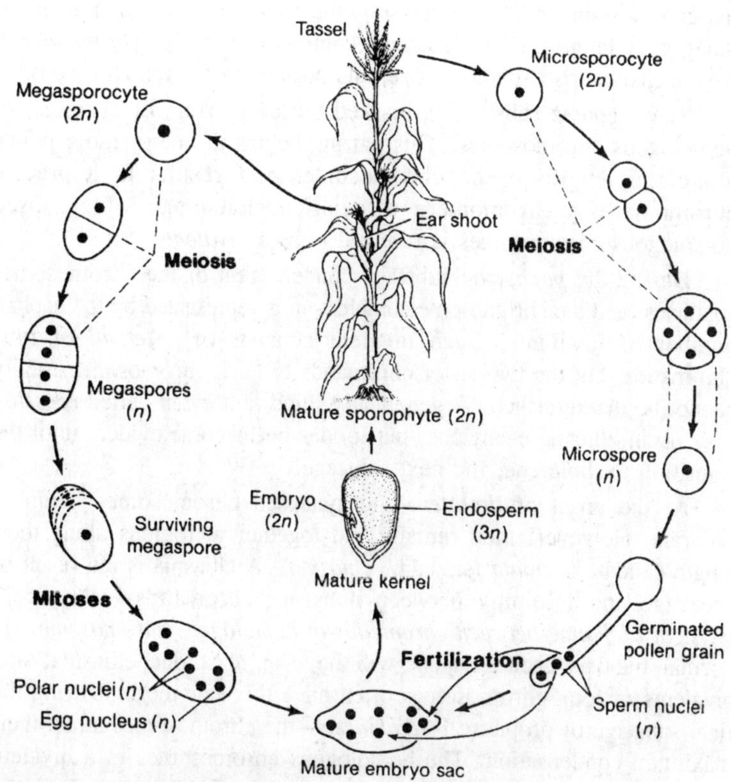

Fig. 6.5. The life cycle of corn Zea mays.

to those used to describe mitosis. The distinctive events of this important process occur during the first division of the nucleus; these events are described in the following section.

First Meiotic Division

The four main stages of the first meiotic division are called *prophase I*, *metaphase I*, *anaphase I*, and *telophase I*. These stages are generally more complex than their counterparts in mitosis.

Prophase I

This is a long stage, lasting several days in most higher organisms and commonly divided into five substages—*leptotene*, *zygotene*, *pachytene*, *diplotene*, and *diakinesis*.

In the *leptotene* substage the chromosomes become visible as long, threadlike structures. The pairs of sister chromatids can be distinguished by electron microscopy. During this initial phase in condensation of the chromosomes numerous dense granules occur at irregular intervals along their length. These localized contractions, called *chromomeres*, have a characteristic number, size, and position in a given chromosome.

The *zygotene* substage is marked by the pairing, or *synapsis*, of homologous chromosomes. This pairing begins at one or more points along the length of the chromosomes and results in a precise chromomere-by-chromomere association. Each pair of synapsed homologous chromosomes is referred to as a *bivalent*.

During the *pachytene* substage condensation of the chromosomes continues, and the chromosome complement is represented by the haploid number of bivalents. Each bivalent consists of a *tetrad* of four chromatids, but the two sister chromatids of each chromosome usually cannot be distinguished. A genetically important event called *crossing over* occurs during pachytene, but it does not become evident until the transition to diplotene, the next substage.

At the onset of *diplotene* the synapsed chromosomes begin to separate. However, they remain held together at regions along their length called *chiasmata* (singular, *chiasma*). A chiasma is the result of breakage and rejoining between nonsister chromatids— that is, *a physical exchange between chromatids of homologous chromosomes*. In normal meiosis each bivalent will have at least one chiasma, and bivalents of long chromosomes often have three or more. During the final substage of prophase I—*diakinesis*—the chromosomes attain their maximum condensation. The homologous chromosomes in a bivalent remain connected by one or two chiasmata, which persist until the first meiotic anaphase. Near the end of diakinesis the formation of a spindle is initiated and the nuclear envelope breaks down.

Metaphase I

There is a general similarity between this stage and a mitotic metaphase. The bivalents become positioned with the centromeres of the two homologous chromosomes on opposite sides of the plane through the middle of the spindle. The co-orientation of the undivided centromeres of each bivalent relative to the two poles of the spindle occurs at *random* and determines the member of each pair of chromosomes that will subsequently move to a particular pole.

Anaphase I

During this stage the homologous chromosomes, each composed of two chromatids joined at an undivided centromere, separate from one another and move to opposite poles of the spindle.

Telophase I

At the completion of anaphase I a haploid set of chromosomes consisting of one homologue from each bivalent is located near each pole of the spindle. During telophase the spindle breaks down and, depending on the species, either a nuclear envelope briefly forms around each group of chromosomes or the chromosomes enter the second meiotic division after only a limited uncoiling.

Second Meiotic Division

In some species the chromosomes pass directly from telophase I to *prophase II* without loss of condensation; in others, there is an interkinesis stage between the two meiotic divisions. *Chromosome replication never occurs between the two divisions*; the chromosomes present at the beginning of the second division are identical to those present at the end of the first division. After a short prophase (prophase II) and the formation of second-division spindles, the centromeres of the chromosomes in each nucleus become aligned on the central plane of the spindle at *metaphase II*. During *anaphase II* the centromeres (replicated in the *first* division) separate and the chromatids of each chromosome move to opposite poles of the spindle. *Telophase II* is marked by a transition to the interphase condition of the chromosomes in the four haploid nuclei accompanied by division of the cytoplasm. Thus, the second meiotic division superficially resembles a mitotic division. However, there is an important difference: *the chromatids of a chromosome are usually not identical sister along their entire length because of the occurrence of crossing over associated with the formation of chiasmata during prophase of the first division.*

Chromosomes and Heredity

The first clear proof that genes are parts of chromosomes was obtained in experiments concerned with the pattern of transmission of the *sex chromosomes*, the chromosomes responsible for the determination of the separate sexes in some plants and in almost all higher animals. These results are examine in this section.

Some Additional Terminology

In this chapter we will be using data from *Drosophila* genetics. We begin by describing the usual notation employed for this organism.

A superscript + (as in a^+) is used to identify the *wildtype* allele of any gene. Wildtype means that the allele is the one found most commonly in natural populations of the organism. A recessive allele is indicated by a small letter, and a dominant allele by a capital letter, so the dominance relations of the allelic forms of the gene are immediately clear. For example, y^+ is the symbol for the wildtype counterpart of recessive allele *y* determining yellow body colour; whereas Cy^+ is the symbols for the wildtype counterpart of another gene for which the dominant allele, *Cy* (Curly), results in wings that are curled. The letters used for gene symbols usually come from a descriptive term for the trait determined by an altered form of the gene. A common practice is to represent the wildtype allele of a gene only with a + sign (omitting the letter), as in the heterozygous genotype *y*/+, when it is clear from the context which gene is being designated. These conventions are also used with many other species, though most plant geneticists elect to represent recessive alleles by small letters and dominant alleles by initial capital letters without designating the wildtype. In this book the notations used will be those in general use for whichever organism is being considered.

Chromosomal Determination of Sex

The sex chromosomes are an exception to the rule that all chromosomes of diploid organisms are present in pairs of morphologically similar homologues. Early microscopic analysis showed that one of the chromosomes in males of some insect species does not have a homologue. This unpaired chromosome, called the *X chromosome*, was found to divide in only one of the meiotic divisions and to be present in only half of the sperm cells produced. The biological significance of these observations become clear when females of the same species were shown to have two X chromosomes. In other species in which the females also have two X chromosomes, it was observed that a morphologically different chromosome, referred to as the *Y chromosome*, is present with the X in males and pairs with it during meiosis. These differences in the chromosomal constitution of males and females were recognized to represent a method for the determination of sex at the time of fertilization. That is, whereas every egg will contain an X chromosome, only half of the sperm will have an X and the other half will, in some organisms, have a Y or, in some organisms, no sex chromosome. Fertilization of an egg by a sperm carrying an X results in an XX zygote, which develops into a female, and fertilization by a sperm having no X produces an XY (if a Y is present) or an X0 zygote (if no Y is present), which develops into a male. The result is a criss-cross pattern of inheritance of the X chromosome, in which a male receives his only X chromosome from his mother and transmits it only to his daughters. In organisms with this type of chromosomal sex determination—which is now known to occur in mammals including humans, many insects and other animals, and some flowering plants—the female is called the *homogametic* sex and the male the *heterogametic* sex because they produce one and two types of gametes, respectively.

Assuming that the union of gametes in fertilization occurs at random, as Mendel had supposed, the inheritance of the sex chromosomes (one X and one Y) nearly explains the 1:1 sex ratio usually observed, However, other genes controlling various processes involved in sexual development occur in the other chromosome pairs. These nonsex chromosomes are collectively called *autosomes*, and they are present equally in the two sexes. Many genes with functions unrelated to sex are also located in the X chromosomes, as will be seen in the next section. In most organisms, including humans, the Y chromosome carries few genes other than those associated with sex determination.

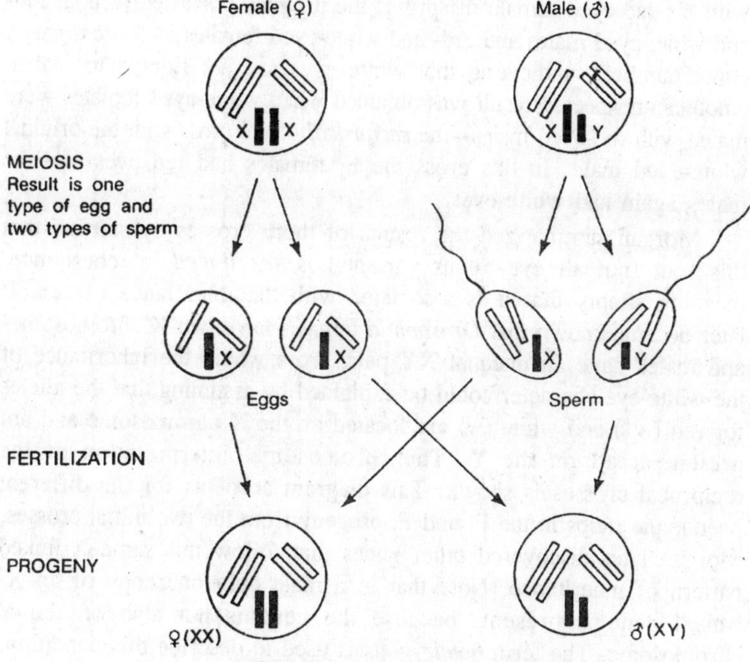

Fig. 6.6. The chromosomal basis of sex determination found in mammals, many insects, and other animals.

X-linked Inheritance

Crosses of the kind considered yield similar progeny when the genotypes of the male and female parents are reversed—this is, in *reciprocal crosses*. One of the earliest observations of an exception to this behaviour was made by Thomas Hunt Morgan in 1910, in an early study of the fruit fly, *Drosophila melanogaster*.

The normal eye colour of the fly is brick red. In a laboratory population that had been maintained for many generations, Morgan found a single male with white eyes. In a mating of this male with red-eyed females, all the F_1 progeny had red eyes, as would be expected if the allele for white is recessive. In the F_2 produced by the mating of F_1 males and females there were 2459 red-eyed females, 1011 red-eyed males, and 782 white-eyed males. These numbers represent a rather poor fit to the expected 3:1 ratio of red- versus white-eyed phenotypes. Note also that all of the white-eyed flies were males, a surprising result. Furthermore, when a white-eyed male was crossed with his red-eyed female offspring, the progeny consisted of both red- and white-eyed males and red- and white-eyed females in approximately equal numbers, indicating that white eyes are not limited to males. Another unexpected result was obtained when white-eyed females were mated with red-eyed males—the *reciprocal* of the cross with the original white-eyed male. In this cross the F_1 females had red eyes and the males again had white eyes.

Morgan summarized the results of these crosses by starting that this trait (not all eye-colour variants) is *sex linked* in inheritance, meaning simply that it is associated with the inheritance of sex. It later became known that *Drosophila* females have two X chromosomes and males have an unequal XY pair, from which the inheritance of the white-eye character could be explained by assuming that the alleles for red (w^+) and white (w) are located on the X chromosome and not present at all on the Y. The chromosomal interpretation of the reciprocal crosses is shown. This diagram accounts for the different phenotypic ratios in the F_1 and F_2 progeny from the two initial crosses. Morgan later discovered other genes that follow this same X-linked pattern of inheritance. Note that in a male only one copy of an X-linked gene is present, because the gene is not also on the Y chromosome. The term *hemizygous* is used to describe this condition.

Let us summarize the signs of sex linked:

1. A heterozygous female will, on the average, transmit the recessive allele to half of her daughters and half of her sons.

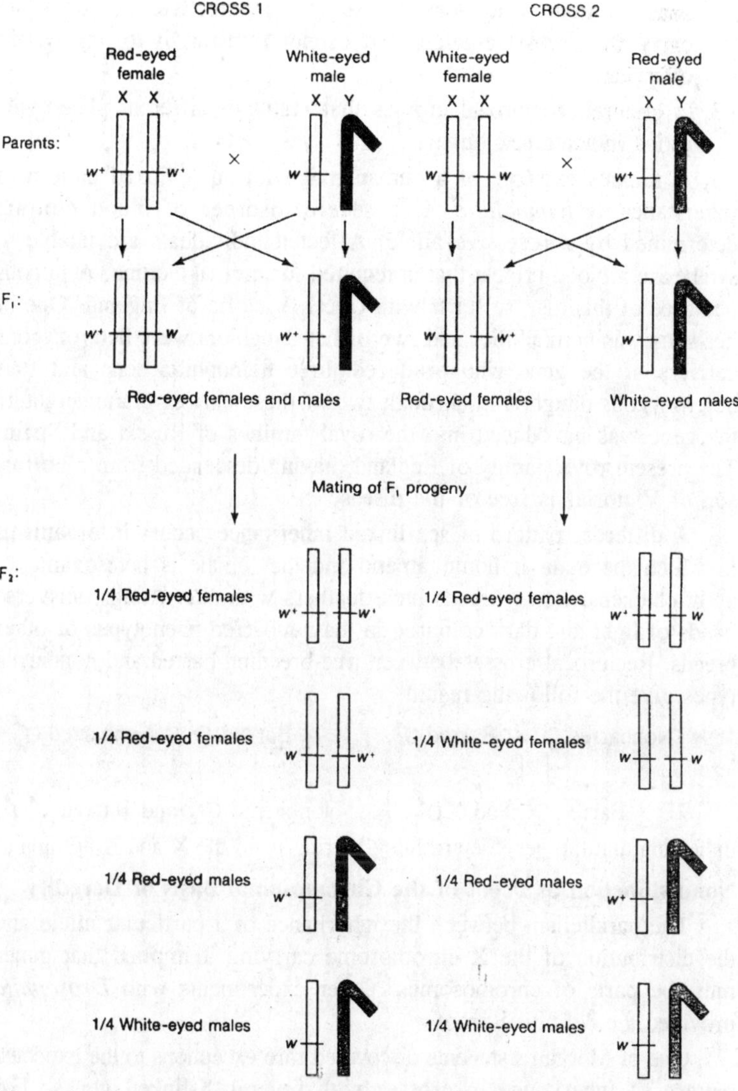

Fig. 6.7. A chromosomal interpretation of the results obtained in F_1 and F_2 progenies when a Drosophila female with red eyes is crossed with a white-eyed male (cross 1) and when the reciprocal cross of white-eyed female with red-eyed male (cross 2) is made.

2. Since a male has only one X chromosome, he will show the defective trait if the recessive allele is present in that chromosome, and will transmit the allele to all of his daughters but none of his

sons. A male who does not show the defective trait does not carry the recessive allele and cannot transmit it to any of his offspring.

3. In general, reciprocal crosses that result in different phenotypic ratios indicate sex linkage.

A famous example of a human trait with an X-linked pattern of inheritance is *hemophilia A*, a severe disorder of blood clotting determined by a recessive allele. Affected individuals are unable to synthesize a blood protein that is required for normal clotting. A famous pedigree of this disease starts with Queen Victoria of England. One of her sons was hemophilic, and two of her daughters were heterozygous carriers of the gene who produced three hemophilic sons and four heterozygous daughters. Through two of these carrier granddaughters the gene was introduced into the royal families of Russia and Spain. The present royal family of England, having descended from a normal son of Victoria, is free of the disease.

A different pattern of sex-linked inheritance occurs in organisms in which the male is homogametic and the female is heterogametic, as in chickens. Some breeds have feathers with alternating transverse bands of light and dark coloured in the nonbarred phenotypes of other breeds. Reciprocal crosses between true-breeding barred and nonbarred types give the following results

Nonbarred ♀ × Barred ♂	Barred ♀ × Nonbarred ♂
↓	↓
Barred ♀♀ and ♂♂	Nonbarred ♀♀ and Barred ♂♂

indicating that the gene determining barring is on the X and is dominant.

Nondisjunction as Proof of the Chromosomal Basis of Heredity

The parallelism between the inheritance of a particular allele and the distribution of the X chromosome carrying it implied that genes must be parts of chromosomes. Other experiments with *Drosophila* provided the definitive proof.

One of Morgan's students discovered rare exceptions to the expected pattern of inheritance in crosses with several X-linked genes. For example, when white-eyed females were mated with red-eyed males, most of the progeny consisted of the expected red-eyed females and white-eyed males. However, about one in every 2000 F_1 flies was an exception, either a white-eyed female or a red-eyed male. It was proposed that these rare exceptional offspring could be accounted for by occasional failure of the two X chromosomes in the mother to

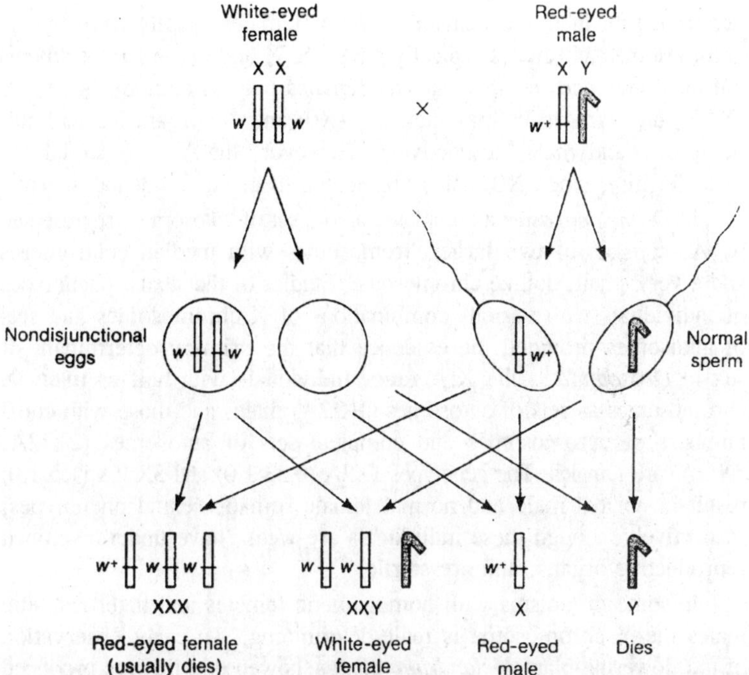

Fig. 6.8. The results of meiotic nondisjunction of the X chromosomes in a Drosophila female.

separate from each other during meiosis—a phenomenon called *nondisjunction*. The consequence of such failure is the formation of some eggs with two X chromosomes and others with none. Four classes of zygotes are expected from the fertilization of these abnormal eggs. Individuals with no X chromosome were not detected because embryos lacking this chromosome are not viable; most progeny with three X chromosomes also die during early development. Examination by microscopy indicated that the exceptional white-eyed females had two X chromosomes and a Y chromosome, and exceptional red-eyed males had a single X but no Y. These X0 males were sterile and could not be used for further genetic analysis.

The results of this and related experiments were of great importance in the development of genetics, because they showed that genes are associated with chromosomes.

Sex Determination in *Drosophila*

The sexual phenotypes resulting from nondisjunction in *D. melanogaster* established that the Y chromosome has no effect on sex determination in this organism. That is, females usually have two X chromosomes and males typically have one X and one Y. Experiments not described showed that unusual females may in addition have a Y (XXY) and rare males may lack a Y (X0) and the organisms will still be female and male, respectively. However, the Y is essential for male fertility, since X0 males are unable to produce normal sperm.

In *D. melanogaster* a complete haploid set of autosomes, represented by A, consists of two large chromosomes with median centromeres and a very small, dotlike chromosome. Studies of the sexual phenotypes of individuals with various combinations of X chromosomes and sets of autosomes provided the evidence that the primary determinant of sex in *Drosophila* is the X:A ratio. Individuals with half as many X chromosomes as sets of autosomes (1X:2A) males and those with equal numbers of chromosomes and complete sets of autosomes (2X:2A, 3X:3A) are females. The genotypes 1X3A (0.33:1.0) and 3X 2A (1.5:1.0) result in normal male and normal female somatic sexual phenotypes, respectively, though these individuals are weak, have underdeveloped reproductive organs, and are sterile.

In some organisms with homogametic females and heterogametic males the Y chromosome is male-determining. An early observation of the flowering plant *Melandrium album* showed that flowers produced by both 2X 2A and 2X 4A plants are pistillate (female), and those of XY 2A and XY 4A plants are invariably staminate (male). When another X chromosome is present, in XXY 2A or XXY 4A plants, most of the flowers are staminate, though an occasional one is bisexual (having both stamens and a functional pistil). This correspondence between the expression of sex and chromosome constitution indicates that in this plant the X is female-determining and, unlike *Drosophila*, the Y is male-determining. A balance between the effects of the two sex chromosomes is approached in XXXXY 4A plants, which produce mostly bisexual flowers, though a few are staminate. Many other variations in the primary sex-determining. The first evidence for this conclusion came from analyses of the chromosome complements associated with two abnormal sexual phenotypes. Males with a condition known as *Klinefelter syndrome* have 47 chromosomes (an extra X) rather than the normal 46 and are XXY. A similar phenotype characterized by underdeveloped testes and sterility has been found to result from XXYY, XXXY, XXXYY, and XXXXY sex-chromosome constitutions. Females with the aberrant condition known as *Turner syndrome* have only 45 chromosomes and are X0. These sexually underdeveloped individuals are sterile.

Humans with an XXX sex-chromosome complement are fertile females, and XYY individuals are fertile males.

The role of the Y chromosome in sex determination in humans and other mammals is to induce a rudimentary embryonic gonad with both male and female potentials to develop into a testis instead of any ovary. A major factor in this induction appears to be related to a substance determined by a gene on the Y chromosome. It is not yet known whether the Y-chromosome gene determines the substance directly or in some way acts to regulate the expression of the gene located on another chromosome. A newly formed testis secretes both a substance that suppresses the development of the primitive oviduct and the hormone testosterone, which is required to induce development of both the genital duct system and secondary sexual characteristics of the male. In the absence of a Y chromosome, the rudimentary gonad differentiates into an ovary and the primitive oviduct into Fallopian tubes and a uterus.

Probability in Genetic Prediction and in Analysis of Genetic Data

Genetic ratios are the result of chance operating in the assortment of genes into gametes and in the combination of gametes into zygotes. Thus, exact predictions are not possible for a particular event. However, it is possible to state in advance that an event has a certain probability of occurring, as we have seen. In this section we consider some of the probability methods used in interpreting genetic data.

Use of the Binomial Distribution in Predicting Genetic Outcomes

The probability that each of three children in a family will be of the same sex is an example of a simple problem that uses both the addition and multiplication rules of probability. The probability that all three will be girls is (1/2) (1/2) (1/2) = 1/8, and the probability that all three will be boys is also 1/8. Since these outcomes are mutually exclusive, the probability of one or the other is the sum of the two probabilities, which is 1/4. The remaining possible outcomes are that two of the children will be girls and the other a boy, or two will be boys and the other a girl. For each of these outcomes only three different orders of birth are possible—for example, GGB, GBG, and BGG—each having a probability of 1/8. Thus, the probability of two girls and a boy, disregarding birth order, is the sum of the probabilities for the three possible orders, or 3/8; likewise, the probability of two boys and a girl is also 3/8. Therefore, the distribution of probabilities for the sex ratio in families with three children is:

GGG	GGB	GBB	BBB
	BGB	BGB	
	BGG	BBG	
$(1/2)^3$	$3(1/2)^2(1/2)$	$3(1/2)(1/2)^2$	$(1/2)^3$
1/8 +	3/8 +	3/8 +	1/8 = 1

This sex-ratio information can be obtained more directly by expanding the binomial expression $(p + q)^n$, in which p is the probability of the birth of a girl (1/2), q the probability of the birth of a body (1/2), and n is the number of children; in our example,

$$(p + q)^3 = \mathbf{1}p^3 + \mathbf{3}p^2q + \mathbf{3}pq^2 + \mathbf{1}q^3$$

in which the bold numbers can be compared to the numbers above. Similarly, the binomial distribution of probabilities for the sex ratios in families of five children is

$$(p + q)^2 = 1p^5 + 5p^4q + 10p^3q^2 + 10p^2q^3 + 5pq^4 + 1q^5$$

Each term tells us the probability of a particular combination. For example, the third term tells the probability of three girls (p^3) and two boys (q^2) in a family having five children—namely,

$$10(1/2)^2(1/2)^2 = 10/32 = 5/16$$

There are $n + 1$ terms in a binomial expansion. The exponents of p decrease from n in the first term to 0 in the last term, and the exponents of q increase by one from 0 in the first term to n in the last term. The coefficients generated by successive values of n can be arranged in a regular triangle known as *Pascal's triangle*. Note that the horizontal rows of the triangle are symmetrical, and that each number is the sum of the two numbers on either side of it in the row above.

In general, if the probability of event A is p and that of event B is q, and the two events are independent and mutually exclusive, the probability that A will occur four times and B two times—in a specific order—is p^4q^2, by the multiplication rule. However, suppose that we are interested in the occurrence of this combination of events: four of A and two of B, irrespective of order. In that case, we multiply the probability that the combination 4A:2B will occur in any one specific order by the number of possible orders. The number of different combinations of six events, four of one kind and two of another, is

$$\frac{6!}{4\ 2} = \frac{1\times2\times3\times4\times5\times6}{1\times2\times3\times4\times1\times2} = 15$$

(The symbol ! is for factorial, or the product of all positive integers from one through a given number.) This calculation provides the coefficient of the p^4q^2 term in the expansion of the binomial $(p + q)^6$. Therefore, the probability that event A will occur four times and event B two times is $15p^4q^2$.

The general rule for repeated trails of events with constant probabilities is: if the probability of occurrence of event A is p, and the probability of the alternative event B is q, the probability that in n trails event A will occur s times and event B will occur t times is

$$\frac{n}{s!\,t!}p^s p^t$$

in which $s + t = n$, and $p + q = 1$. To use this expression, one must remember that 0! is defined to be 1, and that any number raised to the zero power (e.g., 2^0) also equals 1. Note that each individual term in the expansion of the binomial $(p + q)^n$ is the same as the expression just given.

Let us consider a specific example, in which we calculate the probability of a mating between two heterozygous parents yielding a typical 3:1 ratio of the dominant and recessive traits among families of a particular size. The probability p of one child showing the trait determined by the dominant allele is 3/4, and the probability q of one child showing the trait determined by the recessive allele is 1/4. Suppose that we wanted to know how often in such families with eight children would six of the children have the trait determined by the dominant allele and two children have the trait determined by the recessive allele. In this case, $n = 8$, $s = 6$, $t = 2$, and the probability of this combination of events is

$$\frac{8!}{6!\,2!}(p)^6(q)^2 = \frac{6!\times 7\times 8}{6!\times 2!}(3/4)^6(1/4)^2 = 0.31$$

That is, in 31 percent of the families of eight children the offspring will show the ideal 3:1 phenotypic ratio, the other families will deviate in one direction or the other because of chance variation.

Evaluating the Fit of Observed Results to Theoretical Expectations: The Chi-Squared Method

An important question in dealing with any event whose outcome is subject to the rules of probability is whether an observed result is in agreement with the theoretical expectation based on a particular hypothesis. That is, how far can observed results deviate from those expected before they must be considered an indication of a real difference and not just an accident of sampling? For example, suppose that 100 seedlings from a cross between a green F_1 corn plant and a yellow-green plant of one of the parental strains were grown to test for a 1:1 ratio, and that 65 of the seedling were green and 35 yellow-green. Does this mean the expected ratio is wrong and that the genetic basis for the colour difference is more complex than a single pair of alleles, or might the deviation from the 1:1 ratio merely result from chance? The answer depends on how often, in progeny samples of size 100, a deviation as large or larger would be expected to occur by chance if the expectation were correct.

The probability of obtaining results that differ from an expected 50:50 by 15 individuals or more *in either direction* can be determined from the expansion of the binomial $(p + q)^{100}$. This tedious chore would yield 101 terms, representing the probabilities of all possible outcomes—that is, from 100 green and 0 yellow-green to 0 green and 100 yellow-green. Summing the appropriate terms would tell us that the chance of a deviation as large or larger than 15 is about 1 in 300—a probability so small that we would most likely conclude that the 1:1 expectation is wrong.

There is, of course, still the one chance in 300 that the result is due to chance. Two kinds of mistakes are possible in decisions about the agreement between observed and expected results: (1) a correct hypothesis may sometimes be rejected, or (2) an incorrect hypothesis may sometimes be accepted. In this section a method is described that is commonly used to evaluate the fit of observation and theoretical expectation.

Calculation of exact probabilities for assessing goodness of fit between observed values and those predicted from a specific genetic hypothesis is extremely laborious when the number of observations is large. A particularly useful statistical test is the *chi-square* (χ^2) *method.*

The value of χ^2 is given by the expression

$$\chi^2 = \sum \frac{(\text{Observed} - \text{expected})^2}{\text{Expected}}$$

in which Σ means the summation over all the classes. For each class the difference between the observed number and expected number is squared and then divided by the expected number. To illustrate, suppose we have an F_2 generation consisting of 99 wildtype and 45 mutant individuals, and we wish to know whether this set of observed numbers is in satisfactory agreement with the expected 3:1 ratio resulting from the segregation of a single pair of alleles. Calculation of the value of χ^2 is carried out in table. Note that the sum of the deviations of observed from expected numbers equal zero; when there are only two classes, as in this example, the deviations are always equal in magnitude but opposite in sign. Note also that a deviation of 7 would be a very large one if the expected number were 12, but a relatively small one if the expected number were 120. Since sample size is important, the measure of the size of the deviation from expectation—(observed – expected)^2— is exponents of χ^2, and their summation gives a value of 3,00 in the example.

To make use of the χ^2 value, it must be related to the probability of obtaining, by chance, deviations as large or larger than those observed, if in reality the 3:1 expectation is correct. The larger the difference between expected and observed numbers, the larger χ^2 is for a given sample size. The larger χ^2 is, the less likely it is that the deviations are simply a result of chance. In the interpretation of a χ^2 value the number of classes of data must be taken into account. Every χ^2 test has associated with it a number called its *degree of freedom*, which is used in assessing the significance of the χ^2 value. For the type of χ^2 test, the number of degrees of freedom is simply the number of classes of individuals minus 1.

In general terms, degrees of freedom refers to the number of classes that can vary independently. Two classes of individuals, and since the total sample size is fixed at 144, there is only one degree of freedom. Once the number of individuals in either class has been determined, the number in the other class is automatically set. Similarly, when there are four classes of data, three of them can have frequency, but the frequency of the fourth class must equal the difference between the sum of the three and the total. That is, with four classes of individuals there are three degrees of freedom—the total number of classes minus 1.

When a χ^2 value has been calculated and its degrees of freedom determined, the final step in assessing goodness of fit between expected and observed data is straightforward. One uses either a graph or a table of χ^2 values to determine the approximate probability of obtaining, by chance alone, a deviation as great or greater than that actually observed. In the example, $\chi^2 = 3.00$ with one degree of freedom. The top line corresponds to one degree of freedom, and the probability associated with a χ^2 value of 3.00 is somewhere between 0.05 ($\chi^2 = 3.841$) and 0.20 ($\chi^2 = 1.642$). This range implies that a deviation from the 3:1 ratio at least as great as that observed would be expected to occur by chance in more than 5 percent, but fewer than 20 percent, of similar experiments. Some arbitrary conventions are used to interpret these probability values. If the probability is <5 percent, differences between observed and expected values are considered ***significant***, and the hypothesis is rejected. When the probability is >5 percent, differences are considered to be nonsignificant. In this case the agreement between observation and expectation is considered to be satisfactory. This does not mean that the hypothesis is true, but only that it is consistent with observed results.

7

Recombinant DNA Vaccines

Although the development of methods in genetic engineering has been rapid and their subsequent usage has, in part, revolutionized basic biomedical research, to date the number of marketable, industrial applications have been few. The field of veterinary biologics has been the primary benefactor of the limited number of genetically engineered products. Several recombinant DNA (r-DNA) vaccines used to prevent enterotoxigenic *Escherichia coli* disease in livestock animals are currently available world-wide and a vaccine for foot and mouth disease is available in limited markets.

Rationale

The use of r-DNA technology in vaccine development provides several obvious and in some cases necessary advantages over previous methods. In general, bacterial vaccines have been divided into two groups: bacterins and toxoids. As their name implies, bacterins are prepared from whole bacterial cells. The cells may remain viable or treated chemically or physically to kill them. For example, some pertusis vaccines are prepared by chemically killing live, virulent *Bordetella pertusis* cultures. The analogies for viruses include live virus, attenuated live virus, or killed virus vaccines. Toxoid vaccines are based on a single cellular component derived from a relevant toxin. The toxins, such as diptheria toxin, are usually de-toxified by chemical treatments. Toxoids actually fall into a larger more general class of vaccines described as subunit vaccines. Subunit vaccines are prepared from subcellular components of bacteria and therefore do not contain whole cells nor all components of lysed cells. In principle, so long as the appropriate antigen(s) is selected, subunit vaccines may be prepared to protect against virtually any infectious disease. The greatest advantage of a subunit vaccine compared to bacterins is the elimination of cellular components that do not contribute to eliciting of a protective response. Some of these components may actually have inherent toxic activity, and thus their elimination is advantageous. Endotoxin is an integral part of all gram-negative bacteria, being a component of the lipopolysaccharide in their outer membranes. The biological activities of endotoxin are diverse and apparently common to endotoxins from all gram-negative species. Some of the toxic activities of endotoxin include shock, pyrogenicity, vascular permeability, and abortions in pregnant females. Although it is almost impossible to completely exclude endotoxins from products prepared from gram-negative organisms, the complications that result from endotoxemia can be so severe that it is advantageous to reduce endotoxin contamination to a minimum. This may be accomplished by preparing subunit vaccines.

Recombinant DNA techniques may be used to construct bacterial strains that are more amenable to use in bacterins or in the development of subunit vaccines. The following discussion will serve to illustrate some of the advantages of using r-DNA techniques in vaccine development. In some cases the examples are based on real knowledge of microbes and relevant antigens, while other examples represent perceived problems.

Situations where the organisms being used for vaccine production are of high virulence pose a real health threat to individuals in a production operation. Therefore, as an example, growth of *Salmonella typhi* strains, responsible for typhoid fever, would be considered prohibitive in large scale. The hazards associated with this organism and others and their potential for epidemic spread is of such a magnitude that the development of vaccines by traditional techniques would be avoided. A corollary to this problem is if the organisms were virulent and used as a live bacterin, it may very well cause the disease it was meant to control. If the mucosal surface of the small intestine was to be stimulated by live organisms to induce a local immune response, an organism that did not cause disease at the site would obviously be required. However, organisms producing the necessary antigen probably are natural pathogens. The development of a *Shigella*-based vaccine to prevent shigellosis provides an example of this problem. The intent is to induce a local immune response in the gut. However, genetically altered and attenuated strains fail to induce the appropriate antibodies, and suitable wild type strains cause disease. The growth characteristics of the organisms being used also are important. Organisms that do not grow well or that require extraordinary manipulations to promote growth are likely to be poor candidates for vaccines. One extreme of this situation pertains to viruses. The yield for many viruses is probably inadequate for large scale production and requires added time and labor when compared to bacteria. Furthermore, cultivation of viruses can be a relatively expensive operation.

Besides the inherent problems associated with the growth and virulence of some bacteria, issues of gene, and therefore antigen, expression must be considered. Antigens inappropriately displayed on a cell or produced in small amounts pose serious problems in vaccine development. Surface antigens frequently are masked as a result of cellular polysaccharide capsules and thus do not illicit good immune responses. Expression of the *E. coli* pilus antigen K99 exemplifies this phenomenon. Selection of growth conditions frequently is critical in obtaining maximal antigen expression. Growth *in vitro* often times is not appropriate for the expression of antigens produced *in vivo*. Gene instability also may pose a problem. Some genes, especially those that are plasmid encoded, are readily lost by growth *in vitro*. Thus the organism may no longer produce relevant antigen(s).

Probably the biggest problem encountered in vaccine development is determining which antigens must be included. In the case of bacterins, the process may be intuitive since killed, virulent organisms are usually used. However, in order to use gene cloning to construct appropriate vaccine strains, whether used to produce attenuated bacterins or subunit vaccines, the relevant antigens must first be identified.

The development of vaccines to prevent severe *E. coli*-induced diarrhea in neonatal livestock animals was based on the identification of several factors that directly contributed to virulence. Enterotoxigenic *E. coli* (ETEC) differ from commensal strains in at least two aspects. Firstly they produce enterotoxins that cause the small intestine to secrete water. And secondly, they have the capacity to intensively colonize the small intestines of infected animals. ETEC produce two different classes of enterotoxins: heat labile (LT) and heat stable (ST). ST subsequently has been divided into two distinct groups, STa and STb, based on solubility in methanol and spectrum of biologic activity. Both STs are poorly antigenic while LT is highly antigenic. ETEC colonize by adherence to the musocal epithelium of small intestines and subsequently proliferating at that site. Adhesion is facilitated by proteinaceous appendages on surface of the bacterial cell called pili.

Because of the protein nature of pili, they are good antigens. Antibodies produced against specific pilus-adhesions neutralize their adhesive activities and thus prevent colonization of small intestines and therefore disease.

ETEC that induce diarrhea in neonatal pigs produce one or more of three pilus-adhesins termed K88, K99, or 987P. Since all three pilus adhesins occur in nature and since one or all may be present on strains in a specific geographic location all three must be present in a vaccine designed to prevent ETEC induced diarrhea. Alternatively, prevention of symptoms by neutralizing enterotoxin activity also could be exploited in vaccine development. However, based on the known immunogenic properties of the STs this approach would likely be less useful unless the STs could be made more reactive.

The ultimate development of the *E. coli* vaccine was based on a thorough understanding of the mechanism of pathogenesis. Although this is a powerful approach, it is not always possible. Supportive information used to determine the relevant antigens to be included in vaccine may come from the results of serologic analysis. The use of serums from acute and convalescent patients has been particularly useful in determining those antigens that stimulate immune responses. Such antibodies may be protective against disease, and thus their induction would be the desired result of vaccination. Monoclonal antibodies prepared against the pathogenic organisms provide powerful tools in further defining both the important antigens as well as important antigenic determinants (epitopes) on the antigens.

To clone antigen-specifying genes it remains essential to know precisely what antigens are desired. The subsequent selection of the correct recombinant clones usually requires the use of antibodies and thus is closely inked to serologic studies.

METHODS

It is not reasonable to describe all the methods and protocols that could be used in the development of r-DNA vaccines.

Preparation of DNA

The method used to prepare DNA for gene cloning is, in part, dependent upon the location of the specific gene. The methods employed, for example, to obtain plasmid DNA are different than those used for chromosomal DNA. Frequently, information on the gene loci are known thus facilitating the selection of methods for DNA purification. For *E. coli* pilus-adhesins, it was known that some were plasmid encoded.

If the genetic loci are not known, the procedures used to prepare chromosomal DNA should suffice since these are general methods for purifying total cellular DNA.

The procedures for cellular DNA purification can be divided into two general classes. One is based on .the procedures of Marmur which entail cell lysis, deproteinization using proteases and/or phenol extraction, and followed by selective precipitation. Various extraction and precipitation solvents have been used by various investigators. Frequently, these procedures result in material contaminated with RNA. Several enzymes used in the gene cloning process are inhibited by RNA thereby requiring an RNA-free substrate. Treatment of the DNA samples with DNase-free RNase prior to deproteinization will eliminate the vast majority of RNA. However, free ribonucleotides and oligonucleotides may co-precipitate with DNA in the presence of ethanol, thus demonstrating the need for further purification of the DNA.

The second general procedure in cellular DNA purification is based on equilibrium centrifugation. Cessium chloride is commonly used to generate salt gradients at sufficiently high concentrations such that DNA obtained, for example, by the Marmur procedure will reach an equilibrium point equal to its buoyant density. Due to the extreme differences in density of protein, RNA, and DNA these three components are readily separated resulting in highly purified DNA.

Nucleic acids also maybe separated by a variety of chromatography procedures including selective binding to and elution from hydroxyapatite.

The procedures used to purify plasmid DNA exploit two physical properties unique to these DNA molecules: their superhelical structure and their relatively small size. Gentle lysis procedures have been devised that minimized chromosome membrane disruption while maximizing chromosome integrity. These lysates are cleared of most of the large debris including the intact chromosome by centrifugation. The resultant supernatant contains the smaller plasmid DNA, broken, linear chromosomal DNA, RNA, and proteins. Much of the RNA and protein can be eliminated by RNase treatment followed by phenol extractions. Highly purified plasmid DNA can be ultimately prepared by separation in CsCI gradients containing the intercalating dye ethidium bromide. Alternatively the plasmids may be purified from the cleared lysates by procedures such as reversed phase chromatography, agarose gel chromatography, agarose gel electrophoresis, or affinity chromatography (e.g., on columns of acridine tagged agarose).

The yield of plasmid from a culture is partly dependent upon size and copy number. The copy number of some plasmids may be amplified by treating the culture with certain antibiotics such as chloramphenicol or spectinomycin.

DNA Cloning Vectors

The intent of gene cloning is to insert the desired gene into a self-replicating DNA vector. The selection of the vector is dependent upon the source of the DNA to be cloned and the outcome desired from gene cloning. Most cloning vectors possess several common properties. Most are constructed plasmids that have readily selectable gene markers. Antibiotic resistance markers are most commonly used. Sites within the antibiotic resistance markers are used as the site of DNA cloning which results in insertional inactivation of the resistance gene. For example, the vector pBR322 encodes resistance to both ampicillin and tetracycline. Cloning into one of the resistance genes results in the inactivation of that gene. Thus cells possessing pBR322 are resistant to both antibiotics while cells containing a recombinant molecule, with the site of insertion being in the tetracycline resistance gene, are sensitive to tetracycline and resistant to ampicillin. The use of these markers and others facilitates selection of cells containing recombinant DNA molecules.

DNA restriction endonuclease

The capacity to insert DNA into antibiotic resistance genes is based, in part, upon the presence of specific palindromic DNA sequences, such as GGATCC, within the resistance genes. These symmetrical sequences have the same sequence on the complementary strand when read in the opposite direction. The palindromes are recognized by a class of DNases called class two DNA restriction endonucleases and thus have also been termed restriction sites. Most, if not all, bacteria produce one or more of these restriction enzymes. Their function is to restrict the entry and establishment of DNA into bacteria from genetically divergent sources. The enzymes recognize a specific palindrome cutting at a specific-site within or near the palindrome. In the absence of a repair process, such DNA molecules are lost due to their inability to be replicated in bacteria after cleavage. To prevent self-destruction, organisms produce modifying enzymes that alter certain nucelotides within the restriction site, thereby preventing cleavage by their own restriction endonucleases.

Different restriction endonucleases recognize different palindromic sequences. There are more than 200 restriction endonucleases recognizing different palindromes, many of these enzymes being available commercially. Therefore, many sites can be used in gene cloning. Typically, restriction sites within selectable genes are most useful.

Another relevant property of restriction endonucleases is that they cut DNA molecules symmetrically. If the cuts are staggered, the cleavage results in so-called sticky ends, while straight cuts result in blunt ends.

DNA ligation

DNA molecules from any source cleaved with the same restriction endonuclease (that results in staggered cuts) have sticky ends with the same nucleotide sequence. The sticky ends can reanneal and in the presence of the enzyme DNA ligase, a phosphodiester bond between the 3' -hydroxyl and 5' -phosphate is formed (ligation). If the sticky ends were from the same DNA molecule, a circular structure would be formed. On the other hand, if the sticky ends were from two different DNA molecules, a new hybrid molecule would result. The two ends from the hybrid can also be ligated resulting in a recombinant, circular DNA molecule. If one of the DNA molecules had been the cloning vector and the other encoding a gene, the end result would be a "cloned gene." The probability that the vector and insert DNA are ligated together rather than the ligation of several pieces of insert DNA can be increased by selection of appropriate concentrations of the various DNAs or more particularly, the concentration of 5'-ends.

Cloning of fragments cut with restriction endonucleases that create blunt-end can also be accomplished. However, the ligating enzyme must be T4 DNA ligase. It should be noted that when sticky-ended DNA is ligated, the restriction site is recreated and may subsequently be recut. Blunt-end ligations may or may not generate the original restriction sites.

Cloning strategy

Frequently, little is known about the genetic location of specific genes. In such cases the cloning of these genes follows an empirical process (often called "shotgun" cloning). Furthermore, in the absence of information about the number and distribution of restriction sites on and around the genes to be cloned it is difficult to design specific cloning strategies. If a restriction site is within the gene being cloned and if that enzyme is used to generate fragments to be cloned, an active gene probably will not be recovered due to its cleavage prior to cloning. To obtain an intact gene, two fragments would need to be cloned, and they must be arranged in the proper orientation. To increase the likelihood of cloning genes, several techniques have been employed. Each makes use of randomizing the locations of restriction cuts. This may result in the generation of large fragments.

Obviously, the larger the fragment, the greater the probability of cloning a gene and subsequently identifying such clones from a large DNA pool. The randomizing of cut sites decreases the chances that a cut will occur within the gene. Controlled shearing of DNA will result in fragments with a given size range. Partial digestion with restriction endonucleases can also be employed to obtain large fragments. In this regard, enzymes that recognize 4-base pair palindromes, such as Sau 3AI, are particularly useful since such sites, if randomly distributed, would occur every 256 base pairs. Partial digests with Sau 3AI should result in a random distribution of cuts and therefore fragments. In the case of sheared DNA, homopolymer tailing of the fragments and of the cloning vector is necessary. Otherwise, recognition sites for annealing would not be available, resulting in an inability to ligate and thus clone the fragment. Alternately, the ligation of linkers (synthetic DNA sequences containing a restriction site) to fragments may be performed and used as a means for gene cloning. For example, Bam HI linkers may be blunt end ligated to fragments generated by shearing. If the shearing resulted in single stranded regions at the termini, it would be necessary to make them blunt ended prior to ligation by either removing the single stranded region using nuclease S1 or by filling in the single stranded region with DNA polymerase. The fragments could then be inserted and ligated into a Bam HI site in a vector.

In generating random cuts, it often is difficult to restrict the fragments to a narrow size range. The amount of fragments of certain sizes may be increased by separation and elution from molecular sieves or by electrophoresis through agarose.

Recombinant DNA molecules are usually introduced into bacteria by transformation. Most of the best understood cloning systems were developed in *E. coli* and therefore transformation most frequently is into *E. coli*. The procedure of Cohen et al. and its many derivations are used for transformation. Efficiencies of transformation are dependent upon the host strains and the size of the DNA molecules. Several *E. coli* strains have been selected that are transformed more efficiently than other strains. Furthermore, the transforming DNA must be in a covalently closed circular configuration for efficient uptake. Of major consideration in selecting recipient strains are the effects of host restriction and modification enzymes on the incoming DNA. It is certain that host restriction enzymes if functioning properly would eliminate the incoming DNA unless it had been derived from a strain with a functional modification system. Restriction and modification mutants are usually employed for transformation to eliminate this possibility. If significant DNA sequence homology exists between the host and incoming DNA, the risk of homologous recombination can be reduced using *recA* mutants.

Cosmid cloning

As the size of the DNA to be transformed into a strain increases, the efficiency of the process decreases. For the situations described above where fragment sizes are kept large, transformation frequencies are often low. In many cases, the frequency is too low to allow convenient cloning. In these situations, the use of cosmids may solve the problem.

Cosmid cloning vectors are constructed by combining part of a conventional plasmid cloning vector with a bacteriophage lambda sequence called *cos*. *Cos* (or cohesive ends) is a sequence necessary to package DNA into lambda. The procedure proceeds by ligating a fragment into a cosmid cloning vector as usual except that the DNA concentration is kept high to promote concantamer formation. The ligated DNA is then packaged *in vitro* into lambda heads and tails and transduced into a lambda sensitive strain. Since the only lambda sequence in the cosmid cloning vector is *cos*, the only way the DNA can replicate is to use the plasmid-specified functions (also encoded in the cosmid cloning vector). In order to be packaged into lambda heads and tails DNA must, in addition to having *cos*, be approximately 45 kilobases in length. This means that 30 to 40 kb of insert DNA may be cloned. Conventional plasmid cloning due to size limitations in transformation is usually restricted to inserts of approximately 10-15 kb.

The selection of the cloning vector to be used is ultimately dependent upon the DNA source and expected outcome of cloning. A somewhat empirical approach must be used. Several considerations include: what enzymes will be used and the size of the fragments generated; will the cloned DNA be expressed in *E. coli*; will the end product of gene expression be degraded by the host; if it is an intergeneric cloning experiment, will the hybrid DNA molecule be expressed in the other genus? Several expression vectors have been constructed that contain appropriate promoter sequences. Expression of genes in intergeneric cloning may require introduction back to the "natural" host. Shuttle vectors that can replicate in different genera have been constructed for this purpose. The prevention of gene product break down may be prevented in some cases, by fusion of the protein to another of host origin. Cloning vectors that perform this function also are available.

Identification of Recombinants

The success of gene cloning is also dependent upon the ability to detect cells containing the gene. Detection of the desired recombinants have been based on two properties. The first and most desirable property is the synthesis of the desired gene product. Gene products that have enzymatic activity may be detected using an appropriate enzyme assay. However, most gene products that would be used in vaccines are not enzymes and must be detected as antigens. Antigens present on the cell surface may be detected by serum agglutination either in suspension or by precipitation reactions in agar. Colonies also may be transferred to sheets of polyvinyl or nitrocellulose, left intact or lysed, and subjected enzyme linked immunoassays or radioimmunoassays. If DNA probes are available, recombinants can be detected by colony hybridization. One of the most sophisticated approaches is based on the synthesis, in the laboratory, of specific DNA probes. The product of the gene to be cloned is purified and a partial amino acid sequence determined. Based on the amino acid sequence, short oligonucleotides are synthesized that represent the possible coding sequences. In a bacterial system a 10-12 base pair probe would be sufficiently large to eliminate chance that the same sequence would occur in another gene. The DNA probes would then be used in colony hybridization procedures to detect the putative recombinants.

Development of an r-DNA Vaccine to Prevent ETEC Induced Diarrhea

Rationale for Development

As mentioned above, the pathogenesis of ETEC induced diarrhea is depenent upon the expression of two known attributes of virulence: 1. secretion of enterotoxin that induces net water efflux by the small intestine; and 2. colonization of small intestines by adherence to the mucosal surfaces. Adherence of ETEC to small intestines is mediated by protein, surface appendages called pili. In baby pigs, three antigenically different pili are known to facilitate adhesion: K88, K99, 987P.

Initially, it was shown that baby pigs suckling dams that had been vaccinated individually with purified pili were protected against severe, fatal diarrhea after experimental challenge with ETEC producing the same pili as were in the vaccine but not from ETEC producing different pili. The apparent mechanism of protection was believed to be due to the consumption of colostrum containing pilus-specific antibodies. The antibodies neutralized the adhesive activity of the pili, thus preventing the ETEC strain from adhering and colonizing the small intestine. Some evidence indicated that these antibodies also could reverse the adhesion of already adherent bacteria.

Table 7.1. Affect of vaccination with 987P on mortality of pigs challenged with ETEC producing 987P or K99.

Vaccine	*Pilus on challenge ETEC*	*Number of pigs challenged*	*Number that died (%)*
987P	987P	75	1 (1%)
	K99	30	9 (30%)
K99	987P	41	0 (0%)
	K99	80	7.2 (17%)
	987P	35	14 (40%)
Control	K99	78	12.5 (22%)

The next step was to determine whether a vaccine containing all three pilus-types would have a broad-spectrum, protective effect. The results shown in Table 7.2 demonstrate that vaccination with a cell-free mixture of the three pili protected against disease after the pigs were challenged with ETEC producing K88, K99, or 987P.

Table 7.2. Effect of vaccination with EcoBac on mortality of challenged animals.

Vaccine	*Pilus on challenge ETEC*	*Number of pigs challenged*	*Number that died (%)*
	K88	32	5 (15.6%)
	K99	34	1 (2.9%)
EcoBac	987P	44	0 (0%)
	K88	47	29 (61.7%)
Control	K99	48	27 (56.2%)
	987P	44	31 (70.4%)

The rationale behind usage of r-DNA techniques for the development of this vaccine was based primarily on two potential problems: 1. the K88 plasmid was readily lost by cultivation *in vitro*; 2. the yield of

K99, especially compared to K88 and 987P, was very low. Both problems were solved by gene cloning.

Cloning of *E. coli* Pilus Genes

The cloning strategy for two of the three *E. coli* pilus-specifying genes used to produce Eco Bac, a vaccine to prevent ETEC diarrhea in pigs, was based on the observation that these genes were plasmid encoded.

The K99 specifying plasmid (pK99) from a virulent, wild type ETEC strain was moved into an *E. coli* K 12 strain by conjugal mating. The plasmid was subsequently isolated from a 500 ml culture grown in trypticase soy broth at 37°C. A lysate was prepared using the procedure of Ish-Horowicz and Burke. Cells were harvested at 10,000 × g and resuspended in 0.1 volumes glucose (0.1 M), EDTA (0.01 M), tris (0.05M, pH 8). After five minutes at room temperature, two volumes of SDS (1%) in 0.2N NaOH were added to lyse the cells and denature the DNA. After an additional five minutes on ice, 1.5 vols potassium acetate (5M, pH 5) was added and incubated on ice five minutes. The resulting precipitate, including membranes, SDS and chromosome, was removed by centrifugation at 10,000 × g for ten minutes. To the supernatant, 2.5 volumes ice cold ethanol (95%) were added and the mixture stored at least 30 minutes at -20°C. The precipitate was collected, dried in vacu, and resuspended in 10.6 ml Tris (0.01 pH 8), EDTA (1 mM, TE). To the dissolved DNA, 2.1 ml ethidium bromide (1 mg/ml) and 8 gms CsCl were added and the mixture centrifuged to equilibrium. The RNA, covalently closed circular plasmid DNA, linear and open circular DNA (chromosome fragments and plasmids), and proteins and membrane debris separate into four different regions of the gradient based on their bouyant densities in CsCl. The CCC plasmid DNA was removed, extracted with butanol to remove ethidium bromide, ethanol precipitated and dried.

Vector DNA (pBR322) was prepared in the same manner, except that chloramphenicol (150 μg/ml) was added to the log phase cells to amplify plasmid synthesis.

Both vector and pK99 DNA were cleaved with the restriction endonuclease BamHI using the conditions described by the manufacturer. The 5' -phosphate group of pBR322 was removed by addition of calf intestine alkaline phosphatase. Both the digested and diphosphorylated vector and digested pK99 were phenol extracted (two times), ether extracted (three times), and precipitated with ethanol. The precipitates were collected, dried, and resuspended in TE. One μg of pBR322 and 0.5 g pK99 digest were mixed together. Two μl of ligation buffer (0.66 M tris, pH 7.5, 50 mM $MgCl_2$ 50 mM dithiothreitol and 10 mM ATP) were added and the mixture adjusted to 20 μl. T4 DNA ligase was added (two units) and the mixture incubated 18 hours at 10°C. One half of the mixture was used to transform competent *E. coli* cells prepared by the procedure of Cohen, et al. Cells were plated on trypticase-soy agar containing ampicillin (25 μg/ml). After overnight incubation, colonies were replica plated to ampicillin plates and tetracycline (15 μg/ml) plates. Colonies containing ampicillin resistant tetracycline sensitive cells were screened for K99 production by serum agglutination using rabbit antisera prepared against purified K99.

Cloning of K88 genes was identical except that the DNA source (pK88) and antiserum (antiK88) were different, and the restriction enzyme employed was Hind III instead of Barn HI.

Identification of the most useful clones (i.e., most stable and best expression), quantitation and the ultimate formulations for the final vaccine were empirical.

Subcloning

It often is desirable to subclone DNA fragments particularly when the fragment is large as is the case in cosmid cloning. This can be performed using different restriction enzymes or by varying the length of time of digestion (for recombinants derived from partial digests).

For most practical applications of gene cloning, the subcloning process is to remove unnecessary DNA sequences. Frequently, recombinant DNA molecules are unstable, this phenomenon often being associated with the size of insert. Thus plasmids constructed with large inserts may be less stable than those with small inserts demonstrating a need to subclone. Unfortunately, this is not a hard and fast rule. Other properties of the DNA molecules (other than size) are important in determining stability. The hybrid K99 molecule described above is 11.15 kb in size. Upon successive subculturing (three passages), the plasmid was lost at high frequency. This problem can be circumvented, however, by selectively growing cells to maintain the plasmid. For the example of the K99 recombinant, the culture medium is supplemented with ampicillin. Since the K99 recombinant DNA molecule encodes ampicillin resistance (a vector encoded trait), only those cells containing the plasmid grow. Loss of inserted DNA has not been a problem for either K99 or K88 specifying recombinant DNA molecules.

Final Remarks

The ease of development of the *E. coli* vaccine was greatly enhanced since the genes being cloned were from *E. coli* and were to be expressed in *E. coli*. Furthermore, since the gene products being exploited (pilus-adhesins) are surface antigens, their transport to the cell surface in the constructed recombinant strains occurred in an unaltered manner.

Several problems envisaged in construction of strains for bacterial vaccines relate to issues of the cellular location of the antigens and whether they are even expressed in the constructed strains. For example, genes from eucaryotic sources probably will not be expressed in procaryotes. Likewise, animal virus genes probably would not be expressed in procaryotes. However, gene cloning is most readily performed in bacteria and particularly in *E. coli*. Thus, shuttle vectors must be employed. However, there are few shuttle 'vectors available, and therefore the number of eucaryotic hosts (or for that matter other bacterial hosts) that can be used is equally low.

Bacillus subtilis has been used as a procaryotic host for gene cloning. The particular advantage of this organism over *E. coli* is that some proteins are readily secreted simplifying recovery and purification of the proteins in commercial applications. However, the number and diversity of vectors for *B. subtilis* is low greatly reducing its current usefulness in generalized commercial applications. Furthermore, *B. subtilis* produces potent proteases that may degrade the desired material, if composed of protein prior to its isolation.

The recent development of methods for viral gene cloning and expression in vaccinia virus has greatly facilitated the developments of vaccines for viral diseases. As the techniques of gene cloning are improved and refined and the techniques to enhance gene expression and antigen transport are developed, the overall progress in r-DNA vaccine development will accelerate as will all areas of biotechnology using r-DNA techniques.

8

Duplication of DNA

During cell division the chromosomes which are made up of DNA molecules undergo spletting and the sister chromatids move apart to maintain the usual number of chromosomes. Each chromatid has a single chain of DNA. To make it double and to convert chromatid upto chromosome the duplication of DNA becomes essential.

Replication of the DNA Molecule

Kornberg showed that DNA can replicate in a test tube with no cells present. All that is required is a mixture containing DNA, a specific enzyme (which he called DNA-polymerase), and a mixture of the four precursors: the nucleoside triphosphates deoxy-ATP, deoxy-CTP, deoxy-GTP, and deoxy-TTP. If any one of the four nucleoside triphosphates is omitted from the reaction mixture, DNA does not replicate itself. The intact DNA serves as a template for the reaction—a guide to the exact placement of nucleotides in the new strand. Where there is a T in the template, there must be an A in the new strand, and so forth.

Two other models of how the double helix might replicate were suggested after the original paper by Watson and Crick. In the model of conservative replication, the original double helix would serve somehow as a template but would either be reconstituted or, perhaps, would never unwind at all. Thus the new molecule would contain none of the atoms of the original. According to the model of dispersive replication, fragments of the original molecule would serve as templates, assembling two molecules, each containing old and new parts, perhaps at random. In *semiconservative replication* (the model proposed by *Watson* and *Crick*), the original two strands would separate, and each would function as the template for a new partner. Each molecule produced by semiconservative replication would therefore consist of one old and one new strand. After a short time, experimental work confirmed the model of semiconservative replication.

Synthesis of DNA from monomeric units of deoxyribonucleotides using a pre-existing DNA molecule as a template is DNA replication. Three mechanistic models have been proposed regarding DNA replication. These are:

(i) Dispersive

(ii) Conservative

(iii) Semi-conservative.

(i) Dispersive

According to this model, it was proposed that the original DNA is broken into many fragments, which are more or less uniformly distributed among the progenies. Inside the progeny, these fragments grow to form complete DNA segment. However, according to this model one would expect unequal distribution of genes among progenies leading to great variations among them. It would lead to irregular arrangement of nucleotides. However, these expectations are not

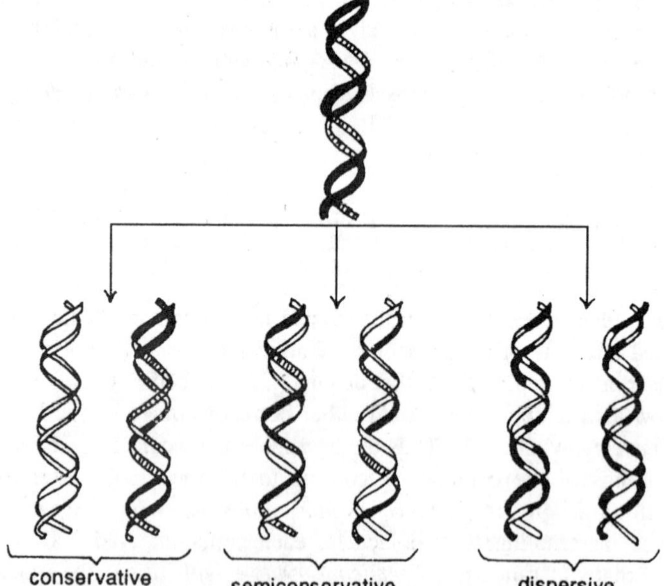

Fig. 8.1. A diagram illustrating three possible modes of DNA replication.

realized and hence the dispersive mechanism of DNA replication does not find support.

(ii) Conservative

According to this model, the original DNA is copied as such and then one copy of the complete DNA is transmitted to one of the progenies. Thus, according to this model, both primary and secondary structures of the DNA are conserved. There is neither breakage nor unwinding of the parental DNA during replication. The experimental evidence however, do not support this model as well, although such a mechanism of DNA replication is possible structurally and chemically.

(iii) Semi-conservative

This mechanism is based on Watson-Crick's model of DNA structure and is now accepted mechanism of DNA replication. The replication of DNA involves the unwinding of double delical DNA molecule but no breakage of the separated strands. Thus, the primary but not the secondary structures of parental DNA are conserved in progenies.

The idea of semiconservative mechanism of DNA replication came from the careful studies of the double helical model of DNA structure. The complimentary strand structure is an indication of the possible strand separation and formation of complimentary strands on each of the free single strands. During semiconservative replication, the hydrogen bonds between the complimentary base pairs are broken and the two separated strands start replicating as soon as a few bonds are broken. The two strands do not separate completely before the new strands are formed. That is why during the process of DNA replication some 'Y' shaped regions in the DNA molecule are observed. Each strand can act as a template or mould for the formation of new complimentary chains. The nucleotides are put together into short pieces of DNA by one of the DNA polymerases. The short segments of the newly synthesized DNA are known as *Okazaki fragments*. The complimentarity of the strands is maintained because of the selective pairing between A and T and G and C. The small Okazaki fragments are joined together to form complete DNA strands by the enzyme polynucleotide ligase. In experiments, where ligase is selectively inhabited, the accumulation of Okazaki fragments take place.

Experimental Proof

A clever experiment by Matthew Meselson and Franklin Stahl convinced the scientific community that semiconservative replication is the correct model. Working at the California Institute of Technology

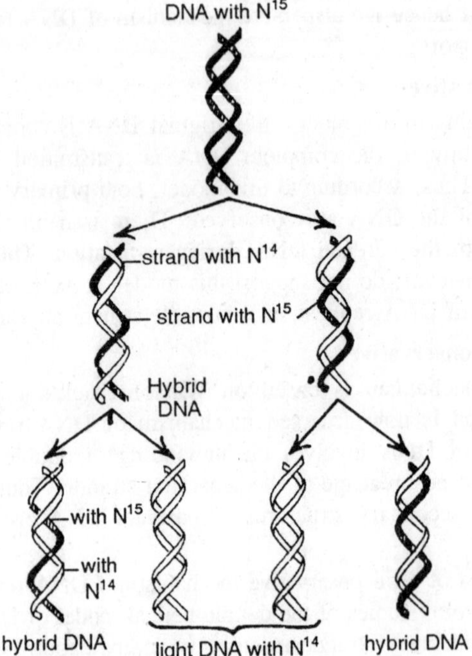

Fig. 8.2. Meselson and Stahl's experiment.

in 1957, they devised a simple way to distinguish old strands of DNA from new ones. The key was to use a "heavy" isotope of nitrogen. Heavy nitrogen (^{15}N) is a rare, nonradioactive isotope that makes molecules more dense than chemically identical molecules containing the common isotope ^{14}N. To study DNA of different densities (that is, DNA containing ^{15}N versus DNA containing ^{14}N), Meselson, Stahl, and Jerry Vinograd invented a new centrifugation procedure that allowed them to determine the density of DNA from a specific sample. At a certain molarity, a solution of cesium chloride (CsCl) has a density very close to that of DNA; at high gravitational forces produced in an ultracentrifuge, cesium ions sediment to some extent, thus establishing a density gradient. When a DNA sample is dissolved in CsCl and centrifuged at about 100,000 times the force of gravity, the DNA gathers in a layer in the centrifuge tube at a position where the density of the CsCI solution equals that of the DNA. To cluster in this way, DNA that is initially lower in the tube, where the density is greater than its own, must rise; DNA that is in a region of lower density must sink.

After developing this method of measuring DNA densities, Meselson and Stahl could begin experimenting. They grew a culture of the bacterium *Escherichia coli* for 17 generations on a medium in which the nitrogen source (ammonium chloride, NH_4Cl) was made with ^{15}N instead of ^{14}N. As a result, all the DNA in the bacteria was "heavy." Another culture was grown on medium with ^{14}N. They extracted DNA from both cultures. When these extracts were combined and centrifuged with CsCl, two separate DNA bands formed, showing that this method would work for separating DNA samples of slightly different densities.

Meselson and Stahl then conducted their main experiment. They grew another culture on ^{15}N medium and *transferred* it to normal ^{14}N medium. *E. coli* reproduces every 20 minutes, and Meselson and Stahl collected some of the bacteria from each generation after the transfer. They extracted DNA from the samples. After DNA was duplicated and the cells divided to produce each new generation, the DNA banding in the density gradient was different from the original banding. Initially, the DNA was uniformly labeled with ^{15}N and hence was relatively dense. After one generation, when the DNA had been duplicated once, all the DNA was of an intermediate density. After two generations, there were two equally large DNA bands: one of low density and one of intermediate density. In samples from subsequent generations, the proportion of low-density DNA increased steadily.

These data can be explained by the semiconservative model of DNA replication. The high-density DNA had two ^{15}N strands, the intermediate-density DNA had one ^{15}N and one ^{14}N strand, and the low-density DNA had two ^{14}N strands. In the first round of DNA replication, the strands of the double helix, both heavy with ^{15}N, separated; during the process of separation, each acted as the template for a second strand, which contained only ^{14}N and hence was less dense. Each double helix then consisted of one ^{15}N and one ^{14}N strand and was of intermediate density. In the second replication, the ^{14}N-containing strands directed the synthesis of partners with ^{14}N, creating low-density DNA, and the ^{15}N strands got new ^{14}N partners.

If the DNA had replicated in accord with either of the other models, the results would have been quite different. Under the conservative model, after one generation there would have been two bands—one for heavy DNA (^{15}N–^{15}N) and the other for light (^{14}N–^{14}N); no DNA of intermediate density would have formed at any time. If dispersive replication had taken place, the first round of replication would have produced DNA of intermediate density, but the density of all the DNA would have decreased with each subsequent replication. The crucial observation proving the semiconservative model was that intermediate-density DNA (^{15}N-^{14}N) appeared in the first generation and continued to appear in subsequent generations.

Cairn's Autoradiography Experiment

J. Cairns demonstrated the semiconservative type of replication by using a technique of autoradiography as given in the following lines:

In autoradiography technique, the material is first supplied with a suitable radioactive material like tritiated thymidine (H^3-TdR; H^3 is heavy isotope of hydrogen and it replaces normal hydrogen in thymidine to give rise to tritiated thymidine). Tritiated thymidine is used since this will selectively label only DNA and will not label RNA, since thymine base is absent in RNA. The tritiated thymidine gets incorporated into DNA and replaces ordinary thymidine. The material is then sectioned or else the cells may be broken down to release the intact bacterial chromosome on slides. These slides are then covered by photographic emulsion and stored in the dark. During this storage the particles emitted by tritiated thymidine will expose the film, which can be developed. This photograph will then show the regions of the presence of tritium and thus indirectly the presence of labelled DNA.

Using this above technique, the replication of DNA could be easily followed by *J. Cairns* and the results were reported in 1963. During incorporation of tritiated thymidine, autoradiographs could be prepared

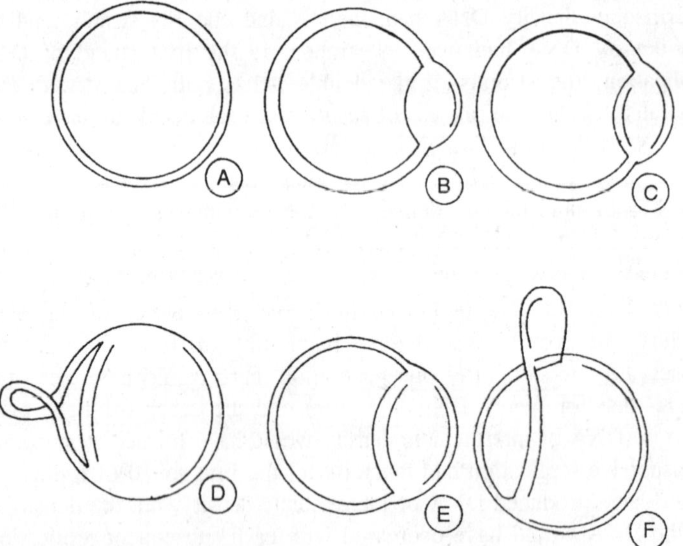

Fig. 8.3. Cairn's model for circular DNA replication.

at regular known intervals. Autoradiographs from this replicating material prepared at regular known intervals demonstrated semi-conservative mode of replication. It would be expected that in regions, where one cycle of replication of DNA has been completed, the density of dots will be higher than in the region where the replication has not taken place. The lighter density of dots is considered to indicate that only one of the two strands is labelled, while a heavier density of dots will indicate that only one of the two strands is labelled, while a heavier density of dots will indicate that both strands are labelled. Such a situation was actually observed. The rate at which the replication proceeds could also be worked out by measuring the length of DNA undergoing replication in a known interval of time. The generation time was worked out by *Cairns* as 30 minutes in *E. coli*. The length of the chromosome was worked out to be about 1 mm. The rate of replication as obvious would be approximately 30μ to 40μ per minute (1 mm = 1000μ).

That after replication, one of the two strands in the daughter DNA molecules is derived from the parent molecule and the other is synthesized afresh. In θ shaped figure, which is obtained in the second cycle of replication is presence of label, the two areas in the split region would never be equally labelled. For instance, one arc would be twice as heavily labelled as the other arc. This is what was actually observed by *Cairns*. Thus the observation support the semiconservative type of replication of DNA.

Taylor's Experiment on Vicia faba Root Tips

J.H. Taylor (1957) and his coworkers also demonstrated the semiconservative method of DNA duplication in the root tip cells of *Vicia faba* by autoradiography. The roots were grown in a medium containing radioactive thymidine, so that the radioactivity is incorporated in the DNA of these cells. The outline of this labelled chromosome on a photographic film appears in the form of scattered black dots of silver grains. When these root tips with labelled chromosomes were transferred to the unlabelled medium containing colchicine and studied for radioactivity, the following observtions were made:

1. In the chromosomes of first generation the radioactivity was found to be uniformly distributed in both the chromatids, because in them the original strand of DNA double helix was labelled with radioactivity and the new one was non-labelled.
2. In the chromosomes of second generation only one of the two chromatids in each chromosome was radioactive.

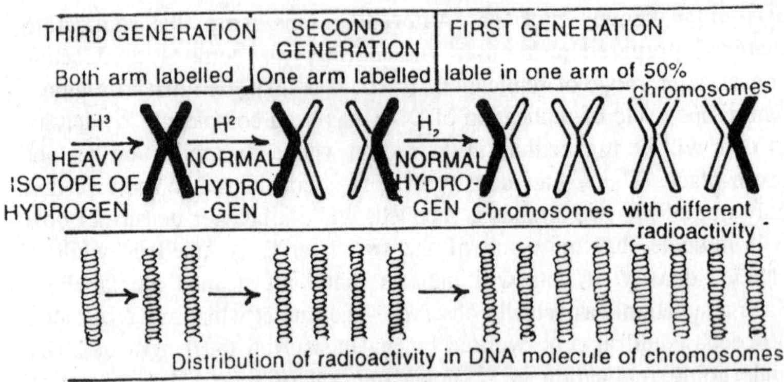

Fig. 8.4. Taylor's experiment on Vicia faba.

This experiment demonstrates the semiconservative method of chromosome replication. But as it is known that each chromatid is composed of only one double helical molecule of DNA, it is the replication of DNA.

ROLE OF ENZYMES AND PROTEINS REPLICATION

The complicated process of DNA replication in *E. coli* requires many different proteins and enzymes, some of which also function in other cellular process such as the repair of DNA damage and genetic recombination. The key enzyme in the synthesis of new DNA is DNA polymerase.

In *E. coli* there are three DNA polymerases. Each catalyzes the DNA template-directed condensation of deoxyribonucleoside 5′-triphosphates. There are significant differences in their activity in DNA synthesis and in the nuclease (DNA or RNA break-down) activities associated with them. All three polymerases catalyze new DNA synthesis in the 5′-3′ direction. The rate of DNA synthesis catalyzed by the three enzymes varies circular prokaryotic DNA molecules replicate bidirectionally (in both directions away from the origin of replication). In these cases there are two replication forks that are mirror images of one another; picture two Ys joined head-to-head by the two pairs of arms. The area of a double-stranded DNA molecule that has denatured for replication is called a *replication bubble*.

In *E. coli* the unwinding of the DNA is catalyzed by an enzyme called helicase, the product of a gene called *rep*. For each 10 base pairs of DNA unwound, one turn of the helix becomes untwisted. As a result, as the helix unwinds a torsional strain is imposed on the DNA. That is, since the DNA molecule is circular, the pulling apart of the helix at one location will cause increased tightening of the molecule elsewhere, much like what happens when the strands of a piece of rope that is fixed at each end are pulled apart. This torsional strain is relieved by the action of enzymes called topoisomerases, which cause single-stranded breaks in DNA away from the replication forks and then allow one strand to rotate relative to the other strand. The unwinding of the DNA produces single-stranded regions, which are stabilized by DNA *single-stranded binding* (SSB) *proteins*. Each SSB protein is a tetramer of a 74,000 dalton subunit, and this tetramer binds to eight bases of single stranded DNA. Over 250 tetramers bind to each replication fork. Once the strands have started to unwind, the internal bases become available for the formation of bonds with bases in the new chain. Initiation of DNA replication next takes place. In *E. coli*, primases bind to the single-stranded DNA of both arms of the replication fork and synthesize short primers. The primers are lengthened by the action of DNA polymerase III, which synthesizes the complementary DNA chains to the template strands. Note that, because of the opposite polarities of the two DNA template strands and the requirement for a 5′- to -3′ direction of new DNA synthesis, the primers are located at different relative positions on the two template strands.

The DNA helix continues to unwind. On the template on the left the new strand continues to be synthesized continuously (in the same

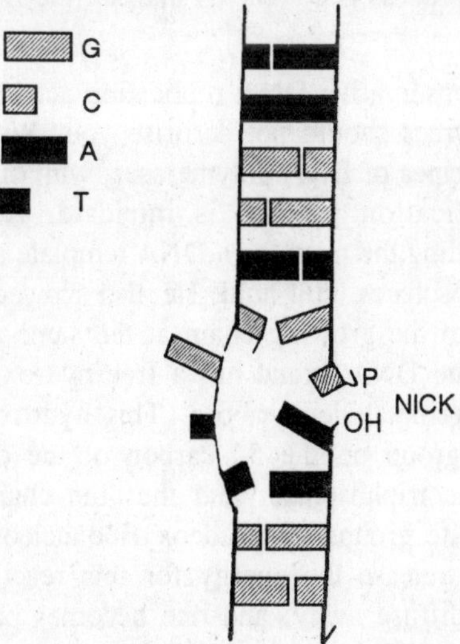

Fig. 8.5. A nicked duplex DNA molecule such as is formed by an endonuclease enzyme.

direction as the direction of unwinding). However, because DNA polymerases can only synthesize DNA in the 5′-to-3′ direction, the new synthesis on the template on the right has gone as far as it can. Thus, a new initiation of DNA synthesis occurs on the right-hand template on the newly exposed single-stranded template close to the site of helicase activity. As before, a primer is made, which is lengthened by DNA polymerase III action. The overall result is that the new DNA strand being made on the left-hand template is synthesized continuously while the new DNA being made on the right-hand template is synthesizd discontinuously so that Okazaki fragments are produced. Eventually the Okazaki fragments are joined together into a continuous DNA strand. This process requires the activities of the enzymes DNA polymerase I and DNA ligase. If we consider two adjacent Okazaki fragments, the 3′ end of the newer DNA fragment is adjacent to but not joined to the primer 5′ end of the previously synthesized fragment. The DNA polymerase III that has just completed the synthesis of the newer fragment now dissociates from the DNA and DNA polymerase I takes its place. This enzyme continuous the 5′-to-3′ synthesis of the newer DNA fragment, at the same time removing the primer of the older fragment through the action of the 5′-to-3′ exonuclease function of DNA polymerase I. When DNA polymerase I is finished, there is a gap between the two new DNA fragments, and this gap is sealed in a reaction catalyzed by DNA ligase, thereby producing a longer DNA strand. The whole process is completed until all the DNA is replicated.

Replication Fork

How is semiconservative DNA replication accomplished? That is depends upon enzymes should not surprise you. We now know that there are different types of DNA polymerases, with different functions, and that the replication process is intricate. Kornberg's basic observations, including the need for a DNA template and for a mixture of nucleoside triphosphates, still hold. He also showed that nucleotides are always added to the growing chain at the same end: the 3′ end, the end at which the DNA strand has a free hydroxyl groups on the 3′ carbon of its terminal deoxyribose. This hydroxyl groups reacts with a phosphate group on the 5′ carbon of the deoxyribose of a deoxyribonucleoside triphosphate, and thus the chain grows. Bonds linking the phosphate groups of the deoxyribonucleoside triphosphate break and thereby release the energy for this reaction. Two of the phosphate groups diffuse away, and one becomes part of the sugar-phosphate backbone of the growing DNA molecule.

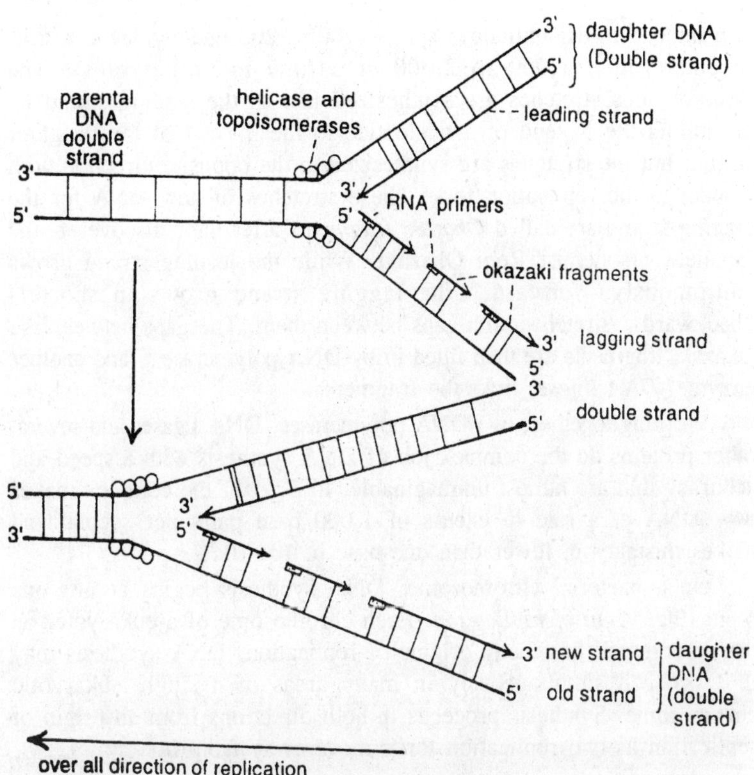

Fig. 8.6. Diagrammatic representation of discontinuous synthesis of DNA.

For double-stranded DNA to replicate, the strands must unwind and separate from each other. Only then can they function as templates for the synthesis of new, complementary strands. The unwinding results in a *replication fork*—a moving, Y-shaped structure that is the region where new DNA strands are being synthesized. Recall that the two original strands are antiparallel, with the 3′end of one strand paired with the 5′ end of the other. As the replication fork moves along the parent DNA molecule, an enzyme, DNA polymerase III, catalyzes the replication of both strands. How can this be accomplished, given that new nucleotides are added only at the 3′ end of a polynucleotide chain?

One parent strand is being exposed beginning at its 3′ end, which presents no problem—its complementary strand is synthesized continuously as the replication fork proceeds. This daughter strand is called the *leading strand*. The other daughter strand, the *lagging strand*, is produced in discontinuous spurts (100 to 200 nucleotides at a time in eukaryotes; 1,000 to 2,000 at a time in prokaryotes). The discontinuous stretches are synthesized just as the leading strand is, by adding the 5′ end of a nucleotide to the 3′ end of the daughter strand, but the stretches are synthesized in the opposite direction with respect to the replication fork. These stretches of new DNA for the lagging strand are called *Okazaki fragments* after their discoverer, the Japanese biochemist Reiji Okazaki. While the leading strand grows continuously "forward," the lagging strand grows in shorter, "backward," stretches with gaps between them. The gaps between the Okazaki fragments are then filled in by DNA polymerase I, and another enzyme, *DNA ligase*, links the fragments.

Working together, two DNA polymerases, DNA ligase, and several other proteins do the complex job of DNA synthesis with a speed and accuracy that are almost unimaginable. In *E. coli*, the complex makes new DNA at a rate in excess of 1,000 base pairs per second and makes mistakes in fewer than one base in 10^8–10^{12}.

On a bacterial chromosome, DNA synthesis begins at just one point, the *origin of replication*. Each chromosome of a eukaryote, on the other hand, has many origins of replication. DNA synthesis may thus proceed simultaneously in many areas of a single eukaryotic chromosome. Synthesis proceeds in both directions from an origin of replication as two replication forks move away from it.

Enzymes used in Replication

The enzymatic synthesis of DNA is a complex process, primarily because of the need for high fidelity in copying the base sequence and for physical separation of the parental strands. The number of steps that must be completed is far too great to be accomplished by a single enzyme and, in fact, about twenty proteins are known at present to be necessary. Thus, in an effort to provide some understanding with a minimum of confusion, each step in the process will be treated separately. We will consider the basic chemistry of polymerization, the source of the precursors, the problems raised by the chemistry of polymerization, the means of initiating and terminating synthesis, and the mechanisms for eliminating replication errors.

Polymerases

In 1957, Arthur Kornberg showed that in extracts of *E. coli* there exists a DNA polymerase (now called *polymerase I* or *pol I*). This enzyme was able to synthesize DNA from four precursor molecules—

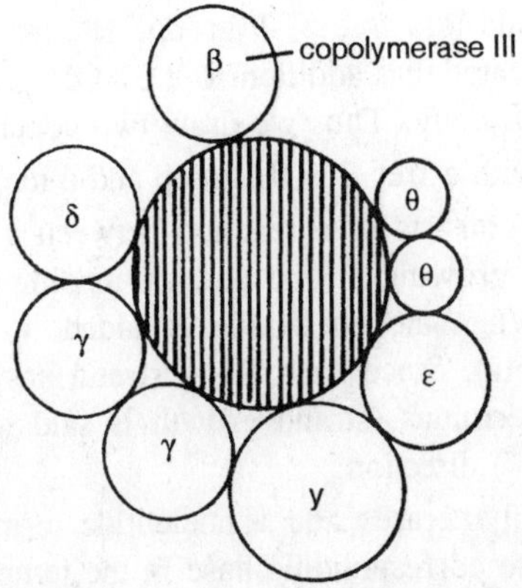

Fig. 8.7. DNA polymerase-III holoenzyme. It consists of subunit α, β, γ, ε θ, δ *and* ε.

namely, the four deoxynucleoside 5´-triphosphates (dNTP), dATP, dGTP, dCTP, and dTTP—as long as a DNA molecule to be copied (a *template* DNA) was provided. Neither 5´-monophosphates nor 5´-diphosphates, nor 3´-(mono-, di-, or tri-) phosphates can be polymerized—only the 5´-triphosphates are substrates for the polymerization reaction; soon we will see why this is the case. Some years later, it was found that pol I, though playing an essential role in the replication process, is not the major polymerase in *E. coli*; instead, the enzyme responsible for advance of the replication fork is polymerase III or pol III. Pol III also exclusively uses 5´-triphosphates as precursors and requires a DNA template before polymerization can occur. Pol I and pol III have many features in common and, in fact, a few types of DNA molecules replicate by using only pol I. The overall chemical reaction catalyzed by both DNA polymerases is:

$$\text{Poly(nucleotide)}_n\text{-3}'\text{-OH} + \text{dNTP} \rightarrow \text{Poly(nucleotide)}_{n+1}\text{-3}'\text{-OH} + \text{PP}$$

in which PP represents pyrophosphate cleaved from the dNTP.

Pol I and pol III have many features in common. Both enzymes only polymerase deoxynucleoside 5´-triphosphates and can do so only while copying a template DNA. Furthermore, polymerization can only occur by addition to a *primer*—that is, an oligonucleotide hydrogen-bonded to the template strand and whose terminal 3´-OH group is available for reaction (that is, a "free" 3´-OH group). The meaning of a primer is already made clear, which depicts six potential template molecules; of these, only three can be said to be active—(c), (e), and (f)—each of which has a free 3´-OH group. The lack of activity with (d) and the direction of synthesis with (e) and (f) indicate that nucleotides do not add to a free 5′-P group. The lack of any synthesis with (a) or (b) indicates that addition to a 3′-OH group cannot occur if there is nothing to copy. Thus we draw two conclusions:

1. Both a primer with a free 3′-OH group and a template are needed.
2. Polymerization consists of a reaction between a 3′-OH group at the end of the growing strand and an incoming nucleoside 5′-triphosphate. When the nucleotide is added, it supplies another free 3′-OH group. Since each DNA strand has a 5′-P terminus and a 3′-OH terminus, strand growth is said to proceed in the 5′→3′ (5′-to-3′) direction.

Occasionally polymerases add a nucleotide terminus that cannot hydrogen-bond to the corresponding base in the template strand. This may be purely a mistake or may result from the tautomerization of adenine and thymine. In any case, it is important that the unpaired base be removable while its incorrectness is recognizable—namely, as an unpaired base at the 3′-OH terminus of a growing strand.

Pol I responds to an unpaired terminal base by terminating polymerizing activity, because the enzyme requires a primer that is correctly hydrogen-bonded. When such an impasse is encountered, a 3′→5′ exonuclease activity, which may be thought of simply as pol I running backwards or in the 3′→5′ direction, is stimulated, and the unpaired base is removed. After removal of this base, the exonuclease activity stops, polymerizing activity is restored, and chain growth begins again. This exonuclease activity is called the *proofreading* or *editing function* of pol I.

Another function of polymerase I is that of a 5′→3′ exonuclease. This activity has the following features:

1. Nucleotides are removed from the 5′-P terminus only, one by one.
2. More than one nucleotide can be removed by successive cutting.
3. The nucleotide removed must have been base-paired.
4. The nucleotide removed can be either or the deoxy-or the ribo-type.
5. Activity can be at a nick as long as there is a 5′-P groups.

The main function of the 5′→3′ exonuclease activity is to remove ribonucleotide primers. The 5′→3′ exonuclease activity at a single-strand break (nick) can occur simultaneously with polymerization. That is, as a 5′-P nucleotide is removed, a replacement can be made by the polymerizing activity. Since pol I cannot form a bond between a

3′-OH group and a 5′-monophosphate, the nick moves along the DNA molecule in the direction of synthesis. This movement is called *nick translation*.

Experimental conditions can be chosen so that polymerization will occur at a single-strand break without concomitant 5′→3′ exonuclease activity. The growing strand then displaces the parental strand. This is thought to be an important step in the mechanism of genetic recombination. Of all *E. coli* polymerases known to date, polymerase I is the only one capable of carrying out an unaided displacement reaction. In other strand displacement reactions, auxilary proteins are required and ATP is cleaved to fuel the unwinding of the helix; this will be discussed when the events at a replication fork are described.

Pol III is a very complex enzyme. In its most active form it is associated with eight other proteins to form the *pol III holoenzyme*, occasionally termed pol III. The term holoenzyme refers to an enzyme that contains several different subunits and retains some activity even when one or more subunits is missing. The smallest aggregate having enzymatic activity is called the *core enzyme*. The activities of the core enzyme and the holoenzyme are usually very different. Genes encoding five of the subunits have been identified; these are called *dnaE*, *dnaN*, *dnaQ*, *dnaX*, and *dnaz*. The *dnaE* protein possesses the major polymerizing activity but each of the subunits, except for the *dnaQ* protein is essential for replication. Pol III shares wtih pol I a requirement for a template and a primer but its substrate specificity is much more limited. For instance, pol III cannot act at a nick nor is it active with single-stranded DNA primed by either a DNA or RNA nucleotide fragment. The principal activity *in vitro* is on gapped DNA in which the gap is less than 100 nucleotides long. Such a gap is akin to the state of the DNA at a replication fork—that is, the parental strands are separated and bear short single-stranded regions ahead of the growing daughter chain.

Pol III cannot carry out strand displacement either, and another system is needed to unwind the helix in order that a replication fork will be able to proceed. The enzyme, like pol I, possesses a 3′→5′ exonuclease activity which performs the major editing function in DNA replication. This function is carried out by the *dnaE* subunit, which is also the major polymerizing subunit, as we have just mentioned. The *dnaQ* subunit plays an important role in editing, also, but the biochemical basis of the role is not yet known; it probably interacts with the *dnaE* subunit. The principal evidence supporting the view that it is involved in providing fidelity to the replication process is that bacteria containing a mutation in the *dnaQ* gene have a somewhat higher mutation frequency. Pol III also possesses a 5′→3 exonuclease activity; however, the enzyme acts only on single-stranded DNA so that it cannot carry out nick translation. The biological role of the 5′→3′ exonuclease activity of pol III is unknown at present. Although pol III holoenzyme is the major replicating enzyme in *E. coli*, much less is known about it than about pol I, because it is a more complex enzyme. Study of pol III is currently an active field of research.

All known polymerases (for both DNA and RNA) are capable of chain growth in only the 5′→3′ direction; that is, the growing end of the polymer must have a free 3′-OH group. It is possible for the following reasons that the enzymes evolved in this way to facilitate editing. If 3′→5′ growth were to occur, the growing strand would also be terminated with a 5′-triphosphate and the 3′-OH group of the incoming nucleotide would react with it. Chemically this is certainly acceptable but since the bonds formed contain only a single phosphate, an editing function would leave a free 5′-monophosphate. In order for chain growth to proceed, an enzymatic system would be needed to enter the replication fork and convert the monophosphate to a triphosphate. There is already a great deal going on in the replication fork, so that it would seem more economical for the cell to require 5′→3′ growth exclusively. However, the observation that chain growth proceeds in only one direction introduces what is probably the greatest complication in the entire replication process; this will be described shortly.

DNA Ligase

Neither replication from a primed circular single strand nor gap filling results in a continuous daughter strand. Discontinuity results because no known polymerase can join a 3′-OH and a 5′-monophosphate group. The joining of these groups is accomplished by the enzyme *DNA ligase*, which functions in replication and other important processes. *E. coli* DNA ligase can join a 3′-OH group as long as both are termini of adjacent base-paired deoxynucleotides—the enzyme cannot bridge a gap.

In the usual polymerization reaction, the activation energy for phosphodiester bond formation comes from cleaving the triphosphate. Since DNA ligase has only a monophosphate to work with, it needs another source of energy. It obtains this energy by hydrolyzing either

ATP or nicotine adenine dinucleotide (NAD); the energy source depends upon the organism from which the DNA ligase is obtained. The *E. coli* DNA ligase uses NAD.

Continuous and Discontinuous DNA Replication

Autoradiographic evidence leads us to believe that replication is occurring simultaneously on both strands. *Continuous* replication is, of course, possible on the 3′→5′ template strand, which begins with the necessary 3′-OH *primer*. (Primer is double-stranded DNA—or, as we shall see, a DNA-RNA hybrid—continuing as single-stranded DNA template. The strand being synthesized has a 3′-OH available). A *discontinuous* form of replication takes place on the complementary strand, where it occurs in short segments, backward, away from the Y-junction. These short segments, called *Okazaki fragments* after *R. Okazaki* who first saw them, average about 1,500 nucleotides in prokaryotes and 150 in eukaroytes. The strand synthesized continuously

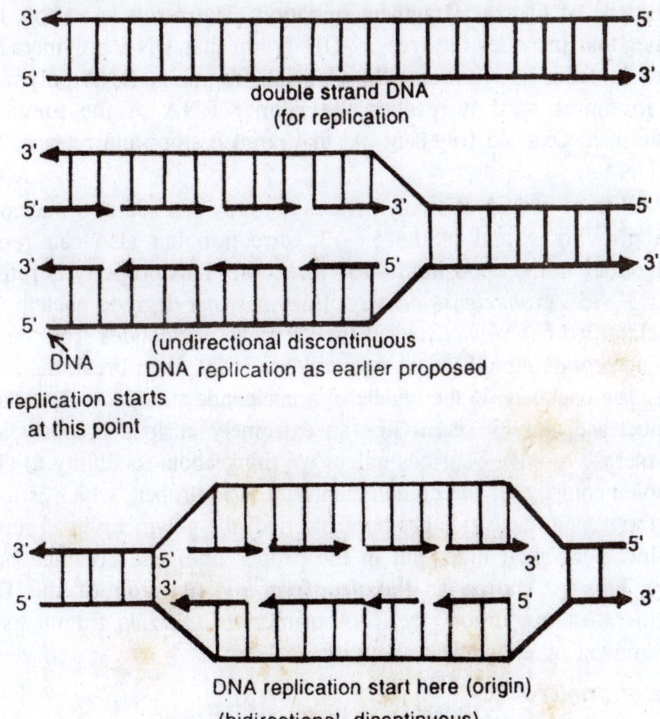

Fig. 8.8. Discontinuous DNA replication.

is referred to as the *leading strand*, and the strand synthesized discontinuously is referred to as the *lagging strand*.

Once initiated, continuous DNA repliction can proceed indefinitely. DNA polymerase III on the leading strand template has what is called high *processivity*: once it attaches, it does not release until the entire strand is replicated. Discontinuous replication, however, requires the repetition of four steps primer synthesis, elongation, primer removal with gap filling, and ligation.

Primer synthesis and elongation

In order for Okazaki fragments to be synthesizd, a primer must be created de novo (Latin, from the beginning). None of the DNA polymerases can create that primer. Instead, one of two enzymes, either *RNA polymerase*, the transcribing enzyme or, more commonly, *primase*, an RNA polymerase coded for by the *dnaG* gene, creates the primer. It is from two to sixty nucleotides, depending on the species, at the site of Okazaki fragment initiation. The result is a short RNA primer that provides the free 3´-OH group that DNA polymerase III needs in order to synthesize the Okazaki fragment. DNA polymerase III continues until it reaches the primer RNA of the previously synthesized Okazaki fragment. At that point it stops and releases from the DNA.

All three prokaryotic polymerases not only can add new nucleotides to a growing strand in the 5´→3´ direction but also can remove nucleotides in the opposite 3´→5´ direction. This property is referred to as *3´→5´ exonuclease activity*. Enzymes that degrade nucleic acids are classified as *exonucleases* if they remove nucleotides from the end of a nucleotide strand or as *endonuclease* if they can break the sugar-phosphate backbone in the middle of a nucleotide strand. At first glance, exonuclease activity seems like an extremely curious property for a polymerase to have—curious unless we think about its ability to check complementarity. If the complementarity is improper, which is to say that the wrong nucleotide has been inserted, the polymerase can remove the incorrect nucleotide, put in the proper one, and continue on its way. This is known as the *proofreading* function of the DNA polymerase. In addition, the RNA primers of Okazaki fragments can be removed by exonuclease activity.

Role of primer

DNA polymerase I is a polymerase when it adds nucleotides, one at a time, and an exonuclease when it removes nucleotide one at a time. To complete the Okazaki fragment, DNA polymerase I acts in both capacities. (Mutants of DNA polymerase I cannot properly connect Okazaki fragments.) DNA polymerase I completes the Okazaki fragment by removing the previous RNA primer and replacing it with DNA nucleotides. When DNA polymerase I has completed its nuclease and polymerase activity, the two previous Okazaki fragments are almost complete. All that remains is a single phosphodiester bond to be made.

Ligation

DNA polymerase I cannot make the final bond to join the Okazaki fragment to the previously synthesized DNA. An enzyme, *DNA ligase*, completes the task by making the final phosphodiester bond in an energy-requiring reaction.

A question of evolutionary interest is why RNA is used for priming of DNA synthesis. Why not use DNA directly and avoid the exonuclease and resynthesis activity? One possible answer is that because priming is inherently more error-prone then regular DNA synthesis it is best for the cell to have the primer nucleotides removed and replaced by DNA synthesized in a less error-prone fashion before DNA synthesis is completed. If the priming nucleotides are RNA, then DNA polymerase I can recognize and remove them in the final stage of Okazaki fragment synthesis, at which time it replaces them with DNA nucleotides inserted with a low error rate.

Another question of evolutionary interest is why DNA synthesis cannot take place in the 3′→5′ direction. Perhaps the answer has to do with proofreading and the exonuclease removal of incorrect nucleotides. When an incorrect nucleotide is found and removed, the next nucleotide brought in, in the 5′→3′ direction, will have a triphosphate end available to provide the energy for its own incorporation. Consider what would happen if the polymerase were capable of adding nucleotides in the opposite direction. The energy for the diester bond would be coming from the triphosphate already attached in the growing 3′→5′ strand. Then, if an error in complementary were detected and the most recently added nucleotide were removed from the 3′→5′ strand by the polymerase, the last nucleotide in the double helix would no longer have a triphosphate available to provide energy for the diester bond with the next nucleotide brought in. Continued polymerization would thus require additional enzymatic steps to provide the energy for the process to continue. This could slow the process down considerably. As it is, the process works at a speed of about four hundred nucleotides incorporated per second with an error rate of about one incorrect pairing per one hundred thousand bases, improved to a rate of only one mistake in ten million by exonuclease proofreading. (Other repair system can improve this error rate another thousandfold, to about one error every 10^{10} times an average base is replicated.

The Initiation of DNA Replication

Each replicon (e.g., the *E. coli* chromosome or a segment of eukaryote chromosome) must have a region in which DNA replication is initiated. In *E. coli* this region is referred to as the genetic locus *oriC*. In order for DNA replication to begin, several steps must occur. First, the specific origin site must be recognized by the appropriate protein. Then the site must be opened and stabilized. And, finally, a replication fork must be initiated in both directions, involving continuous and discontinuous DNA replication. Although many of the proteins involved are known, all the steps at the enzymatic level are not, and hence our understanding is a bit sketchy. Following is a description, most of whose steps are known.

OriC, the origin of replication in *E. coli*, is about 245 base pairs long and is recognized by proteins called *initiator proteins* that open up the double helix. The initiator proteins then take part in the attachment of *primosomes*, a complex of two proteins: a primase, which creates RNA primers, and DNA *helicase*, which unwinds DNA at the Y-junction. As the primosomes move along, they create RNA primers used by DNA polymerase III to initiate Okazaki fragments. At some point, leading-strand synthesis begins and Y-junction activity then proceeds as outlined earlier.

In some phages and plasmids, the initiation of replication is not with an RNA primer but with a protein. This protein provides the primer configuration with an OH group from an amino acid. The generality of this type of priming has not been established. Another interesting protein interaction at the origin of replication involves the reverse of initiation of DNA synthesis, the prevention of the initiation of DNA synthesis. This is accomplished by a newly discovered protein, called an "off switch." That is, this protein binds to the DNA at *oriC* and apparently prevents DNA replication from beginning. It does this by its binding activity that prevents the initiator proteins from opening the DNA. Thus this protein may be a very important component in control of the cell cycle stopping the cell cycle from beginning. Presumably, when the appropriate time comes for the cell cycle to begin, the protein is removed.

Events at the Y-Junction

We now have the image of DNA replication proceeding by a primosome, moving along the lagging strand template, opening up the DNA (helicase activity), and creating RNA primers (primase activity). One DNA polymerase III moves along the leading-strand template generating the leading strand by continuous DNA replication, whereas a second DNA polymerase III moves backward, away from the Y-junction, creating Okazaki fragments. *Single-strand binding proteins* (ssb proteins) keep single-stranded DNA stabilized (open) during this process, and DNA polymerase I and ligase are connecting Okazaki fragments.

This simple picture is slightly complicated by the fact that a single DNA polymerase seems to do the entire lagging strand, rather than dropping off the DNA at the completion of an Okazaki fragment and being replaced by a new one at the newest primer near the Y-junction. In addition, there is evidence that the lagging and leading-strand synthesis is coordinated. The *replisome* model has arisen in which both copies of DNA polymerase III are attached to each other and work in concert with the primosome at the Y-junction. According to this model, a single replisome consisting of two copies of the DNA polymerase III *holoenzyme* (each actually made of seven subunits), a helicase, and a primase, move along the DNA. The leading-strand template is immediately fed to a polymerase, whereas the lagging - strand template is not acted on by the polymerase unit an RNA primer has been placed on the strand, meaning that a long (fifteen hundred base) single strand has been opened up.

As the replisome moves along, another single-stranded length of the lagging-strand template is formed. At about the time that the Okazaki fragment is completed, a new RNA primer has been created. The Okazaki fragment is released and a new Okazaki fragment is begun, starting with the latest primer taking the replisome back to the same configuration, but one Okazaki fragment farther along.

Supercoiling

The simplicity and elegance of the DNA molecule masks an inevitable problem of coiling. Since the DNA molecule is made from two strands that wrap about each other, certain operation, such as DNA replication and its termination, meet topological difficulties. Up to this point, we have seen the circular *E. coli* chromosome in its "relaxed" state. However, there are enzymes in the cell that cause DNA to become overcoiled (positively *supercoiled*) or undercoiled (negatively supercoiled). Positive supercoiling comes about either from too many turns of the DNA in a given length or from the molecule wrapping around itself.

Positive supercoiling comes from having the circular duplex wind about itself in the same direction as the helix twists (right handed), whereas negative supercoiling comes about by having the duplex wind about itself in the opposite direction as the helix twists (left handed). The former state increases the number of turns of one helix around the other side (the *linkage number*, L), whereas the latter decreases it. The three forms of DNA, all have the same sequence yet differ in their linkage number. They are referred to as topological isomers (*topoisomers*). The enzymes that create or alleviate these states are called *topoisomerases*.

Topoisomerases affect supercoiling by either of two methods. Type I topoisomerases break one strand of a double helix and, while binding the broken ends, pass the other strand through the break. The break is then sealed. Type II topoisomerases (e.g., *DNA gyrase* in *E. coli*) do the same sort of thing only instead of breaking one strand of a double helix, they break both and pass another double helix through the temporary gap.

As DNA replication proceeds, positive supercoiling builds up ahead of the Y-junction. This is eliminated by the action of topoisomerases that either create negative supercoiling ahead of the Y-junction in preparation for replication or alleviate positive supercoiling after it has been created. The major components of DNA replication in *E. coli* are summarized in table 8.1.

Termination of Replication

The termination of the replication of a circular chromosome presents no major topological problems. The theta-structure replication finishes with both Y-junctions having proceeded around the molecule. The leading strand on one template closes in on the lagging strand begun in the other direction with the same happening on the other template. The process stops with about twenty-five twists remaining at no particular spot on the chromosome (there is no "termination" locus). A topoisomerase then release the two circles and DNA polymerase I and ligase close them up.

Several different mechanisms have been explored for the termination of the linear chromosomes of some viruses and all eukaryotic genomes. Linear molecules have the problem of completing the last Okazaki fragment. An RNA primer on the very tip of the

Table 8.1. Summary of the Enzymes Involved in DNA Replication in E.coli.

Enzyme (Protein)	*Genetic Locus*	*Function*
DNA polymerase I	*pol A*	Gap filling and primer removal
DNA polymerase II	*pol B*	?
DNA polymerase III		
α subunit	*dnaE* (*polC*)	DNA replication
β subunit	*dnaN*	DNA replication
γ subunit	*dnaX*	DNA replication
δ subunit	?	DNA replication
ε subunit	*dnaQ*	3′→5′ exonuclease
θ subunit	?	DNA replication
τ subunit	*dnaX*	DNA replication
Initiator protein	*dnaA*	Binds to origin of replication
RNA polymerase subunit	*rpoA, B, C, D*	RNA primer in some system
Primase	*dnaG*	RNA primer in some system
DNA ligase	*lig*	Closes nicked DNA strands
Helicase	*rep*	Unwinds DNA for replication
Ssb proteins	*ssb*	Single-strand stability
DNA topoisomerase I	*topA*	Supercoiling of DNA
DNA topoisomerase II		
α subunit	*gyrA* (*nalA*)	ATPase
β subunit	*gyrB* (*cou*)	Cutting, closing of DNA

3′→5′ template cannot be replaced by DNA polymerase I, assuming even that a final primer can be put on the very tip of the molecule. In eukaryotes, an enzyme, telomerase, attaches repeats of a short sequence at each chromosome tip.

Replication Models

The model of DNA replication that we have presented here comes primarily from evidence gathered in *E. coli*, which replicates by way of the *theta*-structure intermediate. However, two other modes of replication occur in circular chromosomes: rolling-circle and D-loop.

Rolling-Circle Model

In the *rolling-circle* mode of replication, a nick (a break in one of the phosphodiester bonds) is made in one of the strands of the circular DNA, resulting in replication of a circle and a tail. This form of replication occurs in the Hfr *E. coli* chromosome, or the F plasmid, during conjugation. The F^+ or Hfr cell retains the circular daughter while passing the linear tail into the F^- cell. This method is also used in several phages, which fill their heads (protein coats) with linear DNA replicated from a circular parent molecule.

In the model for rolling-circle replication, the nick made in one strand creates a free 3´-OH end and a free 5´-PO_4 end. Synthesis of a new circular strand occurs by addition of nucleotides to the 3´ end using the complementary intact strand as a template. No primer is needed because the original break produces a primer configuration (3´-OH). As nucleotides are added to one end of the broken strand in a continuous fashion, the other end is displaced as a 5´-PO_4 tail. As replication of the circular templates occurs, the 5´-PO_4 tail is replicated in a discontinuous manner, and the resulting double helix can be severed from the double-helical circle by a nuclease. DNA ligase closes the replicated circular strand and can join the ends of the replicated tail into a circle in the F^- cell, or can package the linear molecule in a phage head, depending upon which type of circular DNA has been replicated.

D-Loop Model

Chloroplasts and mitochondria have their own circular DNA molecules that appear to replicate by a slightly different mechanism than those described. The origin of replication is at different point on each of the two parental template strands. Replication begins on one strand, displacing the other while forming a displacement loop or *D-loop* structure. Replication continues until the process passes the origin of replication on the other strand. Replication is then initiated on the second strand, in the opposite direction. The result is two circles. Some species have chloroplasts and mitochondria with circular DNAs that have multiple D-loops formed.

Eukaryotic DNA Replication

As we saw earlier, linear eukaryotic chromosome usually have multiple origins of replication resulting in figures referred to as "bubbles" or "eyes." Multiple origins allow eukaryotes to replicate their larger quantities of DNA in a relatively short time, even though eukaryotic DNA replication is considerably showed by the presence of histone proteins associated with the DNA to form chromatin. For example, the *E. coli* replication fork moves about twenty-five thousand base pairs per minute, whereas the eukaryotic Y-junction moves only about two thousand base pairs per minute. The number of replications in eukaryotes varies from about five hundred in yeast to as many as sixty thousand in a diploid mammalian cell.

Much less is understood about eukaryotic DNA replication because of the complexity of eukaryotes and the relatively shorter time during which they have been studied effectively. We presume that eukaryotes have solved the same problems faced by prokaryotes in a similar, but not identical, fashion. For example, eukaryotes have five types of DNA polymerases, named DNA polymerase α, β, γ, δ, and ε. DNA polymerases γ, δ, and ε have exonuclease activity. DNA polymerases α and δ are the major replicating enzymes, with polymerase α replicating the lagging strand and polymerase δ replicating the leading strand. The role of polymerase ε is unclear; it seems capable of regular leading- or lagging-strand replication. DNA polymerase β is the major repair polymerase (like polymerase I in prokaryotes). DNA polymerase γ appears to be concerned primarily with mitochondrial DNA replication.

Table 8.2. Eukaryotic DNA Polymerases

Enzyme	*Function*
DNA polymerase α	Replication of nuclear chromosomes (lagging strand)
DNA polymerase β	Repair of nuclear chromosomes
DNA polymerase γ	Replication of mitochondrial chromosomes
DNA polymerase δ	Replication of nuclear chromosomes (leading strand)
DNA polymerase ε	Probably replication of nuclear chromosomes

9

Manipulation of Genes

The manipulation of gene is the called the genetic engineering. Genetics is also important in manipulating biological systems for scientific or economic reasons, an endeavor that has made possible the creation of organisms having new phenotypes or genotypes. The fundamental techniques for accomplishing this have been mutagenesis or recombination followed by selection for desired characteristics. When using such techniques, geneticists have been forced to work with the random nature of mutagenic and recombination events, which required selective procedures, often quite complex, to find an organism with the required genotype among the many types of organisms produced. Since the 1970s techniques have been developed by which the genotype of an organism can instead be modified in a *directed* and *predetermined* way. This is alternately called *recombinant DNA technology*, or *genetic engineering*.

Genetic engineering involves isolation of DNA fragments and recombination outside of a cell. Selection of a desired genotype is still necessary but the probability of success is usually many orders of magnitude greater than that with traditional procedures. The basic technique is quite simple: two DNA molecules are isolated and cut into fragments by one or more specialized enzymes and then the fragments are joined together *in a desired combination* and restored to a cell for replication and reproduction.

Current interest in genetic engineering centers on its many practical applications, a few examples of which are the following:

1. Isolation of a particular gene, part of a gene, or region of a genome.
2. Production of particular RNA and protein molecules in quantities formerly thought to be unobtainable.

3. Improvement in the production of biochemicals (such as enzymes and drugs) and commercially important organic chemicals.
4. Production of varieties of plants having particular desirable characteristics (for example, requiring less fertilizer or resistance to disease).
5. Correction of genetic defects in higher organisms, and
6. Creation of organisms with economically important features (for example, plants capable of maturing faster or having greater yield).

Some of these examples will be considered later in the chapter.

Separation of Particular DNA Fragments

In genetic engineering the immediate goal of an experiment is usually to insert a *particular* fragment of chromosomal DNA into a plasmid or a viral DNA molecule. This is accomplished by techniques for breaking DNA molecules at specific sites and for isolating particular DNA fragments. DNA fragments are obtained by treatment of DNA samples with a specific class of nuclease. Many nucleases have been isolated from a variety of organisms, and most produce breaks at random sites within a DNA sequence. However, the class of nucleases called *restriction endonucleases*, or, more simply, *restriction enzymes*, consists of sequence-specific enzymes. Most restriction enzymes recognize only one short base sequence in a DNA molecule and make two single-strand breaks, one in each strand, generating 3´-OH and 5´-P groups at each position. Several hundred of these enzymes have been isolated from hundreds of species of microorganisms.

The sequences recognized by restriction enzymes are often *palindromes*—that is, the sequence has symmetry of the form:

A B C	C´ B´ A´		A B	X	B´ A´		A B	B´ A´
A´ B´ C´	C B A	or	A´ B´	X´	B A	or	A´ B´	B A

in which the capital letter represent bases, a´ indicates a complementary base, X is any base, and the vertical line is the axis of symmetry. Most of these sequence have 4-6 bases.

One of the most exciting events in the study of restriction enzymes was the observation by electron microscopy that fragments produced by many restriction enzymes spontaneously circularize. These circles could be relinearized by heating, but if after circularization they were also treated with *E. coli* DNA ligase, which joins 3´-OH and 5´-P groups, circularization became permanent. This observation was the first evidence for three important features of restriction enzymes:

Table 9.1. Some restriction endonucleases, their sources, and their cleavage sites.

Name of enzyme	*Microorganism*	*Target sequence and cleavage sites*
Generates cohesive ends		
EcoRI	*E.coli*	G↓A A ¦ T T C C T T ¦ A A↑G
BamHI	*Bacillus amyloliquefaciens* H	G↓G A ¦ T C C C C T ¦ A G↑G
HaeII	*Haemophilus aegyptius*	Pu G C ¦ G C↓Py Py↑C G ¦ C G Pu
HindIII	*Haemophilus influenza*	A↓A G ¦ C T T T T C ¦ G A↑A
PstI	*Providencia stuartii*	C T G ¦ C A↓G G↑A C ¦ G T C
TaqI	*Thermus aquaticus*	T↓C ¦ G A A G ¦ C↑T
Generates blunt ends		
BalI	*Brevibacterium albium*	T G G↓C C A A C C↑G G T
SmaI	*Serratia marcescens*	C C C↓G G G G G G↑C C C

Note : The vertical dashed line indicates the axis of symmetry in each sequence. Arrows indicate the sites of cutting. The enzyme TaqI yields cohesive ends consisting of two nucleotides, whereas the cohesive ends produced by the other enzymes contain four nucleotides. Pu and Py refer to any purine and pyrimidine, respectively.

1. Restriction enzymes make breaks in palindromic sequences.
2. The breaks are usually not directly opposite one another.
3. The enzymes generate DNA fragments with complementary ends.

Examination of a very large number of restriction enzymes showed that the breaks are usually in one of two distinct arrangements: (1)

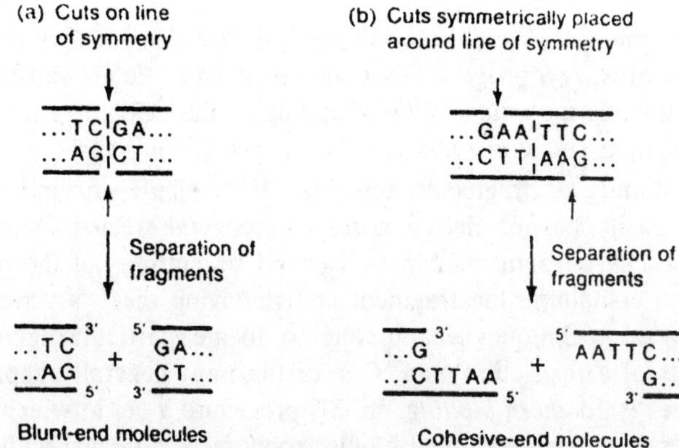

Fig. 9.1. Two types of cuts made by restriction enzymes. The arrow indicate the cleavage sites. The dashed line is the center of symmetry of the sequence.

staggered, but symmetric around the line of symmetry (forming *cohesive ends*) or (2) both at the center of symmetry (forming *blunt ends*). Two types of enzymes produce cohesive ends—those yielding a single-stranded extension with a 5′-P terminus and those yielding a 3′-OH extension. Table 9.1. lists the sequences and cleavage sites for several restriction enzymes, some of which generate cohesive sites and others of which yield blunt ends.

Most restriction enzymes recognize one base sequence without regard to the source of the DNA. Thus,

> Fragments obtained from a DNA molecule from one organism have the same cohesive ends as the fragments produced by the same enzyme acting on DNA molecules from another organism.

Since most restriction enzymes recognize a unique sequence, *the number of cuts made in the DNA from an organism by a particular enzyme is limited.* A typical bacterial DNA molecule, which contains roughly 3×10^6 base pairs, is cut into several hundred to several thousand fragments, and nuclear DNA of mammals is cut into more than a million fragments. These numbers are large but still small compared to the number of sugar-phosphate bonds in an organism. Of special interest are the smaller DNA molecules, such as viral or plasmid DNA, which may have only 1-10 sites of cutting (or even none) for particular enzymes.

Plasmids having a single site for a particular enzyme are especially valuable. Because of the sequence specificity, *a particular restriction enzyme generates a unique set of fragments for a particular DNA molecule.* Another enzyme will generate a different set of fragments from the same DNA molecule. Figure (given below) shows the sites of cutting of *E. coli* phage λ DNA by the enzymes EcoRI and BamHI. A map showing the unique sites of cutting of the DNA of a particular organism by a single enzyme is called a *restriction map*.

The family of fragments generated by a single enzyme can be detected easily by gel electrophoresis of enzyme-treated DNA, and particular DNA fragments can be isolated by cutting out the portion of the gel containing the fragment and removing the DNA from the gel. Several techniques enable one to locate particular genes on fragments of a restriction map. One of the most generally applicable procedures is *Southern blotting*. In this procedure a gel in which DNA molecules have been separated by electrophoresis is treated with alkali to render the DNA single-stranded (denature the DNA) and then the DNA is transferred to a sheet of nitrocellulose in a way that the relative positions of the DNA bands are maintained. The nitrocellulose, to which the single-stranded DNA tightly binds, is then exposed to radioactive RNA or DNA in a way that leads to renaturation.

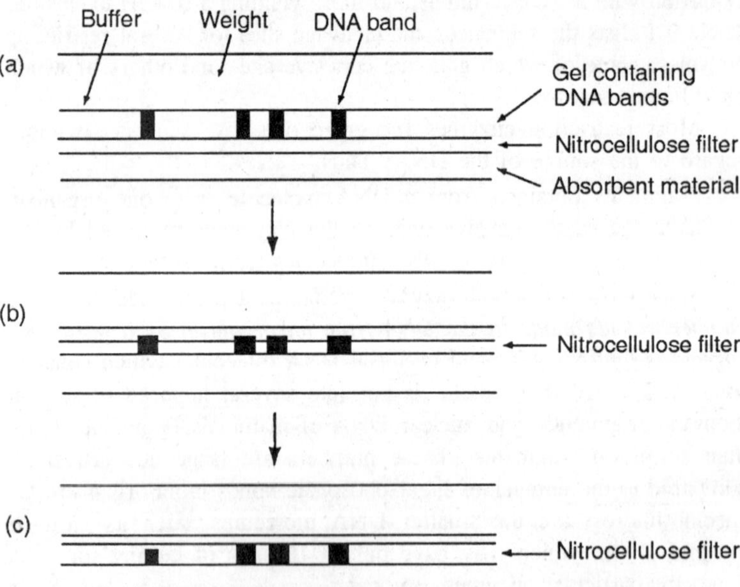

Fig. 9.2. Southern blotting. (a) A weight is placed on a stack consisting of the gel, a nitrocellulose filter, and absorbent material. (b) At a later time the weight has forced the buffer, which carries the DNA, into the nitrocellulose. (c) The lowest layer has absorbed the buffer but the DNA, which remains bound to the nitrocellulose.

Radioactivity becomes stably bound (resistant to removal by washing) to the DNA only at positions at which base sequences complementary to the radioactive molecules are present. The radioactivity is located by placing the paper in contact with x-ray film; after development of the film, blackened regions indicate positions of radioactivity. If a radioactive mRNA species transcribed from a particular gene is used (for example, mRNA isolated from a specialized cell that predominantly makes one type of mRNA), it will hybridize only with the restriction fragment containing that gene.

Restriction Endonucleases

Restriction enzymes are nucleases and as they cut at an internal position of DNA strand (and not at end) they are known as endonucleases. Restriction enzymes fall into following categories:

Type I : These are most complex, bi-functional (i.e. same enzyme possesses both restriction and modification activity) and are made up of 3 subunits. Recognition site is bipartite and asymmetrical. Cleavage site is nonspecific and is atleast 1000 bp from recognition site.

Type II : These are simplest, separate endonuclease and methylase. Recognition site is short sequence (4-6 bp) often palindromic

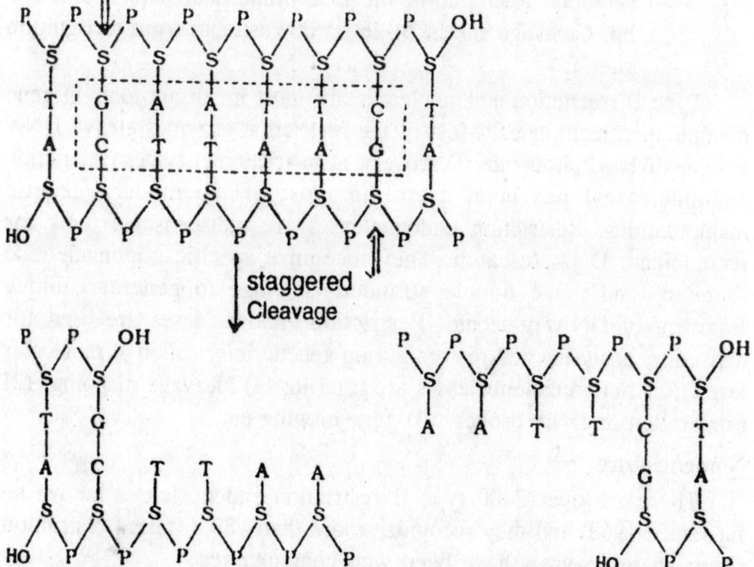

Fig. 9.3. Symmetrical staggered cleavage by Eco RI arrows indicate sites of cleavage.

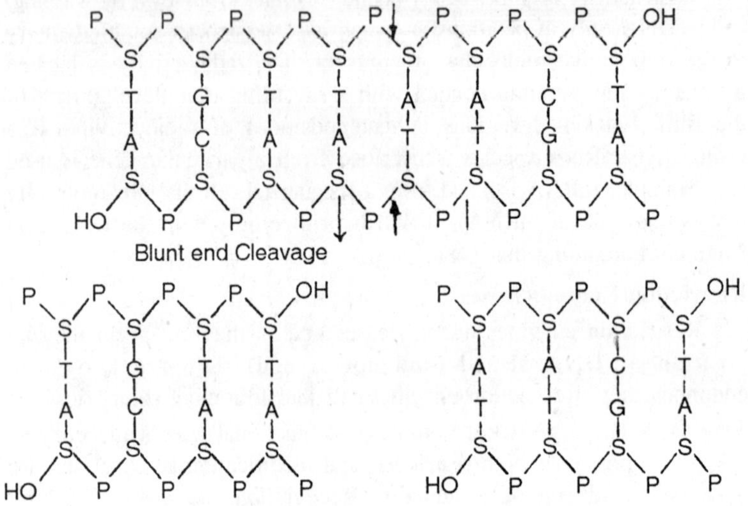

Blunt end Cleavage

Fig. 9.4. Blunt end cleavage by hind III.

(with symmetry). Cleavage site is same as or close to recognition site.

Type III : These are moderate complex, bi-functional, have two subunits. Recognition site is asymmetrical sequence of 5-7 bp. Cleavage site is 24-26 bp downstream from recognition site.

Type II restriction endonucleases are most important tools in gene manipulation techniques. Most of the restriction enzymes cleave DNA at unmethylated target site. Discovery of restriction enzymes (restriction endonucleases) has been useful in most of the modern genetic manipulations. Restriction endonucleases are indispensable tools for recombinant DNA research. They recognize specific oligonucleotide sequence and make double stranded cleavage to generate unique fragment of DNA molecule. Restriction endonucleases are used for dissecting, analysing and re-configuring genetic information at molecular level. Restriction endonucleases are used in: (1) cleavage mapping, (2) preparation of DNA probes, (3) gene cloning etc.

Nomenclature

There are over 2400 type II restriction endonucleases known so far (since 1968) and they recognize more than 188 different restriction sites. These enzymes have been well characterized.

The nomenclature of restriction enzyme is based on various conventions: (1) The generic and specific name of organism in which the enzyme is found are used to provide first part of the designation. Next number in designation indicates serial number or enzyme reported from same organism. Three letter abbrevation is in italic. (2) Strain or type identification is written as a subscript e.g. Eco_k. If restriction and modification system is genetically specified virus or plasmid, the abbreviated species name of host is given and extrachromosomal element as subscript. (3) When a particular host strain has several different restriction and modification system, these are identified by roman numericals I, II, III etc. following 3 letter abbreviation of host strain. (4) The history and incompleteness of the discovery are reflected in nomenclature of restriction endonucleases.

Unidentified bacteria found with these systems receive the genus-species designate *Uba*. As on 1993 hundreds of enzymes were temporarily named as *Uba* (number). In 1973, nomenclature of restriction enzyme was developed and was based on proposals of Smith and Nathan.

Recognition Sites

Restriction enzymes usually recognize a specific DNA sequence of 4, 5 or 6 nucleotides in length and cleave the DNA within this restriction site. There are 4 bases in DNA, randomly distributed. The expected frequency of any particular sequence can be calculated as 4^n where n is the length of recognition sequence. Thus, tetranucleotide sites will occur every 256 basepair, pentanucleotide sites will occur every 1025 basepairs and hexanucleotide sites will occur every 4096 basepairs. Restriction enzymes either cut (1) Straight across the DNA to give blunt ends, or (2) Straight single strand cuts producing short, single stranded projections at each end of the cleaved DNA to produce cohesive or sticky ends.

Table 9.2. Classification of restriction endonucleases based on recognition site.

(I) Tetranucleotides recognizing	—	*Alu I, Hae III, Hpa II, Mbo I, Taq I, Msp I* etc.
(II) Pentanucleotides recognizing	—	*Ava II, Dde I, Eco R II, Hinf I* etc.
(III) Hexanucleotides recognizing	—	*Kpn I, Xma I, Pst I, Hpal, Bam HI, Hind III* etc.
(IV) Heptanucleotides recognizing	—	*Me II* etc.

One unit of enzyme (Restriction endonuclease) is defined as the amount of enzyme required to produce a complete digest of 1 μg of substrate DNA in a reaction volume of 0.05 ml in 1 hour under optimal conditions of salt, pH and temperature. All digestions are performed at 37°C and substrate used is lambda DNA. Optimum temperature for most of the restriction enzymes is 37°C. However, in some cases like *Tha I, Taq I, Bcl I, Bst NI* optimum temperature is 60-65°C. Recognition sequences of *Eco RI* are *palindromic*. Sometimes within a same strand sequence may be palindromic.

Table 9.3. Restriction endonucleases.

	Enzyme	*Source*	*5′ Recognition site 3′*
1.	*Kpn II*	*Klebsiella pneumoniae*	GGTAC↓C
2.	*Bam HI*	*Bacillus amyloliquifaciens*	G↓GATCC
3.	*Sma I*	*Serratia marcesecens*	CCC↓GGG
4.	*Stu I*	*Streptomyces tubercidicus*	AGG↓CCT
5.	*AVA II*	*Anabenia variabilis*	G↓G(A/T) CC
6.	*Pva I*	*Proteus vulgaris*	CGAT↓CG
7.	*Hind III*	*Hemophilus influenzae Rd*	A↓AGCTT
8.	*Rsa I*	*Rhodopseudomonas sphaeroids*	GT↓AC
9.	*Hin C II*	*Hemophilus influenzae C1161*	GTPyPvAC
10.	*ECORI*	*E.coli RY 13*	G↓AATTC
11.	*Xma I*	*Xanthomonas malvacearum*	C↓CCGGG
12.	*Pst I*	*Providencia stuartii*	CTGCA↓G
13.	*Bgl II*	*Bacillus globigii*	A↓GATCT
14.	*Hpa II*	*Hemophilus parainfluenzae*	C↓CGG
15.	*Alu I*	*Arthrobacter luteus*	AG↓CT
16.	*Hae III*	*Hemophilus aegyptius*	GG↓CC
17.	*Hpa I*	*Hemophilus parainfluenzae*	GTT↓AAC
18.	*Mbo I*	*Moraxella bovis*	N↓GATC
19.	*NCo I*	*Nocardia corallina*	G↓CATGG
20.	*Sac I*	*Streptomyces achromogenes*	GAGCT↓C
21.	*Sal I*	*Streptomyces albus G*	G↓TCGAC
22.	*Sau 3AI*	*Staphylococcus aureus 3AI*	N↓GATC
23.	*Taq I*	*Thermus squatius YTI*	T↓CGA
24.	*Hinc II*	*Haemphilus influenzae Rc*	GTPyPuAC
25.	*Acc I*	*Acinetobacter calcoaceticus*	GT↓(AC)(GT)AC
26.	*Dde I*	*Desulfovibrio desulfuricans*	C↓GNAG
27.	*Eco RII*	*E.coli R-245*	↓CC(A/T)GG
28.	*Hinf I*	*Haemophilus influenzae Rf*	G↓ANTC
29.	*Mst II*	*Microcoleus species*	CC↓TNAGG

If cut is not shown it means exact cutting size is not known. Parentheses indicate that either bracketed base will suffice recognition.

Note : Py, Pu indicate non-specific pyrimidines or purines. N indicates any nucleotide base.

Complementary DNA (cDNA)

A second method for obtaining a gene of interest is to create a *complementary DNA (cDNA)* strand from a strand of messenger RNA (mRNA). This procedure is used when specific mRNA molecules can be isolated from cells that produce high levels of a certain protein. For example, a protein called *factor VIII* is necessary for proper blood clotting. It is the protein that most hemophiliacs are missing because

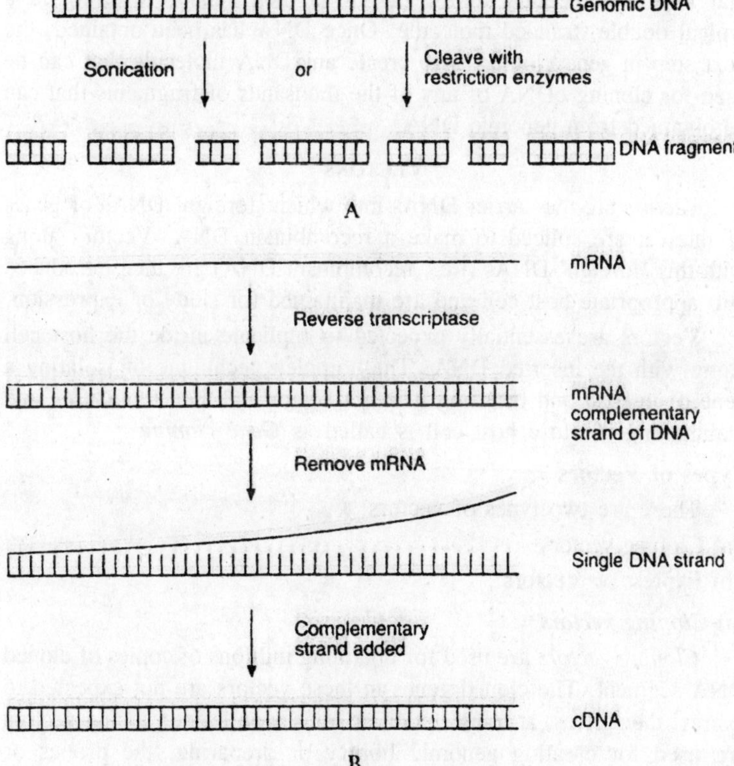

Fig. 9.5. A—To obtain DNA for cloning, large molecules are first broken into smaller segments by sonication or by cleaving them with restriction enzymes. The DNA fragments can then be separated and analyzed to determine which contain a gene of interest. B—If the mRNA encoding a gene of interest can be isolated, a DNA strand that is complementary to this mRNA can be synthesized.

of a mutation in the relevant encoding gene. From several cleaver studies, high levels of this protein were found to be produced in the liver along with the key mRNA molecules directing its synthesis

By separating the factor VIII protein and the affiliated mRNA from other proteins and other mRNAs that were also present, it was possible to identify the gene encoding factor VIII. The major advantage of this method is that a "pure" DNA sequence can be produced that does not contain the extraneous DNA sequences (introns) that occur in genomic DNA. Once the mRNA is isolated and purified, it is used as a template to construct a cDNA strand using the unique enzyme *reverse transcriptase*, which synthesizes DNA from RNA. When the cDNA has been formed, the mRNA strand is removed, and a DNA strand that is complementary to the cDNA is then added, resulting in a typical double-stranded molecule. Once DNA has been obtained, the next step in gene cloning is to create an rDNA molecule that can be used for cloning cDNA or any of the thousands of fragments that can be obtained from genomic DNA.

Vectors

Vectors are the carrier DNAs into which 'foreign' DNAs or genes of interest are spliced to make a recombinant DNA. Vectors along with this 'foreign' DNAs (i.e., recombinant DNA) are then introduced into appropriate host cell and are maintained for study or expression.

Vectors are essentially expected to replicate inside the host cell along with the inserted DNA. The complete technique of isolating a gene of interest and inserting it into a vector and then replicating and maintaining it into a host cell is called as *Gene cloning*.

Types of Vectors

There are two types of vectors:

(a) Cloning vectors.

(b) Expression Vectors.

(a) Cloning vectors

Cloning vectors are used for obtaining millions of copies of cloned DNA segment. The cloned genes in these vectors are not expected to express themselves, at transcription or translation level. Cloning vectors are used for creating genomic library or preparing, the probes or genetic engineering experiments or other basic studies.

(b) Expression vectors

Expression vectors allow the expression of cloned gene, to give the product (protein). This can be achieved through the use of promoters and expression cassettes and regulatory genes (sequences). Expression vectors are used for transformation to generate transgenic plant, animal or microbe where cloned gene expresses to give the product (protein). Commercial production of product of cloned gene may also be achieved by high level expression using the expression vectors.

Promoters in Vectors

Routine manipulations in gene cloning experiments do not require expression of the cloned DNA. But when required, selection of host/vector system plays important role. If cDNA molecule is used for cloning eukaryotic gene in prokaryotic system then problem of post transcriptional modification is obviated. For given host cell, there may be several types of expression vectors, then in addition of other consideration, a key feature of selection of expression vector is the type of promoter used to direct expression of the cloned sequence.

A vector with very efficient promoter is chosen to maximise the expression. Promoters are referred to as strong or weak depending on their efficiency of expression. Weak promoters are used if overexpression and excess product formation is likely to be toxic to host cell. Promoters are regions with a specific base sequence, to which RNA polymerase will bind. In addition to the strength of promoter, it is also desirable to regulate expression. This is done by using promoters that are either inducible or repressible. This exerts some control.

Arrangement of restriction sites immediately downstream from the promoter is critical. Thus expression vector will have following essential features:

(a) Origin of replication that is functional in target host cell.
(b) Antibiotic resistance genes or other genetic selection mechanism.
(c) Promoter (strong or weak as per suitability) along with regulatory control.
(d) Restriction sites immediately downstream form the promoter.
(e) Unique restriction site for cloning located in a position where the inserted cDNA sequence can be expressed effectively.

For expression of cloned gene in plants or animals only plant specific or animal specific promoters work. Promoters can confer the transferred gene a specific pattern of expression.

Expression Cassettes

These are gene constructs which allow the insertion of foreign genes, either as transcriptional or transnational fusions, behind specific promoters. PRT series of plasmids which have been derived from pUC 18/19 are the examples.

Organism	*Gene promoter*	*Induction by*
1. *E.coli*	lac operon, trp operon, λ P_L	IPTG, β-indolylacetic acid, temperature sensitive λ cl protein resp.
2. *A. nidulans*	Glucoamylase	Starch.
3. *S. cerevisiae*	Acid phosphatase Alcohol dehydrogenase galactose utilization Metallothionein	Phosphate depletion glucose depletion galactose heavy metals.
4. *T. reesei*	Cellobiohydrolase	Cellulose
5. Mouse	Metallothionein	heavy metals
6. Human	Heat shock protein	Temperature 740°C
7. Plants (in basal region)	Nopaline synthase (Nos) Mannopine synthase (Mas)	
8. Plants (in leaf tissue)	35 S promoter of Ca MV [high expression (10 fold) at distance of 2 kbp]	
9.	Polyhedrin promoter from baculovirus (used for developing biopesticides or for production of specific chemicals in industry)	

Expression Vectors

Expression vectors are simply designed to express detectable levels of foreign proteins usually at the bench-scale level (i.e. in shake flask cultures).

Production Vectors

Production level vectors are designed for stable large scale production of gene products at economically significant levels. This means production level vectors have additional genetic elements such as *par* sequences, transcription terminators downstream from the cloned genes etc. This is only artificial distinction. Efficients expression vectors are used successfully to express wide variety of proteins. Plasmids like *pASI* are attractive for production purposes as they have strong *pL* transcriptional unit and also a strong transnational unit.

Cloning in *E.coli*

Plasmid DNA used as vector can be cleaved at a site with restriction endonuclease enzyme and foreign DNA segments can be inserted here. Generally, plasmids with two genes for different antibiotic resistance are used—one to identify bacteria that carry plasmid and second to distinguish chimeric plasmid from the parental vector. Insertion of foreign DNA segment is done in site at one antibiotic resistance gene, therefore resistance is lost to one antibiotic.

The chimeric plasmid is resistant to only one antibiotic while the parental plasmid is resistant to two antibiotics. Multicopy plasmids are more suitable since 10-30 copies of one plasmid in a cell along with the inserted gene will give an amplified response. Addition of chloramphenicol often affects new rounds of chromosome replication while plasmid replication is unaffected. This makes isolation of plasmid easier. A common method for cloning DNA is to cleave both plasmid and insert DNA with the same restriction endonuclease. Most of the plasmids which are used as cloning vectors are plasmids derived from natural plasmids so as to posses desirable features.

pBR 322

This is derived from *E. coli* plasmid Col E1. Bolivar and Rodriguez prepared this vector hence the name and 322 are numericals which were significant to these scientists. Alternations and restructuring is done while deriving these plasmids from original natural plasmids. Genes for relaxed replication and genes for antibiotic resistance are inserted in them. pBR 322 is 4362 bp long DNA with genes for resistance to tetracycline and ampicillin. pBR 322 possesses on origin of replication and restriction sites for cleavage by variety of restriction enzymes.

Characteristics of pBR 322

1. It is much smaller in size than natural plasmid. Advantages of small size are : (i) easier to handle, (ii) less susceptible to physical damage, (iii) simpler restriction man, (iv) easy uptake by bacteria during transformation, (v) higher copy number therefore easy detection.
2. Bacterial origin of replication is present and this ensures that plasmid will be replicated in host.
3. Origin of replication is relaxed here, hence activity is not tightly linked to cell division. Resultantly plasmid replication will be initiated for more frequently than chromosomal replication. So more copy number of plasmid per cell is possible.
4. Two genes for different antibiotic resistance.
5. There are single restriction sites for number of enzymes, scattered on plasmid and can be used for cleavage and insertion. Presence

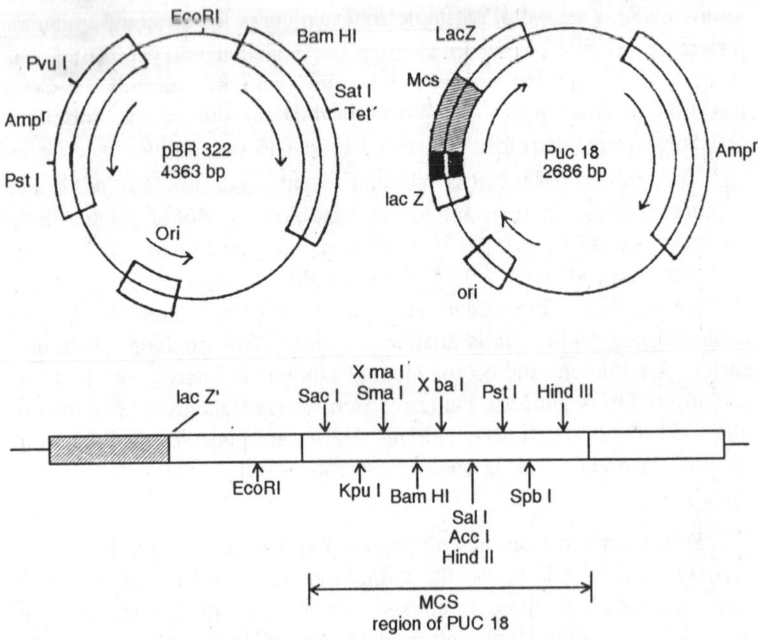

Fig. 9.6. Plasmids pBR 322 and pUC 18.

of such sites within antibiotic resistance gene are important, as they cause respective loss of resistance.

pAT 153 and *pXf 3* are two derivatives of pBR 322 having smaller size and higher copy number.

pBR 327

This is derived from pBR 322, by deletion of nucleotides between 1427 and 2516. These nucleotides are deleted to reduce size of vector and reduce interference. pBR 327 still has genes for resistance against tetracycline and ampicillin.

pUC Vectors

The name is derived from the place of their initial preparation (i.e., University of California). These vectors have 2700 base pairs and ampicillin resistance gene and lac Z gene. Insertion of foreign DNA in lac Z gene causes inactivation of lac Z gene. *E. coli* which is lactose –ve is used for transformation. When culture is grown in presence of IPTG (Isopropyl thiogalactoside) which induces the synthesis of β galactosidase) and X-gal substrate, *E. coli* (transformed with pUC vector with insert) gives white colonies while *E. coli* (transformed with pUC vector but without insert) gives blue colonies. (Substrate is chromogenic hence coloured product is formed if enzyme is active). Vectors of pUC family have a region that contains several unique restriction sites in a short stretch of DNA. This region is known as a polylinker or multiple cloning site (MCS).

Bacteriophages Vectors

Phage has a linear DNA molecule so a single break creates two fragments. Foreign DNA can be inserted between them and two fragments can be joined. Such phages when undergo lytic cycle in host will produce more chimeric DNA. Wild type λ phage could accommodate only 2.5 kb of foreign DNA. Phage vectors are restructured by removing nonessential genes and making vector DNA smaller so that larger insert can be accommodated in phage head during packing. Lambda phage (λ) such prepared has one Eco Rl site and accommodates 20-25 kb of foreign DNA. They are used for preparing genomic library of eukaryotes. One other λ phage devised has two Bam HI sites that flank the I/E region.

Insertion vectors

Insertion phage vector has single restriction site which is used for insertion of foreign DNA. Smaller foreign DNA can be packed here.

Replacement vectors

Replacement vector has two restriction sites which flank a region known as stuffer fragment. Larger fragment of foreign DNA can be replaced between these two sites.

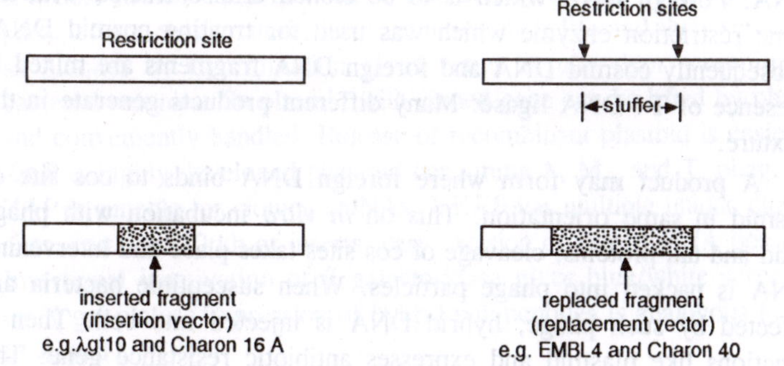

Fig. 9.7. Insertion and replacement vectors.

Cosmids

Cosmids are the novel cloning vectors which possess properties of both plasmid and λ phage. Cosmids first were developed in 1978 by

however may be required having necessary replicon for maintenance in different combinations of unrelated hosts. Shuttle vectors have potential importance in the genetic manipulations of industrially important species. Shuttle vectors can exploit gene manipulative procedures of different hosts for example when *E. coli* amplification will be possible. Shuttle vectors exist for *E. coli* yeast cells combination. *E. coli—Agrobacterium* combination and *E. coli—B. subtilis*. *E. coli—Streptomyces lividans* and *E. coli—mammalian* cells combinations.

Vehicles

A DNA molecule needs to display several features to be able to act as a vehicle for gene cloning. Most important, it must be able to replicate within the host cell, so that numerous copies of the recombinant DNA molecule can be produced and passed to the daughter cells. A cloning vehicle also needs to be relatively small, ideally less than 10 kilobases (kb) in size, as large molecules tend to break down during purification, and are also more difficult to manipulate. Two kinds of DNA molecule that satisfy these criteria can be found in bacterial cells: plasmids and bacteriophage chromosomes. Although plasmids are frequently employed as cloning vehicles, two of the most important types of vector in use today are derived from bacteriophages.

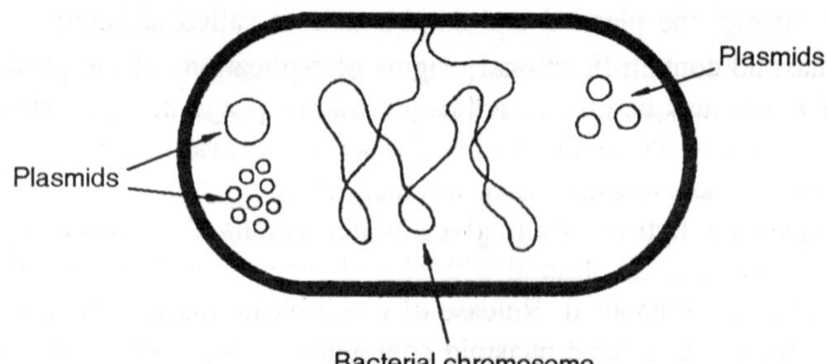

Fig. 9.8. Plasmids independent genetic elements found in bacterial cells.

Plasmids

Basic features of plasmids

Plasmids are circular molecules of DNA that lead an independent existence in the bacterial cell. Plasmids almost always carry one or more genes, and often these genes are responsible for a useful characteristic displayed by the host bacterium. For example, the ability to survive in normally toxic concentrations of antibiotics such as

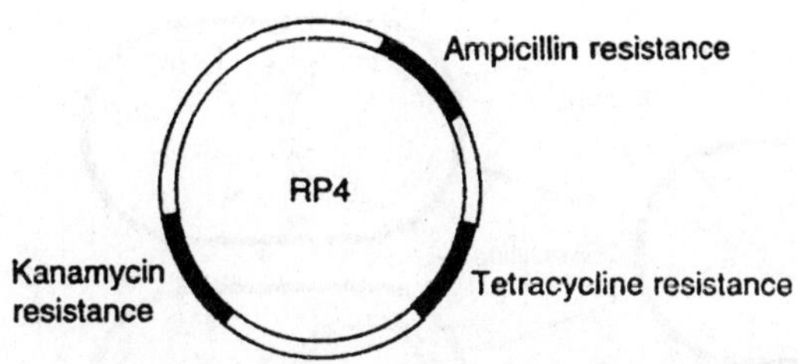

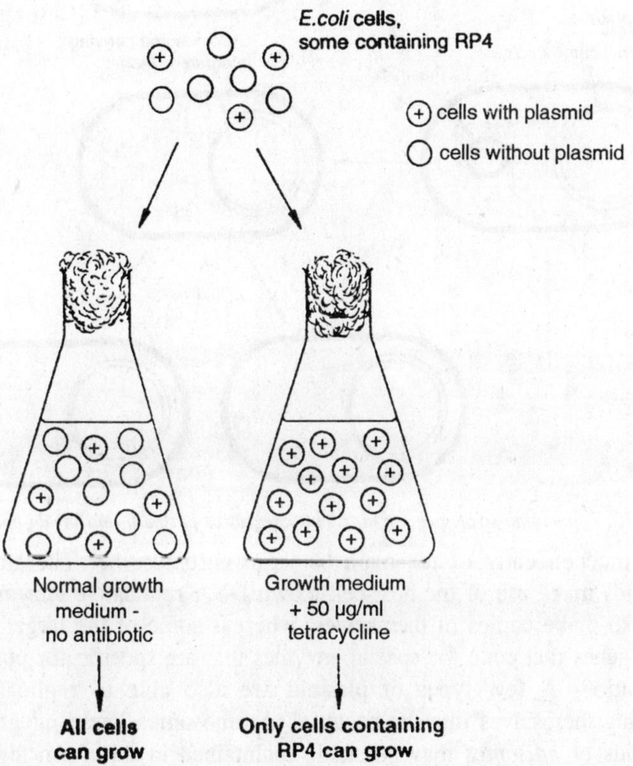

Fig. 9.9. The use of antibiotic resistance as a selectable marker for a plasmid.

chloramphenicol or ampicillin is often due to the presence in the bacterium of a plasmid carrying antibiotic resistance genes. In the laboratory antibiotic resistance is often used as a *selectable marker* to ensure that bacteria in a culture contain a particular plasmid.

All plasmids possess at least one DNA sequence that can act as an *origin of replication*, so they are able to multiply within the cell

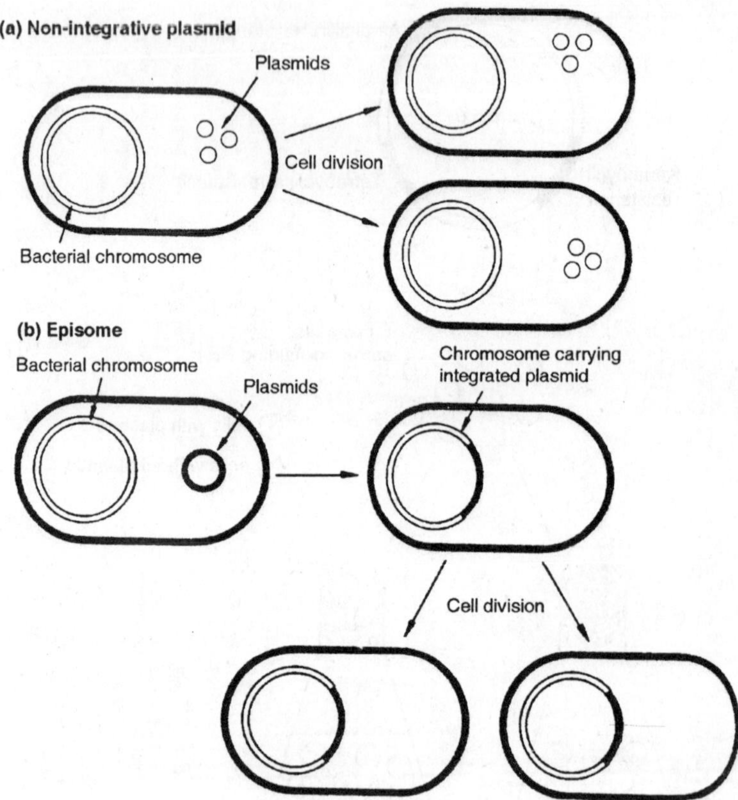

Fig. 9.10. Replication strategies for (a) a non-integrative plasmid, and (b) an episome.

quite independently of the main bacterial chromosome. The smaller plasmids make use of the host cell's own DNA replicative enzymes in order to make copies of themselves, whereas some of the larger ones carry genes that code for special enzymes that are specific for plasmid replication. A few types of plasmid are also able to replicate by inserting themselves into the bacterial chromosome. These integrative plasmids or *episomes* may be stably maintained in this form through numerous cell divisions, but will at same stage exist as independent elements. Integration is also an important feature of some bacteriophage chromosomes and will be described in more detail when these are considered.

Size and copy number

These two features of plasmid are particularly important as far as cloning is concerned. We have already mentioned the relevance of plasmid size and stated that less than 10 kb is desirable for a cloning vehicle. Plasmids range from about 1.0 kb for the smallest to over 250 kb for the largest plasmids, so only a few will be useful for cloning purposes. However, larger plasmids may be adapted for cloning under some circumstances. The *copy number* refers to the number of molecules of an individual plasmid that are normally found in a single bacterial cell.

The factors that control copy number are not well understood, but each plasmid has a characteristic value that may be as low as one (especially for the large molecules) or as many as 50 or more. Generally speaking, a useful cloning vehicle needs to be present in the cell in multiple copies so that large quantities of the recombinant DNA molecule can be obtained.

Table 9.5. Sizes of representative plasmids

Plasmid	*Size*		*Organism*
	Nucleotide length (kb)	*Molecular wt (MDa)*	
pUC8	2.1	1.8	*E.coli*
ColEI	6.4	4.2	*E.coli*
RP4	54	36	*Pseudomonas* + others
F	95	63	*E.coli*
TOL	117	78	*Pseudomonas putida*
pTiAch5	213	142	*Agrobacterium tumefaciens*

Conjugation and compatibility

Plasmids fall into two groups conjugative and non-conjugative. Conjugative plasmids are characterized by the ability to promote sexual *conjugation* between bacterial cell, a process that can result in a conjugative plasmid spreading from one cell to all the other cells in a bacterial culture. Conjugation and plasmid transfer are controlled by a set of transfer or *tra* genes, which are present on conjugative plasmids but absent from the non-conjugative type.

A non-conjugative plasmid may, under some circumstance, be cotransferred along with a conjugative plasmid when both are present in the same cell. Several different kinds of plasmid may be found in a single cell, including more than one different conjugative plasmid at any one time. In fact, cells of *E. coli* have been known to contain up to seven different plasmids at once. To be able to coexist in the same cell, different plasmids must be *compatible*. If two plasmids are incompatible then one or the other will be quite rapidly lost from the

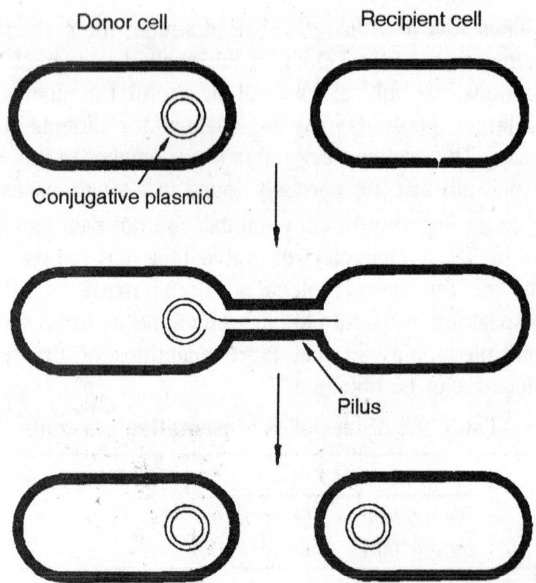

Fig. 9.11. Plasmid transfer by conjugation between bacterial cells.

cell. Different types of plasmid can therefore be assigned to different *incompatibility groups* on the basis of whether or not they can coexist, and plasmids from a single incompatibility groups are often related to each other in various ways. The basis of incompatibility is not well understood, but events during plasmid replication are thought to underlie the phenomenon.

Plasmid classification

There are five main types of plasmid which are as follows.

1. *Fertility* or *'F' plasmids* carry only *tra* genes and have no characteristic beyond the ability to promote conjugal transfer of plasmids, e.g., F plasmid of *E. coli*.
2. *Resistance* or *'R' plasmids* carry genes conferring on the host bacterium resistance to one or more antibacterial agents, such as chloramphenicol, ampicillin and mercury, R plasmids are very important in clinical microbiology as their spread through natural populations can have profound consequences in the treatment of bacterial infections; e.g., RP4, commonly found in *Pseudomonas*, but also occurring in many other bacteria.
3. *Col plasmids* code for colicins—proteins that kill other bacteria; e.g., ColE1 of *E. coli*.

4. *Degradative plasmids* allow the host bacterium to metabolize unusual molecules such as toluene and salicylic acid; e.g., TOL of *Pseudomonas putida*.
5. *Virulence plasmids* confer pathogenicity on the host bacterium; e.g., *Ti plasmids* of *Agrobacterium tumefaciens*, which induce crown gall disease on dicotyledonous plants.

Plasmids in organisms other than bacteria

Although plasmids are widespread in bacteria they are by no means so common in other organisms. The best characterized eukaryotic plasmid is the *2μm circle* that occurs in many strains of the yeast *Saccharomyces cerevisiae*. The discovery of the 2 μm plasmid was very fortuitous as it has allowed the construction of vectors for cloning genes with this very important industrial organism as the host. However, the search for plasmids in other eukaryotes (e.g., filamentous fungi, plants and animals) has proved disappointing and it is suspected that many higher organisms simply do not harbour plasmids within their cells.

Bacteriophages

Basic features of bacteriophages

Bacteriophages, or phages as they are commonly known, are viruses that specifically infect bacteria. Like all viruses, phages are very simple in structure, consisting merely of a DNA (or occasionally RNA) molecule carrying a number of genes, including several for replication of the phage, surrounded by a protective coat or *capsid* made up of protein molecules.

The general pattern of infection, which is the same for all types of phage, is a three-step process.

1. The phage particle attaches to the outside of the bacterium and injects its DNA chromosome into the cell.
2. The phage DNA molecule is replicated, usually by specific phage enzymes coded by genes on the phage chromosome.
3. Other phage genes direct synthesis of the protein components of the capsid, and new phage particles are assembled and released from the bacterium.

With some phage types the entire infection cycle is completed very quickly, possibly in less than 20 minutes. This type of rapid infection is called a *lytic cycle*, as release of the new phage particles is associated with lysis of the bacterial cell. The characteristic feature of a lytic infection cycle is that phage DNA replication is immediately

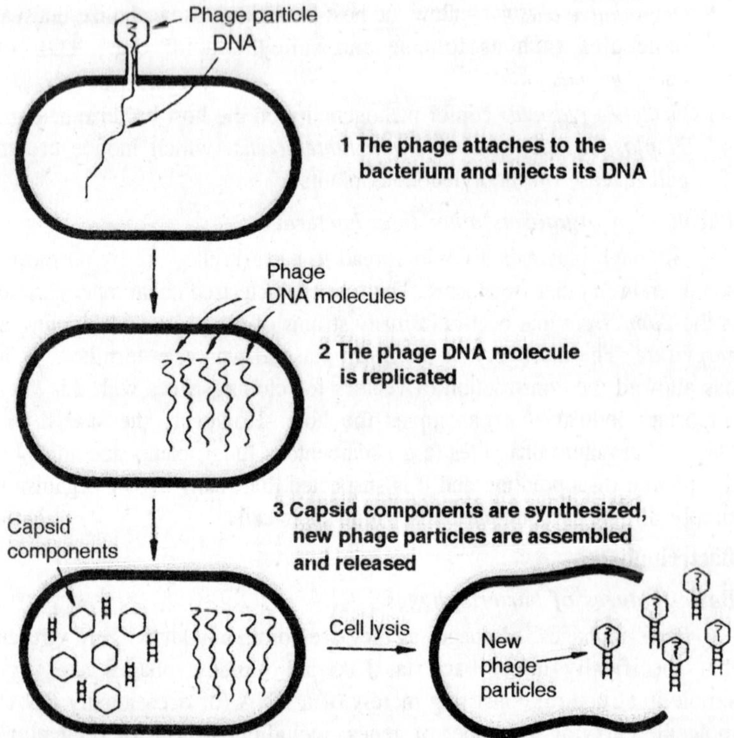

Fig. 9.12. The general pattern of infection of a bacterial cell by a bacteriophage.

followed by synthesis of capsid proteins, and the phage DNA molecule is never maintained in a stable condition in the host cell.

Lysogenic phages

In contrast to a lytic cycle, *lysogenic* infection is characterized by retention of the phage DNA molecule in the host bacterium, possibly for many thousands of cell divisions. With many lysogenic phages the phage DNA is inserted into the bacterial genome, in a manner similar to episomal insertion. The integrated form of the phage DNA (called the *prophage*) is quiescent, and a bacterium (referred to as a *lysogen*) which carries a prophage is usually physiologically indistinguishable from an uninfected cell.

The prophage is eventually released from the host genome and the phage reverts to the lytic mode and lyses the cell. The infection cycle of λ, a typical lysogenic phage is of this type. A limited number of lysogenic phages follow a rather different infection cycle. When M13, or a related phage, infects *E. coli*, new phage particles are continuously

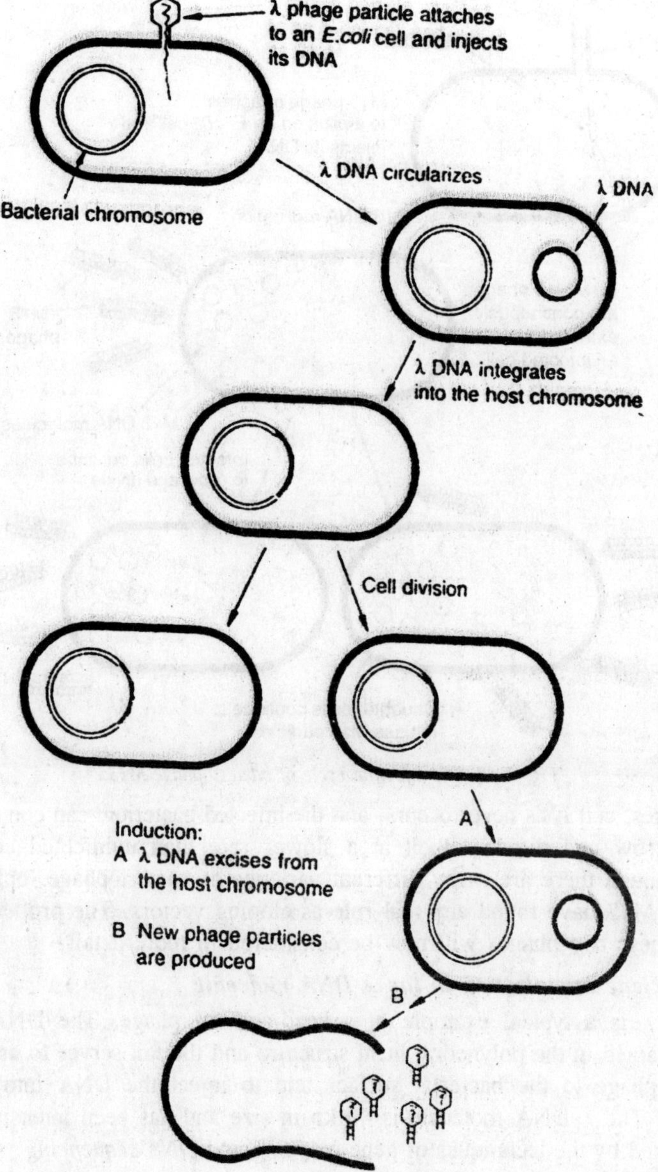

Fig. 9.13. The lysogenic infection cycle of bacteriophage λ.

assembled and released from the cell. The M13 DNA is not integrated into the bacterial genome and does not become quiescent. With these

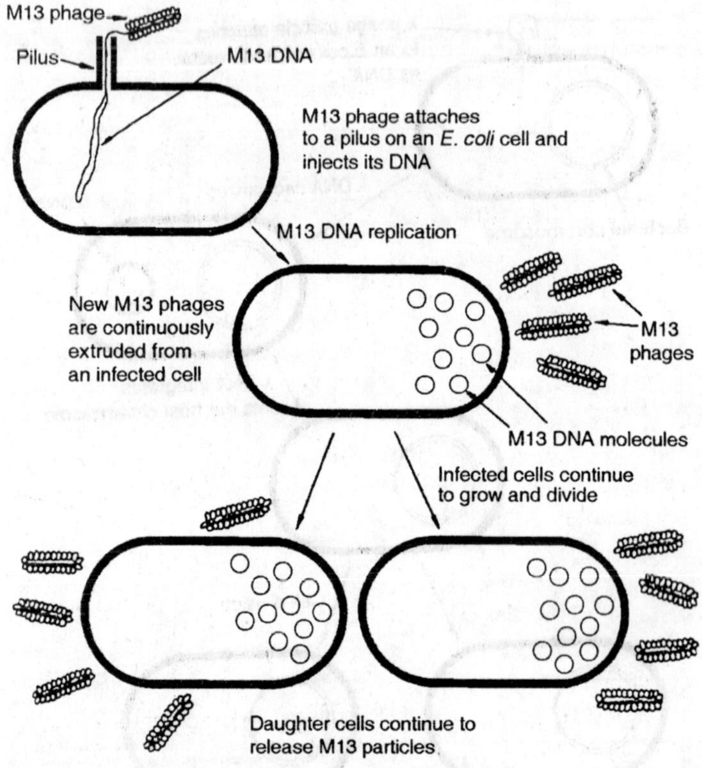

Fig. 9.14. The infection cycle of bacteriophage M13.

phages, cell lysis never occurs, and the infected bacterium can continue to grow and divide, albeit at a slower rate than uninfected cells. Although there are many different varieties of bacteriophage, only λ and M13 have found any real role as cloning vectors. The properties of these two phages will now be considered in more detail.

(a) Gene organization in the λ DNA molecule

λ is a typical example of a head-and-tail phage. The DNA is contained in the polyhedral head structure and the tail serves to attach the phage to the bacterial surface and to inject the DNA into the cell. The λ DNA molecule is 49 kb in size and has been intensively studied by the techniques of gene mapping and *DNA sequencing*. As a result the positions and identities of most of the genes on the l DNA molecule are known.

A feature of the λ genetic map is that genes related in terms of function are clustered together on the genome. For example, all of the genes coding for components of the capsid are grouped together in the left-hand third of the molecule, and genes controlling integration of the prophage into the host genome are clustered in the middle of the molecule. Clustering of related genes is profoundly important for controlling expression of the λ genome, as it allows genes to be switched on and off as a group rather than individually. Clustering is also important in the construction of λ-based cloning vectors.

(b) The linear and circular forms of λ DNA

A second feature of λ that turns out to be of importance in the construction of cloning vectors is the conformation of the DNA molecule. The molecule is linear, with two free, ends, and represents the DNA present in the phage head structure. This linear molecule consists of two *complementary* strands of DNA, base-paired according to the *Watson-Crick rules* (that is, double-stranded DNA). However, at either end of the molecule is a short 12-nucleotide stretch, in which the DNA is single-stranded.

The two single strands are complementary, and so can base-pair with one another to form a circular, completely double-stranded molecule. Complementary single strands are often referred to as *'sticky' end* or 'cohesive' ends, because base-pairing between them can 'stick' together the two ends of a DNA molecule (or the ends of two different DNA molecules). The λ cohesive ends are called he *cos sites* and they play two distinct roles during the λ infection cycle.

First of all, they allow the linear DNA molecule that is injected into the cell to be circularized, which is a necessary prerequisite for insertion into the bacterial genome. The second role of the *cos* sites is rather different, and comes into play after the prophage has excised from the host genome. At this stage a large number of new λ DNA molecules are produced by the rolling circle mechanism of replication in which a continuous DNA strand is 'rolled off' of the template molecule. The result is a catenane consisting of a series of linear λ genomes joined together as the *cos* sites. The role of the *cos* sites is now to act as recognition sequences for an *endonuclease* which cleaves the catenane at the *cos* sites producing individual λ genomes.

The endonuclease (which is the product of gene A on the λ DNA molecule) creates the single-stranded sticky ends, and also acts in conjunction with other proteins to package each λ genome into a phage head structure. The cleavage and packaging processes recognize just the *cos* sites and the DNA sequences to either side of them. Changing the structure of the internal regions of the λ genome, for example by inserting new genes, has no effect on these events so long as the overall length of the λ genome is not altered too greatly.

(c) M13—a filamentous phage

M13 is an example of a filamentous phage and is completely different in structure from λ. Furthermore, the M13 DNA molecule is much smaller than the λ genome, being only 6407 nucleotides in length. It is circular, and is unusual in that it consists entirely of single-stranded DNA. The smaller size of the M13 DNA molecule means that it has room for fewer genes than the λ genome. This is possible because the M13 capsid is constructed from multiple copies of just three proteins, whereas synthesis of the λ head-and-tail structure involves over 15 different proteins. In addition, M13 follows a simpler infection cycle than λ and does not need genes for insertion into the host genome. Injection of an M13 DNA molecule into an *E. coli* cell occurs via the *pilus*, the structure that connects two cells during sexual conjugation.

Once inside the cell the single-stranded molecule acts as the template for synthesis of a complementary strand, resulting in normal double-stranded DNA. This molecule is not inserted into the bacterial genome, but instead replicates until over 100 copies are present in the cell. When the bacterium divides, each daughter receives copies of the phage genome, which continues to replicate, thereby maintaining its overall numbers per cell. New phage particles are continuously assembled and released, about 1000 new phages being produced during each generation of an infected cell.

(d) The attraction of M13 as a cloning vehicle

Several features of M13 make this phage attractive as the basis for a cloning vehicle. The genome is less than 10 kb in size, well within the range that we stated was desirable for a potential vector. In addition, the double-stranded *replicative form* of the M13 genome behaves very much like a plasmid and can be treated as such for experimental purposes. It is easily prepared from a culture of infected *E. coli* cells and can be reintroduced by *transfection*.

Most importantly, genes cloned with an M13-based vector can be obtained in the form of single-stranded DNA. Single-stranded versions of cloned genes are useful for several techniques, notably DNA sequencing and *in vitro* mutagenesis. Using an M13 vector is an easy and reliable way of obtaining single-stranded DNA for this type of work.

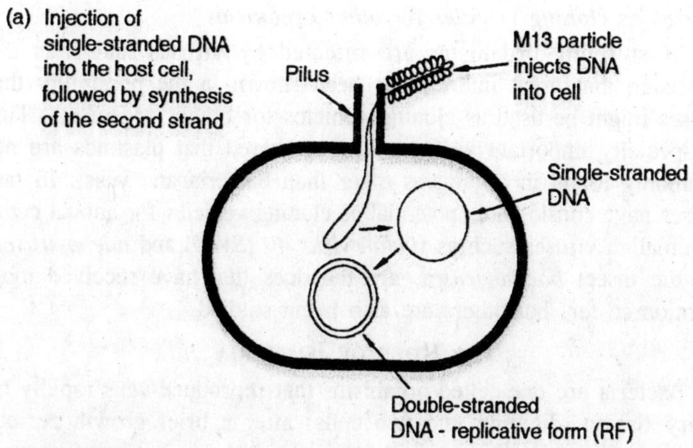

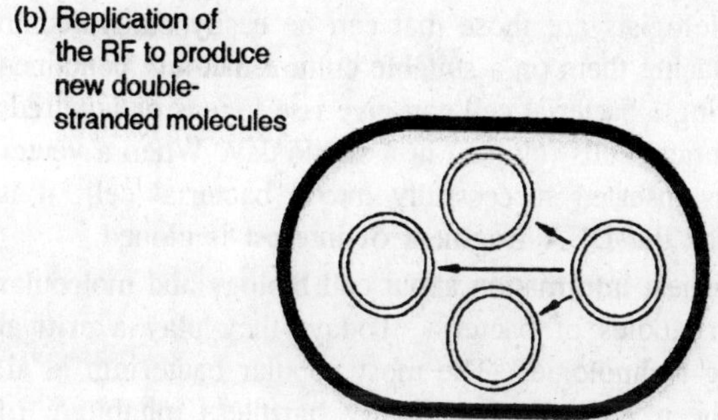

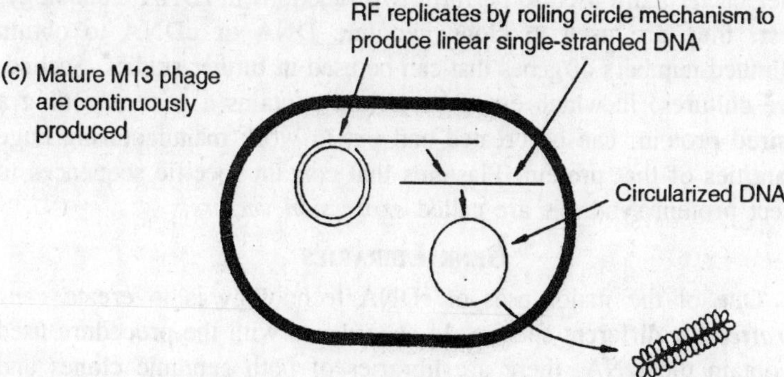

Fig. 9.15. The M13 infection cycle showing the different types of DNA replication.

Viruses as cloning vehicles for other organisms

Most living organisms are infected by viruses and it is not surprising that great interest has been shown in the possibility that viruses might be used as cloning vehicles for higher organisms. This is especially important when it is remembered that plasmids are not commonly found in organisms other than bacteria and yeast. In fact viruses have considerable potential as cloning vehicles for animal cells. Mammalian viruses such as *simian virus 40 (SV40)* and *adenoviruses*, and the insect *baculoviruses*, are the ones that have received most attention so far, but others are also being studied.

The Role of Bacteria

Bacteria are one-celled organisms that reproduce very rapidly by binary fission (dividing into two cells) after a brief growth period. They live in practically every environment, but the species of greatest use to molecular biologists are those that can be easily cultivated in the laboratory by placing them on a suitable culture medium contained in a small dish. A single bacterial cell can give rise to tens or hundreds of thousands of identical cells (clones) in a single day. When a vector containing rDNA is inserted successfully into a bacterial cell, it is replicated, and hence the DNA fragment of interest is cloned.

Much of the earliest information about cell biology and molecular biology came from studies of bacteria. Today, they play a critical role in gene-cloning technologies. The most popular bacterium in all of this work has been *E. coli*, a normally harmless inhabitant of mammalian digestive tracts. More is known about the molecular biology of *E. coli* than about any other species on Earth. *E. coli* and some other bacteria are used to perform two functions in rDNA technology. First, they are used to clone genomic DNA or cDNA to obtain unlimited numbers of genes that can be used in further studies. Second, pure cultures, in which every bacterium contains a gene encoding a desired protein, can be created and put to work manufacturing huge quantities of that protein. Plasmids that contain specific sequences to direct protein synthesis are called *expression vectors*.

Gene Libraries

One of the major uses of rDNA technology is to create *gene libraries* for different species. In accordance with the procedure used to obtain the DNA, there are libraries of both genomic clones and cDNA clones. The number of "volumes" (DNA fragments or genes) required in the library to ensure that all gene sequences are represented is generally related to the size and complexity of the species. For *E. coli*, 1,500 fragments are necessary; for yeast, 4,600; fruit flies, 48,000; and humans, 800,000.

Researchers create a *gene library* for a particular species in the following sequence of events:

1. Genomic DNA is cut into thousands of fragments with appropriate restriction enzymes.
2. Each fragment, representing approximately one gene, is spliced into a phage-cloning vector.
3. The phages are replicated by the "host" bacteria, thus replicating (cloning) the inserted gene fragments.

Such a library is a repository of the entire genome where every gene of a species is represented. The space required for a typical gene library is a test tube or a set of small culture dishes that can be conveniently kept in a refrigerator pending further analysis. Molecular biologists have developed specialized techniques to "screen" gene libraries, that is, to determine where specific genes are located in the pieces of DNA. This involves isolating a gene and identifying the protein it encodes. Once a gene has been isolated and cloned, it is possible quickly to determine its entire base sequence. As you might expect, the greatest emphasis is currently on research related to the human gene library, and remarkable progress has been made in isolating human genes of medical importance. Nevertheless, the human gene library is so immense, and many genes are so difficult to locate, that a complete reading of our library probably lies one or more decades in the future.

Applications of Genetic Engineering

Recombinant DNA technology has revolutionized biology in the past decade. At present, its main uses are (1) facilitating the production of useful proteins, (2) creating bacteria capable of synthesizing economically important molecules, (3) supplying DNA and RNA sequences as a research tool, (4) altering the genotype of organisms such as plants, and (5) potentially correcting genetic defects in animals (gene therapy). Some examples of these applications follows:

Commercial Possibilities

Bacteria with novel phenotypes can be produced by genetic engineering sometimes by combining the features of several other bacteria. For example, several genes from different bacteria have been inserted into a single plasmid that has then been placed in a marine bacterium, yielding an organism capable of metabolizing petroleum; this organism has been used to clean up oil spills in the oceans.

Many biotechnology companies are at work designing bacteria that can synthesize industrially important chemicals. Bacteria have been designed that are able to compost waste more efficiently and to fix nitrogen (to improve the fertility of soil), and an enormous effort is currently being expended to create organisms that can convert biological waste to alcohol. A human insulin, synthesized in *E. coli*, is already commercially available. Altering the genotype of plants is an important application of recombinant DNA technology. Of great use is the bacterium *Agrobacterium tumefaciens* and its plasmid Ti, which produces crown gall tumors in dicotyledonous plants. These tumor result from disruption of the bacterium inside plant cells, release of bacterial DNA, and integration of a segment of the plasmid DNA into the plant chromosome.

It is possible by genetic engineering to introduce genes from one plant into this plasmid and then, by infecting a second plant with the bacterium, transfer the genes of the first plant to the second plant. (Actually genes are first cloned in an *E. coli* plasmid and then recloned in Ti). Attempts are being made to perform plant breeding in this way. An example is the attempted alteration of the surface structure of the roots of grains such as wheat, by introducing certain genes from legumes (peas, beans), in order to give grains the ability of the legumes to establish root nodules of nitrogen-fixing bacteria. If successful, this would eliminate the need for the addition of nitrogenous fertilizers to grain-growing soils.

The first engineered recombinant plant of commercial value was developed in 1985. An economically important herbicide (weed killer) is glyphosate, which inhibits a particular essential enzyme in many plants. However, most herbicides cannot be applied to fields growing crops because both the crop and the weeds would be killed.

The target gene of glyphosate is also present in the bacterium *Salmonella typhimurium*. A resistant form of the gene was obtained by mutagenesis and growth of *Salmonella* in the presence of glyphosate; the gene was cloned in *E. coli*, and then recloned in *Agrobacterium*. Infection of plants with purified Ti containing the glyphosate-resistance gene has yielded varieties of maize, cotton, and tobacco that are resistant to glyphosate. Thus, fields of these crops can be sprayed with glyphosate at any stage of growth of the crop. All weeds are killed, and the crop is totally unharmed. An interesting bacterium is a strain of *Pseudomonas fluorescens*, which lives in association with maize and soybean roots. A lethal gene from *Bacillus thuriengis*, a bacterium pathogenic to the back cutworm, has been engineered into this bacterium. The black cutworm causes extensive crop damage and is usually combatted with noxious insecticides. In preliminary studies inoculation of soil with the engineered *Ps. fluorescens* resulted in death of the cutworm.

Uses in Research

Recombinant DNA technology is extraordinarily useful in basic research. Several examples were already given in mutant bacteria in which particular genetic systems have been altered in an effort to make them more amenable to study (for example, the numerous mutants in the *lac* operon). Such mutants have usually been derived through standard genetic techniques. For simple mutants this procedure is straightforward, but for mutants required to have many genetic markers (which may be very closely linked) the frequency of production of mutants can be so low that their isolation becomes very tedious.

Recombinant DNA techniques can simplify mutant construction, since fragments containing desired genetic markers can be purified, altered, and combined in test tube, and then introduced into another cell. This saves time and labour, and often enables mutants to be constructed that cannot in practice be formed in any other way. An example is the formation of double mutants of animal viruses, which undergo crossing over at such a low frequency that mutations can rarely be recombined by genetic crosses.

The greatest impact of the new technology on basic research has been in the study of eukaryotes, in particular, of eukaryotic regulation. Experiments of the type, which study the regulation of operons in bacteria, have been made possible by the use of mutations in promoters, operators, and structural genes. However, this approach has not been feasible with eukaryotes, because eukaryotes are diploid and, hence, mutants are difficult to isolate.

Furthermore, except for the unicellular eukaryotes such as yeast, there is no simple and rapid way to do multiple genetic manipulations with eukaryotic cells. Cloning techniques have made it possible to study regulation of gene expression in eukaryotes by direct assays of mRNA molecules produced by particular genes. The approach is based on the success with which it has become possible to understand gene activity in bacteria by studying the synthesis of mRNA in a variety of conditions.

An excellent assay is DNA-RNA hybridization, in which either a primary transcript or mRNA is detected by hybridization to a DNA sample that has been enriched for the gene being studied. In microbial experiments specialized transducing particles would be the source of the DNA—for example, *E. coli* gal mRNA is assayed by hybridization of radioactivity labeled intracellular RNA with DNA of *l gal* transducing particles. However, transducing particles do not exist for eukaryotic systems, so recombinant DNA techniques have been used to clone a gene whose regulation is to be studied.

The usual procedure is to clone the gene first in a plasmid (or a phage) vector and then allow the cell containing the plasmid to multiply. Purified plasmid DNA provides the researcher with a large supply of the DNA of that gene. The vector containing that DNA sequence is called a DNA *probe*, because its DNA, in denatured form, can be added to a cell extract containing mRNA, to probe for a particular mRNA by renaturation. Since no natural genes of the vector are present in eukaryotic cells, the vector DNA can be regarded as a source of pure DNA copies of a particular eukaryotic gene because no other genes in the vector will participate in the renaturation.

A further useful technique is to apply denatured probe DNA to a nitrocellulose filter and then to use a filter-binding assay for the specific RNA species. DNA but not RNA binds to these filters. Radioactive mRNA can then be incubated with a filter containing the DNA, using conditions such that renaturation will result. Except where the mRNA has renatured to the denatured DNA, the RNA can be removed by washing the filter. Thus, presence of radioactivity in the filter after washing indicates that complementary mRNA has been bound to the DNA on the filter. This simple technique and variations such as Southern blotting have revolutionized the study of eukaryotic gene regulation.

Production of Eukaryotic Proteins

One of the most valuable applications of genetic engineering is the production of large quantities of particular proteins that are otherwise difficult to obtain. The method is simple in principle. The gene encoding the desired protein is cloned in a vector adjacent to a bacterial promoter, and tests are performed to ensure that the gene is oriented such that its coding strand is linked to the strand containing the promoter. A plasmid for which many copies are present in each cell, or occasionally an actively replicating phage (such as λ), is used as a vector; in both cases, cells can be prepared that contain several hundred copies of the gene, and this can result in synthesis of a gene product to reach a concentration of about 1 to 5 percent of the cellular protein. In practice, production of large quantities of a prokaryotic protein in a bacterium is straightforward. However, if the gene is from a eukaryote, special problems exist:

1. Eukaryotic promoters are not usually recognized by bacterial RNA polymerases.
2. The mRNA transcribed from eukaryotic genes may not be translatable on bacterial ribosomes.
3. Introns may be present, and bacteria are unable to excise eukaryotic introns.
4. The protein itself often must be processed (for example, insulin); and bacteria cannot recognize processing signals from eukaryotes.
5. Eukaryotic proteins are often recognized as foreign material by bacterial protein-digesting enzyme and are broken down.

Several approaches to solving these problems have been taken with some degree of success. One procedure uses a genetically engineered plasmid, called a *shuttle vector*, that consists of both *E. coli* and yeast DNA and replicates in both organisms.

Cloning is done in *E. coli*, which for a variety of technical reasons is easier than cloning in yeast, and then the plasmid is isolated and transferred to yeast for expression of the cloned gene. In yeast the five problems just stated are avoided, or at least substantially reduced. Another approach uses a plasmid containing the *E. coli lac* region, cleaved in the *lacZ* gene by a restriction enzyme making blunt ends. Either c-DNA or a synthetic DNA molecule whose sequence is known from the amino acid sequence of the protein product is inserted into the *lacZ* gene, so that the eukaryotic protein is synthesized as the terminal region of β-galactosidase, from which it can be cleaved.

The first example of this approach resulted in a synthetic gene capable of yielding a 14-residue polypeptide hormone—somatostatin—which is synthesized *in vivo* in the mammalian hypothalamus. The procedure, applicable to most short polypeptides, was the following.

By chemical techniques, a double-stranded DNA molecule was synthesized containing 51 base pair; the base sequence of the coding strand was

TAC-(42 base encoding somatostatin)-ACTATC

The base sequence of the corresponding mRNA molecule was

AUG-(42 bases encoding somatostatin)-UGAUAG

The AUG codon of the mRNA was not used for initiation, but specified methionine. The mRNA terminated with two stop codons UGA and UAG.

The vector was the plasmid pBR322 modified to contain the *lac* promoter-operator region and a portion of the *lacZ* gene encoding the amino-terminal segment of β-galactosidase. The vector was cleaved at a site in the *lacZ* segment by a restriction enzyme that leaves blunt ends; the synthetic DNA molecule was then blunt-end-ligated to the cleaved plasmid. When the *lac* operon was induced, a protein was made consisting of the amino-terminal segment of β-galactosidase *coupled by methionine* to somatostatin, and terminated at the repeated stop codons of the synthetic DNA. This protein was purified and treated with cyanogen bromide, a reagent that cleaves proteins only at the

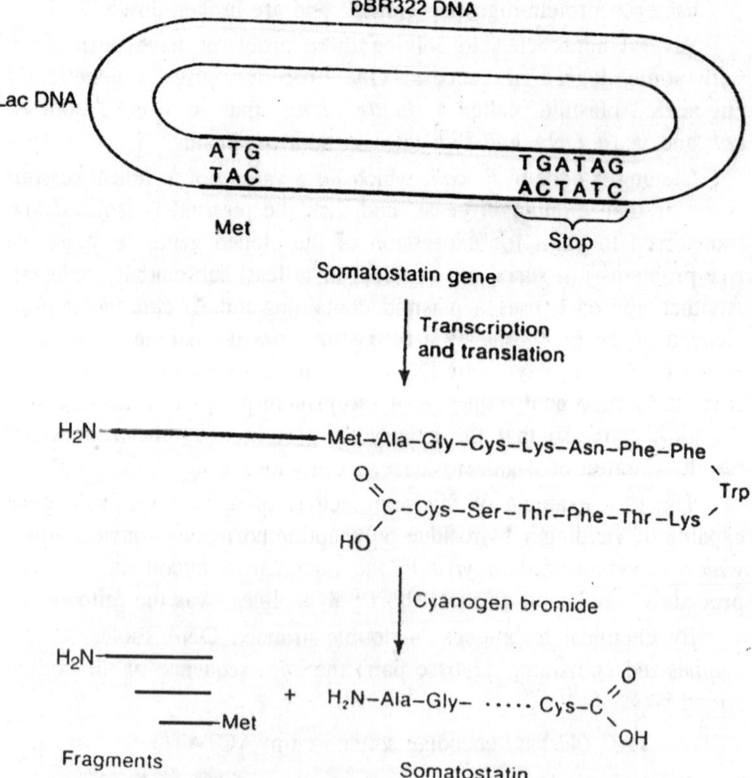

Fig. 9.16. Synthesis of somatostatin from a chemically synthesized gene joined to the plasmid pBR22 lac.

carboxyl side of methionine. In this way, the methionine linker remains attached to a β-galactosidase fragment and somatostatin is released. Use of methionine-coupling followed by cleavage with cyanogen bromide is a useful technique for separating any polypeptide from a bacterial protein to which it is fused, as long as the polypeptide itself does not contain methionine. Another more generally useful linker is $(Asp)_4$-Lys. In a sequence (Asp_4)-Lys-X, in which X is another amino acid, the enzyme enterokinase cleaves between lysine and X. Since such a sequence is not common, this linker is of more value than methionine.

Genetic Engineering with Animal Viruses

Retroviruses are RNA-containing animal viruses that have an unusual life cycle. These viruses, which contain the enzyme reverse transcriptase in their protein coats, use this enzyme to synthesize a double-stranded DNA copy of the viral RNA shortly after infection. As a normal part of the viral life cycle, this double-stranded DNA becomes inserted into the animal-cell chromosome, apparently at one of an enormous number of potential sites. Transcription occurs only after the DNA copy is inserted.

The infected host cell survives the infection, retaining the retroviral DNA in its chromosome. One of the best-understood retroviruses is *Rous sarcoma virus*, which causes tumors in chickens. Retrovirus DNA, either isolated from a cell or prepared synthetically, is useful vector with animal cells. Genetic engineering with retroviruses allows the possibility of altering the genotype of an animal cell. Since a wide variety of retroviruses are known, including two that can grow on human hosts, genetic defects may be correctable by these procedures in the future.

Many retrovirus species contain a gene that produces uncontrolled growth of a cell containing the retrovirus, thereby causing tumors in animals. If a retrovirus vector is to be used to change a genotype, the tumor-causing ability must be removed. This is possible by removal of the tumor-causing gene, which also provides the space needed for incorporation of foreign DNA. The recombinant DNA procedure employed with retroviruses consists of synthesis in the laboratory of double-stranded DNA from the viral RNA, through use of reverse transcriptase. The DNA is then cleaved with a restriction enzyme and, by the techniques already described, foreign DNA is inserted and selected.

Treatment of mutant cells in culture with the recombinant DNA, and application of a transformation procedure suitable for animal cells, yields cells in which recombinant retroviral DNA has been permanently inserted into an animal-cell chromosome. In this way, the genotype of the cell can be altered. Experiments have been done in which human cells deficient in the synthesis of purines have been obtained from patients with Lesch-Nyhan syndrome and grown in culture; these cells have been converted to normal cells by transformation with recombinant DNA. The exciting potential of this technique lies in the possibility of correcting genetic defects—for example, restoring the ability of a diabetic individual to make insulin or correcting immunological deficiencies. This technique has been termed *gene therapy*. However, it must be recognized that retroviruses are not well understood and are potentially dangerous. Gene therapy is not yet a practical technique, for major problems exist; for example, there is no reliable way to ensure that a gene is inserted in the appropriate target cell or target tissue. In addition, some means is needed to regulate the expression of the inserted genes. A major breakthrough in disease prevention has been the development of synthetic vaccines. Production of certain vaccines such as anti-hepatitis B has been difficult because of the extreme hazards of working with large quantities of the hepatitis B virus. The danger would be avoided if the viral antigen could be cloned and purified in *E. coli* or yeast, because the pure antigen could be given as a vaccine.

Several viral antigens have been cloned, but because of either thermal instability or poor antigenicity when pure, attempts to make vaccines in this way have generally been unsuccessful. However, the use of vaccinia virus (the anti-smallpox agent) as a carrier has been fruitful. The procedure makes use of the fact that viral antigens are on the surface of virus particles and that some of these antigens can be engineered into the coat of vaccinia. By 1985 vaccinia hybrids with surface antigens of hepatitis B, influenza virus, and vesicular stomatitis virus (which kills cattle, horses, and pigs) have been prepared and shown to be useful vaccines in animal tests. A surface antigens of *Plasmodium falciparum*, the parasite that causes malaria, has also been placed in the vaccinia coat; this may lead to an antimalaria vaccine.

Diagnosis of Hereditary Diseases

Restriction mapping has also been used in prenatal diagnosis—for example, in detecting sickle-cell anemia. This disease is caused by possession of an altered hemoglobin. The base change causing the mutation generates a cleavage site for a particular restriction enzyme. Cleavage of DNA from a sickle-cell patient produces a restriction map that differs from that of the DNA of a normal patient. The different restriction pattern can be detected by using Southern blotting and, as a probe, radioactive RNA prepared by transcribing cloned globin genes. The technique is so sensitive that the sickle-cell mutation can be detected in a small sample of cells obtained from the amniotic fluid surrounding a fetus, and detection is possible at a very early stage of development. Possibly in the future, gene therapy might be used to produce a normal child.

10

TRANSCRIPTIONAL CONTROL

Jacob and Monod proposed the operon model in 1961 for the co-ordinate regulation of transcription of genes involved in specific metabolic pathways. The operon is a unit of gene expression and regulation which typically includes.

- The *structural genes* (any gene other than a regulator) for enzymes involved in a specific biosynthetic pathway whose expression is co-ordinately controlled.
- Control elements such as an *operator sequence*, which is a DNA sequence that regulates transcription of the structural genes.
- *Regulator gene(s)* whose products recognize the control elements, for example a repressor which binds to and regulates an operator sequence.

The Lactose Operon

Escherichia coli can use lactose as a source of carbon. The enzymes required for the use of lactose as a carbon source are only synthesized when lactose is available as the sole carbon source. The lactose operon (or lac operon) consists of three structural genes: lacZ, which codes for b-balactosidase, an enzyme responsible for hydrolysis of lactose to balactose and glucose; lacY which encodes a galactoside permease which is responsible for lactose transport across the bacterial cell wall; and lacA, which encodes a thiogalactoside transacetylase. The three structural genes are encoded in a single transcription unit, lacZYA, which has a single promoter P_{lac}. This organization means that the three lactose operon structural proteins are expressed together as a polycistronic mRNA containing more than one coding region under

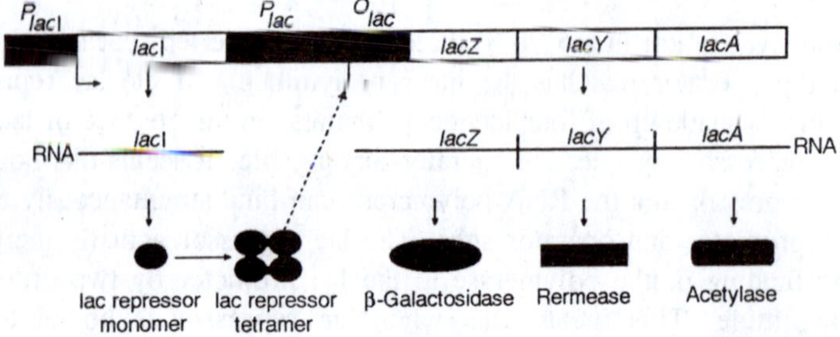

Fig. 10.1. Structure of the lactose operon.

the same regulatory control The lacZYA transcription unit contains an operator site o_{lac} which is positioned between bases —5 and +21 and the 5'-end of the P_{lac} promoter region. This site binds a protein called the lac repressor which is a potent inhibitor of transcription when it is bound to the operator. The lac repressor is encoded by a separate regulatory gene lacI which is also a part of the lactose operon lacI is situated just upstream from P_{lac}.

The Lac Repressor

The lacI gene encodes the lac repressor which is active as a tetramer of identical subunits. It has a very strong affinity for the lac operator-binding site, O_{lac}, and also has a generally high affinity for DNA. The lac operator site consists of 28 bp which is palindromic. (A palindrome has the same DNA sequence when one strand is read left to right in a 5' to 3' direction and the complementary strand is

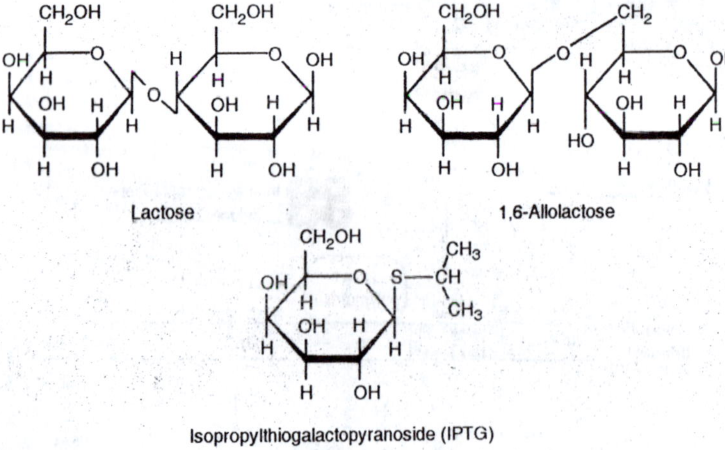

Lactose

1,6-Allolactose

Isopropylthiogalactopyranoside (IPTG)

Fig. 10.2. Structure of lactose, allolactose and IPTG.

read right to left in a 5' to 3' direction. This inverted repeat symmetry of the operator matches the inherent symmetry of the lac repressor which is made up of four identical subunits. In the absence of lactose, the repressor occupies the operator-binding site. It seems that both the lac repressor and the RNA polymerase can bind simultaneously to the lac promoter and operator sites. The lac repressor actually increases the binding of the polymerase to the lac promoter by two orders of magnitude. This means that when lac repressor is bound to the O_{lac} operator DNA sequence, polymerase is also likely to be bound to the adjacent P_{lac} promoter sequence.

Induction

In the absence of an inducer, the lac repressor blocks all but a very low level of transcription of lacZYA. When lactose is added to cells, the low basal level of the permease allows its uptake, and β-galactosidase catalyzes the conversion of some lactose to allolactose.

Allolactose acts as an inducer and binds to the lac repressor This causes a change in the conformation of the repressor tetramer, reducing its affinity for the lac operator. The removal of the lac repressor from the operator site allows the polymerase (which is already sited at the adjacent promoter) to rapidly begin transcription of the lazZYA genes. Thus, the addition of lactose, or a synthetic inducer such as isopropyl-β-D-thiogalactopyranoside (IPTG) very rapidly stimulates transcription of the lactose operon structural genes. The subsequent removal

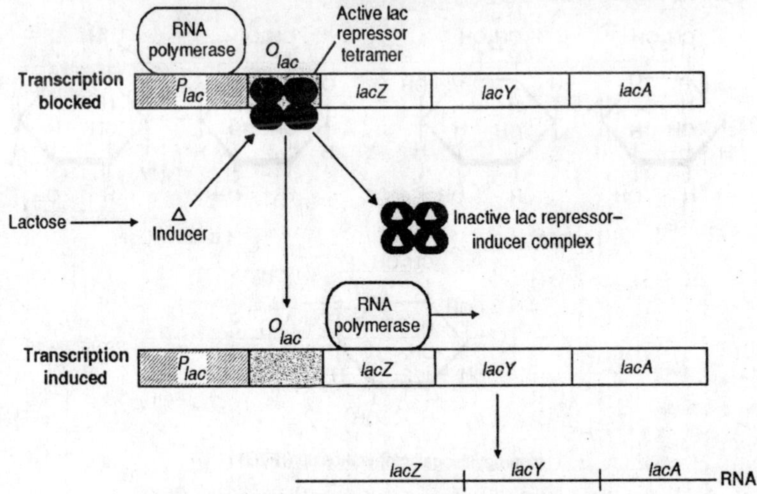

Fig. 10.3. Binding of inducer inactivates the lac repressor.

of the inducer leads to an almost immediate inhibition of this induced transcription, since the free lac repressor rapidly re-occupies the operator site and the lacZYA RNA transcript is extremely unstable.

cAMP Receptor Protein

The P_{lac} promoter is not a strong promoter. P_{lac} and related promoters do not have strong-35 sequences and some even have weak-10 consensus sequences. For high level transcription, they require the activity of a specific activator protein called cAMP receptor protein (CRP). CRP may also be called catabolite activator protein or CAP. When glucose is present, *E. coli* does not require alternative carbon sources such as lactose. Therefore, catabolic operons, such as the lactose operon, are not normally activated. This regulation is mediated by CRP which exists as a dimer which can not bind to DNA on its own, nor regulate transcription. Glucose reduces the level of cAMP in the cell. When glucose is absent, the levels of cAMP in E. coli increase and CRP binds to cAMP. The CARP-cAMP complex binds to the lactose operon promoter P_{lac} just upstream from the site for RNA polymerase. CRP binding induces a 90° bend in DNA, and this is believed to enhance RNA polymerase binding to the promoter, enhancing transcription by 50-fold. The CRP-binding site is an inverted repeat and may be adjacent to the promoter (As in the lactose operon), may lie within the promoter itself, or may be much further upstream from the promoter. Differences in the he CRP-binding sites of the promoters of different catabolic operons may mediate different levels of response of these operons to cAMP *in vivo*.

The TRP Operon

The Trypotophan Operon

The trp operon encodes five structural genes whose activity is required for trypotophan synthesis. The operon encodes a single transcription unit which produces a 7 kb transcript which is synthesized downstream from the trp promoter and trp operator sites P_{trp} and O_{trp} Like many of the operons involved in amino acid biosynthesis, the trp operon has evolved systems for co-ordinated expression of these genes when the product of the biosynthetic pathway, trytophan, is in short supply in the cell. As with the lac operon, the RNA product of this transcription unit is very unstable, enabling bacteria to respond to changing needs for tryptophan.

The Trp Repressor

A gene product of the separate trpR operon, the trp repressor, specifically interacts with the operator site of the trp operon. Thc

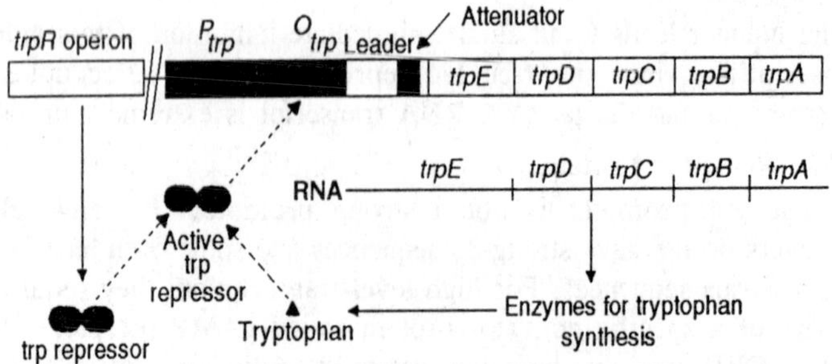

Fig. 10.4. Structure of the trp operon and function of the trp repressor.

symmetrical operator sequence, which forms the trp repressor-binding site, overlaps with the trp promoter sequence between bases -21 and +3 and +3. The core binding site is a palindrome of 18 bp. The trp repressor binds tryptophan and can only bind to the operator when it is complexed with tryptophan. The repressor is a dimer of two subunits which have structural similarity to the CRP protein and lac repressor. The repressor dimer has a structural with a central core and two flexible DNA-reading heads each formed from the carboxyl-terminal half of one subunit. Only when tryptophan is bound to the repressor are the reading heads the correct distance apart, and the side chains in the correct conformation, to interact with successive major grooves of the DNA at the trp operator sequence. Tryptophan, the end-product of the enzymes encoded by the trp operon, therefore acts as a co-repressor and inhibits its own synthesis through end-product inhibition. The repressor reduces transcription initiation by around 70-fold. This is a much smaller transcriptional effect than that mediated by the binding of the lac repressor.

The Attenuator

At first, it was thought that the repressor was responsible for all of the transcriptional regulation of the trp operon. However, it was observed that the deletion of a sequence between the operator and the trpE gene coding region resulted in an increase in both the basal and the activated (depressed) levels of transcription. This site is termed the attenuator and it lies towards the end of the transcribed leader sequence of 162 not that precedes the trpE initiator codon. The attenuator is a rho-independent terminator site which has a short GC-rich palindrome followed by eight successive U residues. If this sequence is able to form a hairpin structure in the RNA transcript, then it acts as a highly efficient transcription terminator and only a 140 bp transcript is synthesized.

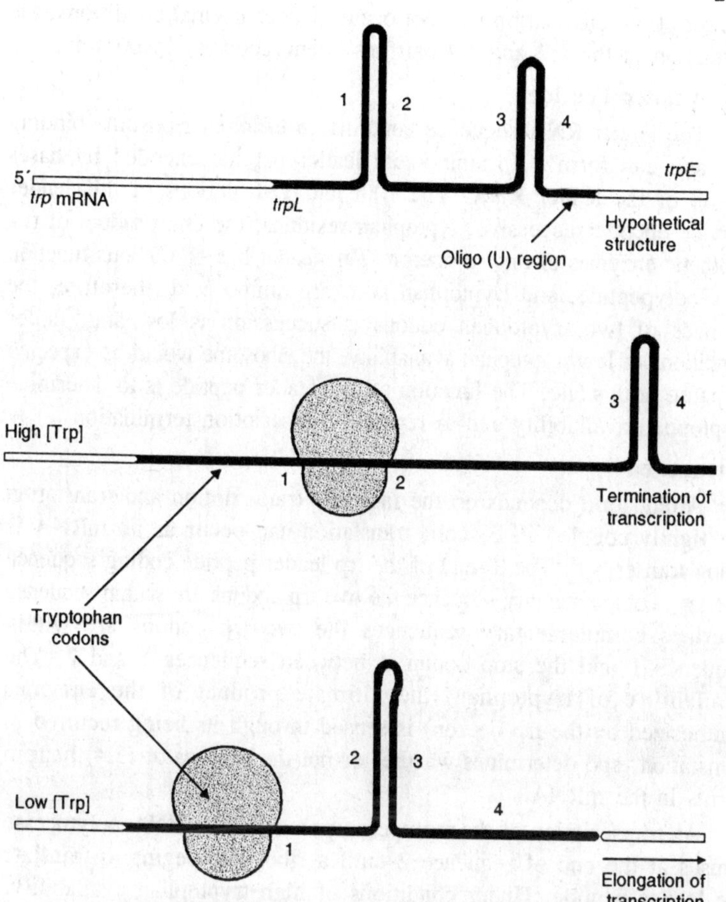

Fig. 10.5. Transcriptional attenuation in the trp operon.

Leader RNA Structure

The leader sequence of trp operon RNA contains for regions of complementary sequence which can form different base-paired RNA structures. These are termed sequences 1, 2, 3, and 4. The attenuator hairpin is the product of the base pairing of sequences 3 and 4. The attenuator hairpin is the product of the base pairing of sequences 3 and 4 (3:4 structure). Sequences 1 and 2 are also complementary and can form a second 1:2 hairpin. However, sequence 2 is also complementary and can form a second 1:2 hairpin. However, sequence 2 is also complementary to sequence 3. If sequences 2 and 3 form a 2:3 hairpin structure, the 3:$ attenuator hairpin cannot be formed and transcription termination will not occur. Under normal conditions, the formation of the 1:2 and 3:4 hairpins is energetically favourable.

The Leader Peptide

The leader RNA sequence contains an efficient ribosome-binding site and can form a 14-amino-acid leader peptide encoded by bases 27–68 of the leader RNA. The 10th and 11th codons of this leader peptide encode successive tryptophan residues, the end-product of the synthetic enzymes of the trp operon. This leader has no obvious function as a polypeptide, and tryptophan is a rare amino acid; therefore, the chances of two tryptophan codons is succession is low and, under conditions of low tryptophan availability, the ribosome would be expected to pause at this site. The function of this leader peptide is to determine tryptophan availability and to regulate transcription termination.

Attenuation

Attenuation depends on the fact that transcription and translation are tightly coupled in E. coli; translation can occur as an mRNA is being transcribed. The 3'-end of the trp leader peptide coding sequence overlaps complementary sequence the two trp codons are within sequence overlaps complementary sequences the two trp codons are within sequence 1 and the stop codon is between sequences 1 and 2. The availability of tryptophan (the ultimate product of the enzymes synthesized by the trp operon) is sensed through its being required in translation, and determines whether or not the terminator (3:4) hairpin forms in the mRNA.

As transcription of the trp operon proceeds, the RNA polymerase pauses at the end of sequence 2 until a ribosome begins to translate the leader peptide. Under conditions of high tryptophan availability, the ribosome rapidly incorporates tryptophan at the two trp codons and thus translates to the end of the leader message. The ribosome is then occluding sequence 2 and, as the RNA polymerase reaches the terminator sequence, the 3:4 hairpin can form, and transcription may be terminated. This is the process of attenuation.

Alternatively, if tryptophan is in scarce supply, it will not be available as an aminoacyl tRNA for translation, and the ribosome will tend to pause at the two trp codons, occluding sequence 1. This leaves sequence 2 free to form a hairpin with sequence 3, known as the anti-terminator. The terminator (3:4) hairpin cannot form, and transcription continues into trpE and beyond. Thus the level of the end product, tryptophan, determines the probability that transcription will terminate early (attenuation), rather than proceeding through the whole operon.

Importance of Attenuation

The presence of tryptophan gives rise to a 10-fold repression of trp operon transcription through the process of attenuation alone. Combined with control by the trp repressor (70fold), this means that tryptophan levels exert a 700-fold regulatory effect on expression from the trp operon. Attenuation occurs in at least six operons that encode enzymes concerned with amino acid biosynthesis. For example, the His operon has a leader which encodes a peptide with seven successive histidine codons. Not all of these other operons have the same combination of regulatory controls that are found in the trp operon. The His operon has no repressor-operator regulation, and attenuation forms the only mechanism of feedback control.

THE ARABINOSE OPERON OF E. COLI

The arabinose operon is another example of glucose-sensitive operon. As with lactose, when arabinose is absent, only a few molecules of the enzymes needed for arabinose catabolism are present in the cell. When arabinose is added (provided glucose is absent), there is a vary rapid increase in the number of arabinose catabolic enzymes. This is controlled by a different mechanism from that described for the lactose operon.

The gene governing the metabolism of arabinose comprise what is called a regulation, which is composed of at least three operons. The *araBAD* operon contains the gene for the enzymes involved with the conversion of L-arabinose to D-xylulose 5-phosphate. The controlling sites for this operon are located adjacent to it. Two operons control the transport of arabinose into the cell: *areE*, which is the structural gene for the L-arabinose binding protein, and *araF*. The regulator gene for the system, *araC*, is located between the *araBAD* controlling site region and the leu operon. The *araC* gene controls the expression of *araBAD* by its positive and negative action in the controlling site region. Since *araBAD*, *araE*, and *araF* are inducible by L-arabinose and controlled coordinately by *araC*, it is assumed that the structures of the three controlling site regions are similar. The following discussion will focus on the *araBAD* operon.

The operon is thought to be controlled as follows. The *araC* gene codes for a P1 protein, which has repressor function and exerts its effect by binding to that adjacent *araO* (operator) controlling site and

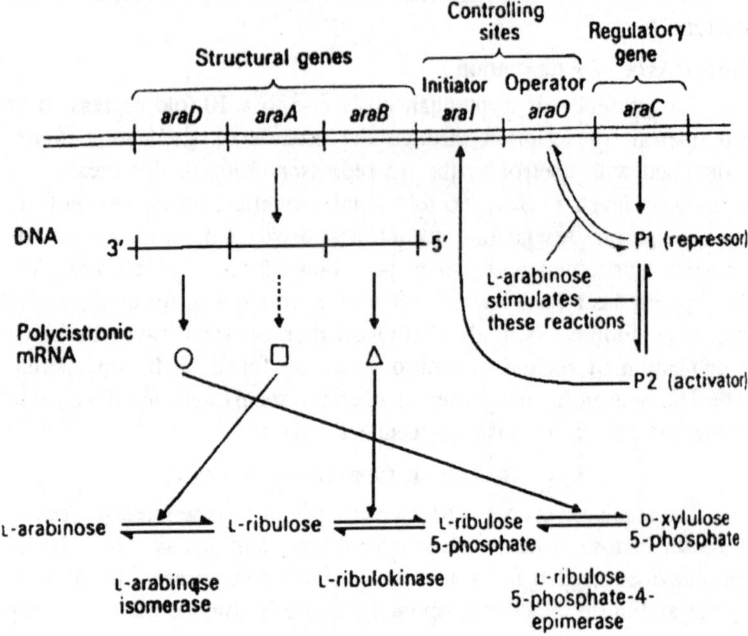

Fig. 10.6. The araBAD (arabinose) operon E. coli and the associated controlling sites and regulatory gene.

preventing the RNA polymerase from binding. Thus the operon is under negative control by P1. When L-arabinose is present, it stimulates the release of P1 from the DNA and the conversion of P1 to P2, which is an activator of the operon. The P2 binds to the *araI* (initiator) site, facilitating RNA polymerase binding and the initiation of transcription (All this, of course, requires the prior binding of cAMP-CGA complex, and this is thought to occur in the same vicinity.) The three structural genes are transcribed on a single polycistronic mRNA. The operon, then, is under positive control by P2.

Much of this is hypothesis, but there is good evidence for some parts of it. There is genetic evidence for the existence of the aral site, and it is thought that part of this has promoter activity and another part is involved with cAMP-CGA binding. The function of the *araC* gene has been demonstrated by the study of genetic mutants. Mapping experiments with these mutants have shown that *araC* consists of only one cistron. Moreover, *araC* nonsense mutants have been shown to have no cis effect on the *araBAD* operon, thus indicating that *araC* is not in the BAD operon.

Three classes of *araC* alleles are known:

1. *araC*$^+$. The wild type allele renders the operon inducible; the three enzymes are induced by L-arabinose.
2. *araD*$^+$. These occur with high frequency and result in a pleiotropoically L-arabinose-negative phenotype. In other words, the enzymes are not inducible by L-arabinose.
3. *araC*^{c}. These are quite rare and give a pleiotropically constitutive phenotype in that the enzymes are produced even in the absence of the inducer.

As with the regulatory mutants of the lactose operon, diploid studies have been used to obtain an understanding of the function of the *araC* gene. From these studies it was sown there *araC*$^-$ is recessive to C$^+$ in either the cis or trans arrangement. The C$^-$ alleles are complemented by A$^-$, B$^-$, and D$^-$ alleles, and thus the pleiotropic negative phenotype is not the result of a polarity effect on the araBAD operon. The conclusion from studies of C$^-$ and C$^+$C$^-$ strains was the C$^+$ produces a protein that, in the presence of L-arabinose, is necessary for the expression of the region. This suggested some positive control in the system and contrasts with the lactose operon C$^-$ mutants, which are constitutive because of the loss of negative control

The Cc alleles are cis- and trans-dominant to C$^-$, suggesting that they produce the activator P2 in the absence of L-arabinose; this activator is able to turn on an operon that is either cis or trans to the Cc allele. On the other hand, *C*$^+$ is dominant to *C*^{c}. This suggests that there is some negative control of the operon. Based on the model for regulation of the operon that was presented (which was, of course, proposed on the basis of the data now being discussed), in the absence of arabinose, P1 acts as a repressor, preventing expression of the operon by araCc activator. That is, when P1 is on the operator, transcription ceases even if P2 is present.

There is some biochemical evidence to support the regulatory model. Studies of heat-sensitive *araC–* mutants have shown that both the repressor and activator functions are heat-labile, thereby indicating that arac produces a protein that can serve both functions. The *araC* protein has been purified and it has been shown to have both repressor and activator activity. Indeed there is some evidence that the P2 form of the protein is a dimer of the P1 form. There is also evidence that L-arabinose interacts directly with the *araC* protein to bring about the necessary conversion. Also, there is direct evidence that the activator form of *araC* protein, P2, is required for transcription of the operon. In an *in vitro* system, synthesis of ara mRNA shows an absolute requirement for *araC* protein.

In conclusion, the L-arabinose regulation is under both positive and negative control, with the *araC* protein playing a pivotal role in the regulatory process. The exact nature of the controlling sites remains to be worked out. In contrast to the lactose operon where inhibition must be removed for the genes to be expressed, the arabinose operon requires activation for transcription to begin.

11

Gene Expression in Prokaryote

Transcriptional control requires a three-component system : a regulatory gene, a control sequence of DNA (the operator) where the regulatory gene product binds and the regulated structural gene or operon (group of adjacent structural genes governed by the same transcriptional regulatory system). The whole transcriptional control system can be either *negative* or *positive*. The system in which the regulatory gene product binds to inhibit transcription are called *negative transcriptional control* systems. The regulatory gene in a negative transcriptional regulatory system is coding for a repressor protein. The positive transcriptional control system is one in which the regulatory gene codes for an expressor protein, one necessary for the gene to be expressed. There is evidence that a number of operons are controlled by both positive and negative transcriptional control systems. In such cases, the two different regulatory genes produce their repressor and expressor products. There are separate control sequence for each of these products. The genes under such control are expressed only when both systems are "go"—repressor is not bound to its control sequence and expressor is bound to its control sequence.

The simple distinctions that have been given are each further subdivided into inducible and repressible operons within each category. An inducible operon is normally not expressed but can be induced (turned on, make to begin transcription of mRNA) when the right environmental signal is received. A repressible operon is normally expressed but can be repressed (turned off, stopped from transcribing mRNA) when the right environmental signal is received. Repressible operons can sometimes be depressed (expressed more than usual) as well as being repressed. A small molecule usually serves as the environmental signal; this small molecule is called the *inducer* for inducible operons and the corepressor for repressible operons. The role of the small molecule signal is to bind to the regulatory gene product (the repressor protein or expressor protein), changing the ability of the regulatory protein to bind to its control sequence or operator.

Small molecule signals are almost always input or output compounds from the biochemical reactions catalyzed by the proteins coded by the operon. Occasionally they are compounds belonging to other pathways, for example, pathways whose use is preferred to those of the regulated operon, or compounds whose concentration must be balanced with those in the pathway.

Negative Transcriptional Control

The negative transcriptional system is one in which the regulatory gene produces a repressor protein, and the system may be either inducible or repressible. The *E. coli* lactose i (lac i) gene interaction with the lactose (lac) operon is the classic example of a negative inducible system.

Lactose operon of *E. coli*

Function in the wild type

The sugar lactose cannot be used as an energy source of *E. coli* unless it is first broken down into its glucose and galactose components. This reaction is catalyzed by the enzyme β-galactosidase, which has a tetrameric structure of identical 135,000-daltons polypeptides.

In a wild type cell growing in a medium that does not contain lactose, there are only a few molecules of *β-galactosidase*. However, if the cell is growing in a medium containing lactose as the sole carbon and energy source, there are about 3000 copies of the enzyme. In other words, induction of enzyme synthesis occurs as a result of the presence of lactose. Lactose itself is not the actual inducer in this system, but rather induction is brought about by the action of *allolactose* which is produced from lactose by the enzyme activity of the few molecules of β-galactosidase will remain high, but once the lactose runs out the enzyme level will diminish rapidly. This

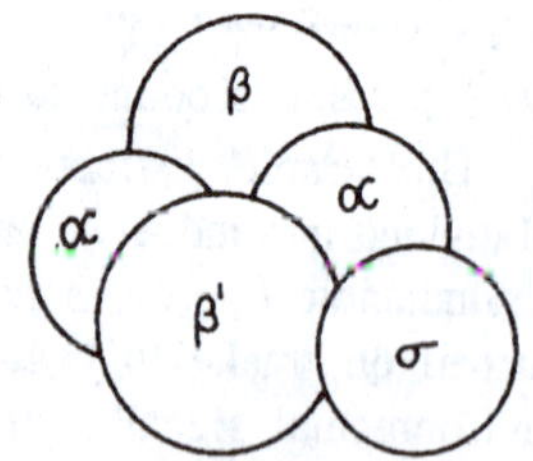

Fig. 11.1. A model for the structure of prokaryotic RNA polymerase showing association of five polypeptides ($\alpha_2\beta\beta'\sigma$).

is not an all-or-none phenomenon; the amount of enzyme produced is directly proportional to the amount of inducer present (up to the maximum amount found in the cell).

The addition of lactose to the cell not only brings about the increase in β-galactosidase levels has but also a rapid increase in the synthesis of *β-galactoside permease* and *thiogalactoside transacetylase*, two enzymes also needed for lactose breakdown. Genetic experiments have shown the genes for all of the three proteins are linked on the chromosome and adjacent to them are two regulatory sites, the operator and the promoter. A short distance away is the *i* gene, which codes for a repressor gene involved in the regulation of the system. As a result of the phenotypic properties of regulatory mutations, *F. Jacob* and *J. Monod* proposed their classical operon model for the control of gene express bacteria, a more up-to-date description of which will be given here. By definition, an operon is a genetic unit consisting of contiguous genes that function coordinately under the joint control of an operator and a repressor.

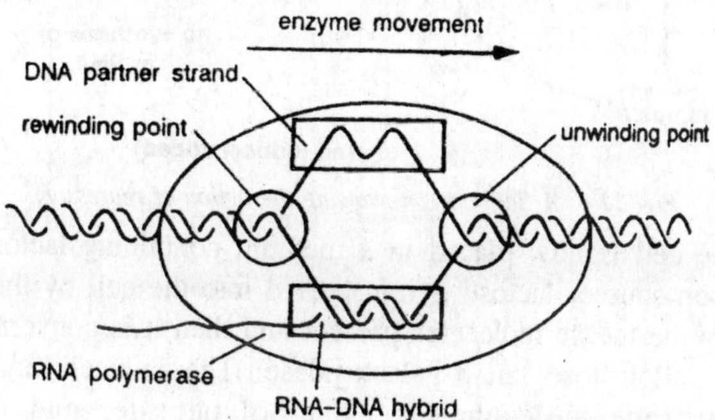

Fig. 11.2. Active centres in bacterial RNA polymerase enzyme.

The z^+, y^+, and a^+ structural genes code for *β-galactosidase*, *permease*, and *transacetylase*, respectively. Genetic experiments with mutant strains lacking enzyme activity showed that the three genes are adjacent on the chromosome. The i^+ gene (repressor gene) codes for a repressor molecule, which is a protein. The expression of this gene is constitutive; that is, the product is synthesized all the time. Control of the amount of gene product, then, is a function of the rate of which RNA polymerase can bind to the i gene promoter (pi^+) and initiate message produces a polypeptide that aggregates to produce the functional repressor tetramer. If the cell is growing on a medium lacking lactose, the repressor will bind to the operator (o^+) region,

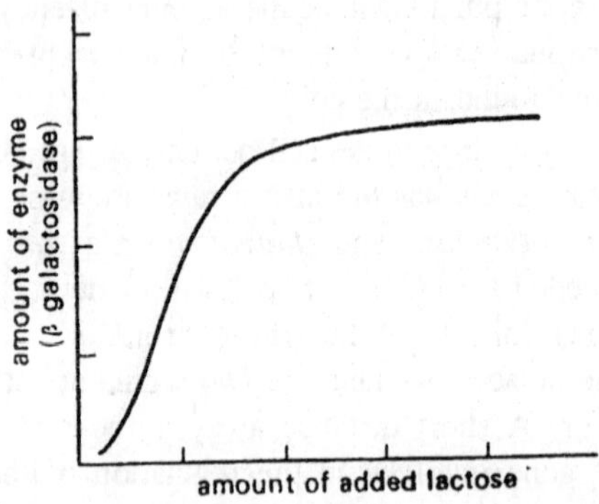

Fig. 11.3. Induction of β galactosidase synthesis by lactose.

which is located adjacent to the z^+ gene. When this complex is formed, RNA polymerase cannot bind to the structural gene promoter (p^+) region, and thus transcription of the gene is prevented.

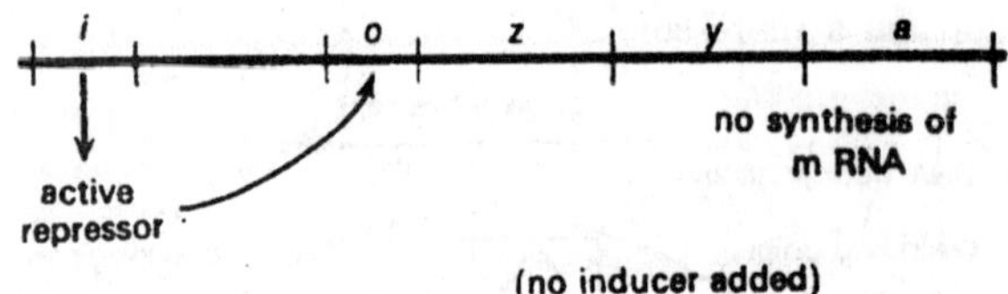

Fig. 11.4. A 'lac' operon showing the action of repressor.

If the cell is now placed in a medium containing lactose as the sole carbon source, lactose is transported into the cell by the activity of the few permease molecules present and then it is converted to the inducer, allolactose, an β-galactosidase. The inducer binds to the repressor, one molecule for each polypeptide, and causes a conformational change in the repressor such that it no longer has affinity for the DNA. As a consequence, the repressor falls of the operator and once that has occurred. RNA polymerase can bind to the structural gene promoter and initiate transcription of the operon. The lactose operon is a single transcriptional unit so that the RNA polymerase transcribe the z, y, and a genes, in that order, into a single polygenic (polycistronic) mRNA. This mRNA is translated by ribosome attaching to the 5' end (z gene end) and moving down the molecule. Thus the β-galactosidase is made fit and, after the stop codon of that region, the ribosome continues moving toward the 3' end of the message. Then after recognizing the initiation sequence for the permease gene, the ribosome begins to translate the part of the message to produce permease. The process is repeated at the boundary between the permease

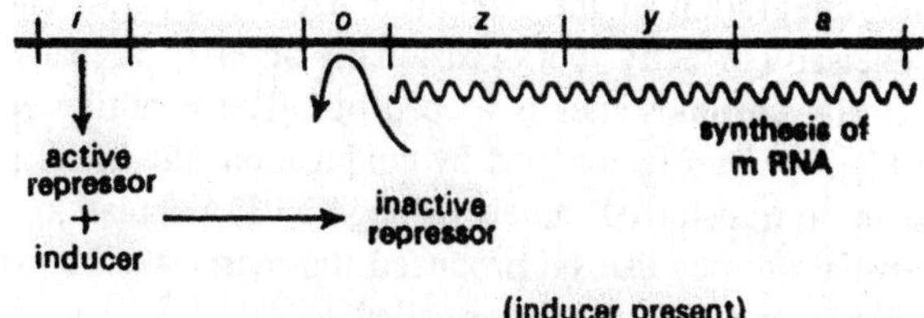

Fig. 11.5. A 'lac' operon showing induction by inducer.

and transacetylase gene sequences. The general principle that pertains here is that ribosomes can only initiate translation at the end of a polycistronic mRNA. (This makes the coordinate production of proteins of related function easy to control.) It is not possible for ribosomes to bind and initiate translation at the start regions of the permease or transacetylase sequences, presumably because the correct initiation sequence only found at the 5' end of the message.

Elucidation of the Regulation of the Lactose Operon by Studies of Mutants

The Jacob-Monod operon model of control of gene expression was based on studies of a number of regulatory mutants in which the control of the expression of the lactose operon was abnormal. An important part of the Jacob and Monod's mutant studies involved partial diploid strains, and the construction of these strains will be described before the properties of the mutants themselves are discussed.

In the discussion of Hfr strains of *E. coli*, it was shown that Hfr strains could revert to the F+ state if there was a reversal of the

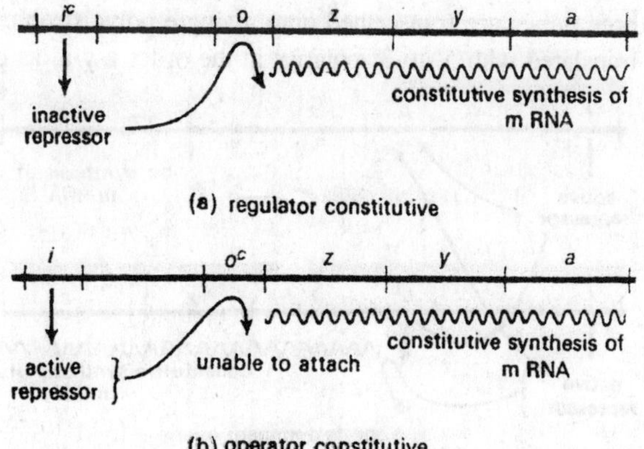

Fig. 11.6. Constitutive strains resulting due to mutations in regulator gene and operator gene.

process for integration of the circular F factor. In most cases the F factor is excised correctly, but occasionally an error is made and part of the bacterial chromosomes is looped out. The resulting episome is called an F' (*F-prime*) factor, and by conjugation, the genes picked up by the F can be transferred rapidly through an F^- population. Thus, by this process, episomes can be produced that carry the lactose region of the chromosome, and these are called *F' lac*.

Mutants of the structural genes

Both missense and nonsense mutants have been isolated for the three structural genes. Mapping these mutants provided evidence for the order z-y-a on the chromosome. A missense mutation in one of the genes results in the loss of activity of the enzyme for which the gene codes but does not affect the activities of the other two enzymes in the system.

On the other hand, the *nonsense mutations* have some interesting properties. For example, a nonsense mutation in the z gene (z^-), providing it is not too near the end of the gene, will not only abolish z gene activity but also reduce or abolish the expression of the y and a genes. This is called *polarity*, or the polar effect, and thus the nonsense mutations are sometimes called *polar mutations*. Indeed there is a gradient of polarity of the effect of chain-terminating mutants in the z gene on the expression of the other two genes. The closer the mutation is to the operator, the more severe the polar effect; that is, the less likely permease and transacetylase will be produced. Polar mutations in the y gene effect the activity of the y gene and a gene but not the activity of the z gene. The interpretation of these data was that the three genes are transcribed onto a single polycistronic mRNA, which is translated with 5'-to-3' polarity in the order z-y-a, as discussed

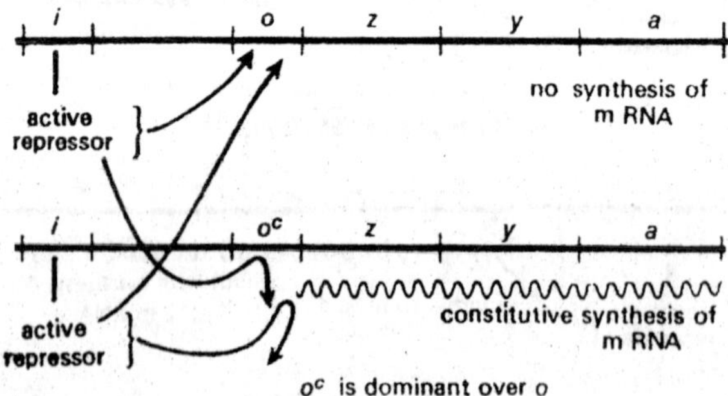

Fig. 11.7. Check on enzyme synthesis in a partial diploid, heterozygous for regulator gene (I/I^c).

previously. The polar effect of chain-terminating mutations is the consequence of the distance ribosomes have to travel before recognizing a new translation initiation sequence–the farther it is, the more likely the ribosome will fall off the message.

Operator mutants

A series of mutants were isolated that were constitutive for enzyme production. In other words they had lost regulatory control such that the enzymes were produced whether or not the inducer was present. When the mutations were mapped, it was found that several were clustered next to the z gene in a region that is now called the operator. These so-called operator-constitutive (o^c) mutants have lost the ability to bind the repressor protein such that transcription cannot be prevented. This latter point is supported by genetic and biochemical evidence. We will discuss the former here.

The effect of oc mutations on the contiguous structural genes and on z, y, and a genes located on a different piece of DNA was examined by haploid and partial diploid strains. In these studies, mutant strains were constructed with various combinations of operator and structural gene alleles, and the production of the enzymes was monitored in the absence and presence of the inducer, lactose.

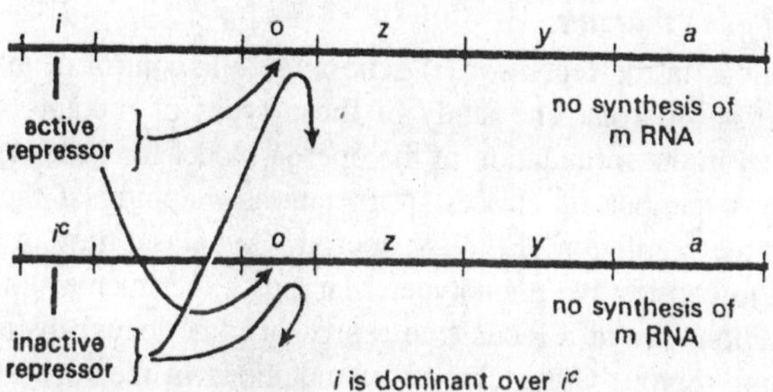

Fig. 11.8. Constitutive enzyme synthesis in a partial diploid, heterozygous for operator gene (O/O^c).

Class 1 is the wild type where neither enzyme is produced until lactose is added as the inducer. *Class 2* is a haploid strain carrying on oc mutation that leads to constitutive production of the enzymes. *Class 3* is a partial diploid carrying oc and y^- mutations on one chromosome and the o^+ and z^- alleles on the other chromosome. In this case β-galactosidase (z gene product) was produced constitutively, but permease (y gene product) was produced only in the presence of lactose. Class 4 is a similar case, with oc and z^- on one piece of DNA and the o^+ and y^- alleles on the other DNA. Here the permease is produced constitutively and the β-galactosidase is inducible. The two latter cases show that the effect of the oc mutation is limited to those genes adjacent to it on the same piece of DNA ; that is, it is a cis-dominant mutation. Using class 3 as an example, z^+ is on the same DNA as oc and hence is transcribed constitutively. The y^- mutation results in a nonfunctional permease. In the strain, the wild type y^+ allele is on the chromosome that also carries the o^+ allele. This DNA, then, is under normal regulatory control, and hence the y^+ is inducible. All of these data are consistent with the thesis that the operator region does not produce a diffusible product. Further, biochemical experiments showed that DNA form o^c strains will not bind the repressor, thus showing that the operator is the site of repressor action. The operator, then, is a controlling region that in the wild type can bind the repressor, which then prevents transcription. In the o^c mutants, the region has been altered (e.g. nucleotide pair change or deletion) so that the required protein—nucleic acid interaction does not occur and transcription cannot be stopped. Finally, since the o^c mutants do not affect the properties of β-galactosidase, it has been concluded that the operator is most probably a region distinct from the *z* cistron.

Repressor gene mutants

Mutations in the repressor (i) gene affect the control of expression of the lactose operon. The study of these types of mutants was also instrumental in the formulation of the operon model for gene regulation. There are a number of classes of mutations mapping at the *i* locus that affect the regulation of the *lac* operon. One class, the i^- mutations, gives rise to constitutive phenotype. Mapping experiments showed that these mutations are at a locus (the repressor gene locus) distinct from the operator region. Other classes of mutations in the repressor gene include the is (super repressed) and the i^{-d} (trans-dominant).

Class 1 is the wild type as before and *class 2* shows the constitutive nature of i^- mutations. *Classes 3* and *4* involve partial diploids in which a z^- mutation is on one DNA any *a* y^- mutation. In both cases normal, inducible production of the enzymes is the case showing that the i^+ is dominant to i^- for genes either on the same or a different DNA. This is called *trans-dominance* and is interpreted to mean that the i^+ gene codes for a diffusible product that acts to prevent transcription in the absence of lactose. This we now know to be the repressor protein that binds to the operator. The i^- mutations produce an inactive repressor that cannot bind to the operator, and this results in constitutive enzyme production in haploid cells. In the i^+/i^- partial diploids, there are enough repressor molecules produced from the i^+ gene to bind to the two operators present and thus normal regulation is in effect.

An i^s mutations results in a complementely negative phenotype with respect to *lac* enzyme production. In partial diploids, i^s is trans-dominant to the i^+ allele. The interpretation is that i^s mutations result in the production of an altered repressor molecule (a super-repressor), which binds to the operator normally but is not capable of recognizing the inducer molecule, lactose. Hence transcription cannot be initiated since the defective repressors remain stuck on the operators.

The last class of *i* mutations are the i^{-d} mutations. Like the i^- mutations, these result in constitutive enzyme production in haploids, but unlike the i^- alleles, they are transdominant to the wild-type i^+ allele. These i^{-d} mutants are very rare, and their phenotype is believed to be related to the tetrameric nature of the repressor. That is, the wild type produces a repressor protein of four identical polypeptides. This repressor molecule has only one binding site for the operator. In i^{-d} mutants the subunits appear nor to combine normally so that no operator-specific binding site results. In the i^+/i^{-d} strains there is a mixture of normal and defective repressor subunits and the trans dominant of i-^{d} is thought to be because repressor made from combinations of the two cannot bind to the operator. Only purely wild-type repressors could bind, and statistically, they would be quite rare. With this in mine, the i^- mutations discussed previously must either be nonsense mutations, which produce short polypeptides, or missense mutations, which result in polypeptides that do not participate in tetramer formation.

In summary, the *i* mutations provide evidence that the *i* gene produces a diffusible product—the repressor–which prevents transcription of the *lac* operon. The site of action of the repressor is the operator region. Also, the *i* mutations show that the lactose repressor has three recognition reactions coded into its structure :

1. With the inducer, lactose ; this presumably alters its shape causing it to dissociate from the DNA.
2. With the operator region.
3. With itself in that it acts as a tetramer.

Promoter mutants

These mutants map to the left of the operator region and are characterized by the lack of mRNA production in the presence or absence of lactose. The date show that p^- mutations are cis-dominant to p^+ in that their effect is limited only to the genes on the same piece of DNA. This is shown especially clearly by classes 5 and 6, where inducible enzyme production resulted for the wild-type gene on the p^+ DNA whereas no enzyme was produced from the wild-type structural gene on p^- DNA. Classes 3 and 4 also indicate that the properties of p^- mutations are not affected by oc or i^- constitutive mutants. These facts are interpreted if we assume the p^+ region is the binding site for RNA polymerase. If it is altered so that polymerase cannot bind, no transcription will occur.

The trp regulatory gene and tryptophan operon

The *E. coli* tryptophan operon is controlled negatively by the *trp* regulatory (*trp R*) gene. This regulatory system is negative and repressible. Notice that this operon codes for catalysts of the biosynthesis of tryptophan, a vital amino acid needed to make proteins. We can expect that biosynthesis of tryptophan will usually be needed; only when there is an influx of tryptophan into the bacterial environment will transcription of these genes by superfluous. The transcription is thus usually occurring, but shut off when conditions permit. Making mRNA and proteins requires a lot of energy, so the cell spares itself the synthesis when the enzymes being coded are not needed.

The *trp R* gene is not near the trp operon in *E. coli*. It codes for a repressor protein that will not bind to the *trp* operator (*trop O*) unless the end product, tryptophan, is bound to it. Tryptophan is the corepressor, and serves as a signal for turning off transcription of these genes. This system is clearly beneficial, since the bacteria need not make catalysts for tryptophan biosynthesis if tryptophan is already in excess. Notice that, in this case, unlike the case of the *lac i/lac* operon interaction, the promoter (RNA polymerase site) is between the operator and the structural genes. The promoter is the site adjacent to the site for mRNA initiation, so the *trp O* region is not transcribed into RNA, whereas the *lac O* region is transcribed. The *trp O* and *trp* promoter (*trp P*) regions overlap, like the analogous regions in the *lac* operon, and the repressor function *in vitro* by blocking the access of RNA polymerase to the transcription initiation site, as in the *lac* operon.

A feature of this repressible operon that was unexpected is the attenuator region. The *attenuator* is a region of the mRNA leader near gene E where transcription usually terminates in the absence of regulatory input. This regulatory input probably involves the transfer

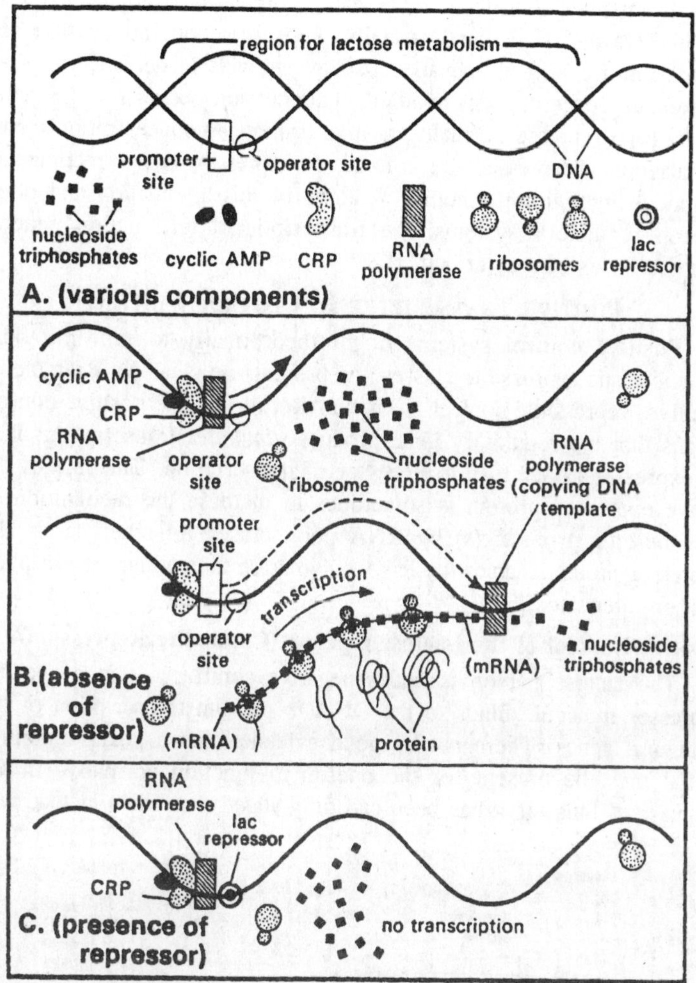

Fig. 11.9. Mechanism involved in the positive control system for the regulation of gene activity in E. coli lac operon.

RNA (tRNA) for tryptophan. If there is insufficient tryptophan to keep the cognate tRNA[trp] acylated for protein synthesis, the RNA polymerase continues past the attenuator to transcribe the entire *trp* operon. Otherwise, transcription ends in a block of G-C pairs, a block of A-T pairs, and a region of two-fold symmetry (palindrome), having produced 160 base pairs of the *trp* operon leader but no trp structural gene mRNA coding sequences. Thus, even thc relatively coarse mechanism of transcriptional control can be graded according to tryptophan demand. At high levels of demand (usually the case), both attenuation and repression are held in abeyance. At lower levels of demand (medium tryptophan concentrations available), attenuation occurs but the genes are still not repressed. Finally, at high tryptophan concentrations, both attenuation and repression are in force, the repression being predominant. A model highlighting common features of this attenuation and other operon regulatory systems, the transcriptional termination general hypothesis, will be described.

POSITIVE TRANSCRIPTIONAL CONTROL SYSTEMS

Positive control systems ought theoretically to fall into both inducible and repressible classes. At present, no clear cut example of positive, repressible control is known. Recall that the positive control means that the regulatory gene product *stimulates* transcription; it is an expressor rather than a repressor of transcription. The role of the active expressor protein is sometimes to increase the recognition of the operon's promoter(s) by RNA polymerases; in other cases, the expressor protein enables the RNA polymerase to continue transcription past an attenuator.

Positive Control of the Lactose Operon (Catabolite Repression)

The lactose person is under negative control in that a specific repressor molecule binds to the DNA to prevent transcription of the structural genes. There is also good evidence that a positive control signal must be present for the operon to function normally. In the discussions thus far it has been carefully stated that when lactose was

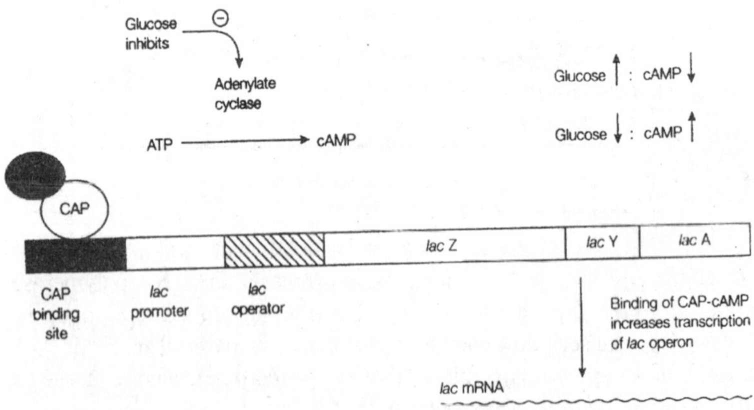

Fig. 11.10. Catabolite repression of the operon.

present, it was the *sole* carbon and energy source. If both glucose and lactose are present in the growth medium, the cells will preferentially catabolize glucose and the lactose operon in not transcribed. A similar situation applies to a number of other operons involved in the catabolism of other sugars, for example arabinose and galactose. These are called *glucose-sensitive operons*, and the phenomenon is commonly called *catabolite repression* (the *glucose effect*).

The effect of glucose on transcription of the operon in question is the result of the action of a breakdown product of that sugar to lower the intracellular amount of cyclic AMP (cAMP : 3', 5'-cyclic adenosine monophosphate). This molecule is made from ATP in a reaction catalyzed by adenyl cyclase and is broken down with the aid of the enzyme phosphodiesterase. Thus the (unknown) catabolite of glucose could bring about the decrease in cAMP levels either by inhibiting adenyl cyclase or by stimulating phosphodiesterase or perhaps by doing both.

The importance of cAMP to the transcription of the lactose and other glucose-sensitive operons is that a complex of cAMP with a *catabolite gene activator* (GCA) protein (a dimer of molecular weight 44,000) must bind to the promoter before RNA polymerase can bind and initiate transcription. Schematically the events shown are needed for transcription of the operons. If insufficient cAMP is present to make the complex, transcription is blocked at the various operons.

Molecular Details of Lactose Operon Regulation

Information has been obtained about the sequence of nucleotide pairs involved in the promoter and operator regions of the lactose operon.

The actual sequencing involved nucleic acid fingerprinting techniques. The repressor binding site was determined by sequencing the stretch of DNA protected from DNase digestion when the repressor protein is bound to the operator. The exact limits of the cAMP-CGA binding site and RNA polymerase interaction sites are not known, but genetic mutations that affect the level of expression of the lactose operon, and which presumably are in the promoter, show the promoter region to be about 80 nucleotide pairs long. It is noteworthy that the end of the *i* gene is immediately adjacent to the promoter sequence. Comparison of the sequence of the lactose operon mRNA with the DNA sequence shows that transcription begins in the operator region within the region protected by the repressor. This is shown in more detail which also shows the region of the mRNA protected by the ribosome (the ribosome binding site) during the initiation step.

As can be seen, the first 38 bases of the mRNA are not translated, the codon for fmet starting at position 39. The first part of the mRNA is a copy of most of the operator region. The ribosome binding site was determined by sequencing the part of the mRNA remaining after RNase digestion while the ribosome is bound in its initiation configuration. This latter arrangement was achieved in vitro by having only fmet-tRNA in the reaction mixture so initiation but not elongation could occur. The ribosome binding site of the lactose operon covers 50 nucleotides of the mRNA and includes the codons for the first seven amino acids of the β-galactosidase. The boxed nucleotides in the diagram indicate sequences that have also been found in ribosome binding sites of other prokaryotic mRNAs. Specifically there is a nonsense codon (UAA here), a purine-purine-UUU-X-purine (where X is usually a purine), an AGGA sequence, and the start codon AUG.

The ara Gene and the Arabinose Operon

This operon (genes *ara B, A* and *D*) was first described as a simple, positive transcriptional control, regulatory system, inducible by the alternative sugar arabinose. The evidence obtained by *Lee*, *Englesberg*, and colleagues showed that deletions or nonsense mutations in the regulatory gene, *ara* C^-, shut off expression in the operon, rather than making it constitutive. Thus, the *ara C* product could be viewed as an expressor rather than a repressor. These are C^- mutations were found to be recessive in merodiploids. This finding makes sense because a mutant expressor protein, which no longer bound the inducer arabinose, would not bind to the operator. Then, the alternative *ara* C^+ expressor could bind to both arabinose and the operator, and its function would dominate the phenotype.

Later, the model was revised in the light of a more complete study of possible mutant phenotypes. The *ara E* gene, not in the operon, was found to be co-regulated with the operon. A class of *ara C* constitutive mutants was also found. By itself, this finding need not have disturbed the model, since the constitutive version of the expressor protein could simply bind to its control regions even when arabinose was lacking. The *ara* C^c mutants should be dominant if so, but the *ara* C^c mutants were recessive to *ara* C^+ in merodiploids. This result was probably found because the *ara* C^+ expressor acted as a repressor when arabinose was absent. When arabinose bound to the repressor, it converted the repressor into a positive control expressor.

COMPLEXITIES OF TRANSCRIPTIONAL REGULATION IN PROKARYOTES

The prokaryotic systems that operate via transcriptional control often utilize different control systems that respond to different signal molecules in the environment. This level of complexity can yield very sophisticated responses in the bacterial cell, as in the case of the catabolite repression system. Nevertheless, there are even more complex transcriptional control systems acting in bacterial. The example presented can be considered a *regulatory cascade*—that is, one regulatory event precipitates a chain reaction on more than one set of genes.

Flagellar Phase Variation

In many species of *Salmonella*, there are two nonallelic structural genes for *flagellin*, the main flagellar protein. These genes are called H1 and H2 (from the German *hauk*). Although both H1 and H2 genes are present in a cell, only one is usually expressed—flagellin in phase 1 is produced from transcription and translation of gene H1, whereas in phase 2, the flagellin is produced by transcription and translation of gene H2. Cells change from phase 1 to phase 2 at a measurable rate on the order of one change per 1000 cells per bacterial generation. The change can be detected using whole cells, by means of antibodies to H1 and H2 flagellin, or the flagella can be isolated and their proteins characterized.

The structure of the regulatory system involved. First, look at the H1 gene regulation. When the *ah* 1 (operator) is open, the mRNA is produced and H1 flagellin is made. The *ah* 1 operator can be turned off or blocked by a repressor coded by the gene *rh* 1 in the operon with the H2 gene. This finding was confirmed by *in vitro* translation of isolated mRNA. The complexity lies in the fact that the repressor of H1 is not regulated by a small molecular signal. Instead, the repressor itself is regulated by transcriptional control. Whenever the repressor is made, it evidently acts as a repressor of H1 without being activated or inactivated by signal molecules. But, when the repressor is made, H2 flagellin is made at the same time, since the H2 gene and the *rh* 1 gene are in the same transcription unit (operon). Thus, the two types of flagellin are alternatives : either one or the other is produced.

The next problem is not the H2, rh 1 operon is regulated. The operator (or perhaps promoter) for this operon is ah 2. The ah 2^- mutants lose the ability to make H2 flagellin and also lose the ability to repress H1 flagellin simultaneously. Their phenotype is stable H1;

they are no longer diphasic. The region labelled *vh* 2 is the region that regulates expression of the H2-rh 1 operon. There is no evidence that vh 2 codes for a protein. Instead, *vh* 2 may be a sequence that undergoes spontaneous inversion at the measurable rate. The region *vh* 2 may allow H2-*rh* 1 expression when it is inserted in one direction but disallow transcription when it is inserted backwards. This idea is proposed because H2 DNA from a diphasic *Salmonella* stock has been subjected to molecular cloning attached to l phage. The DNA was denatured and a heteroduplex analysis showed what appeared to be an inversion loop adjacent to H2 on the molecule. The proportion of these inversion loops correlated with the ratio of *Salmonella* in the two different phases. This evidence suggest (but does not prove) that *vh* 2 may control the expression of the H2 operon depending on its own insertion polarity.

Transcriptional Termination Control Via mRNA Alternative Conformations

General Transcriptional Termination Hypothesis

Several bacterial operons are subject to a control mechanism, transcriptional termination control, that takes advantage of the tight coupling between transcription of mRNA and its translation into protein. In these operons, transcription is terminated and the new RNA is released before the structural genes have been transcribed, unless the newly formed mRNA assumes a particular conformation which allows transcription to continue. This permissive conformation is achieved by interaction with the translation apparatus. Both ribosome binding and a subnormal level of a particular aminoacyl tRNA are necessary in order for transcription to continue into the structural genes.

A general hypothesis is to explain this phenomenon has been presented by *Keller* and *Calvo*. This hypothesis involves two stable mRNA leader conformations. Figure shows schematically the two conformations for the leader of the *leucine* (*leu*) operon mRNA. The first, when transcription continues and the operon is on, involves translation of a short peptide, called the *leader peptide*, by a ribosome. When this ribosome arrives at a group of leucine codons, it becomes arrested due to a limited supply of leucyl tRNA. This arrested state allows the formation of the mRNA conformation containing the preemptor helix and loop. This preemptor helix blocks the formation of the terminator helix that shuts off transcription. The entire operon is transcribed under these conditions. Notice that the supply of leucyl tRNA in normal cells is a very good indication of the supply of the end product, leucine, formed by enzymes encoded by the *leu* operon. When leucine is low, the transcription will occur so more leucine can be made.

When leucine is high, the transcription is terminated, since a ribosome will not be frozen at the leucine codons of the leader peptide. The mRNA leader will form the most stable base-paired interactions it can, which are those of terminator and terminator protector helices. These helices will result in the termination of RNA transcription in the leader, without continuation into the *leu* operon structural genes. Since these helices, the termination helix and its protector helix, are the most stable conformation for the leader mRNA, only a block to their formation can allow transcription.

Notice that this mechanism cannot apply to eukaryotes, because transcription and translation in eukaryotes are separated into different cell compartments by the nuclear envelope. It is a major control on pathways for synthesis of the amino acids leucine, threonine, phenylalanine, tryptophan and histidine in bacteria. In mammals, these pathways are not found. In lower eukaryotes such as fungi, these genes are for the most part widely separated in the genome.

Histidine Operon Control

The *histidine* operon of *Salmonella* has long been a regulatory enigma. There is an apparent operator, his *O,* with cis dominant constitutive mutations. Five unlinked genes also can be mutated to cause his operon constitutive expression : *his U* (tRNA^{hts}), *his R* (histidyl tRNA synthestase), his *W, his T,* and *his S* (genes for enzymes which modify bases of tRNA^{his}). The operon is expressed except in histidine excess, so it was viewed as repressible operon with histidyl tRNA as its small-molecule signal. But where was its repressor or expressor protein? Exhaustive searching did not reveal any additional regulatory genes, though it did reveal that RNA polymerase recognizes the two promoters of this operon better when ppGpp (3'-diphospho- 5'-diphosphoguanosine) is bound to it. This strange nucleotide is known to be high when the general level of amino acids is low, probably being the product of some type of idling reaction at the ribosome.

The transcriptional termination mechanism may be the only control responding specifically to histidine levels in this operon. *Johnston* and collaborators have characterized events at the his O region and found that this region is translated, since it contains nonsense mutants responsive to suppression. The leader peptide contains 16 amino acids,

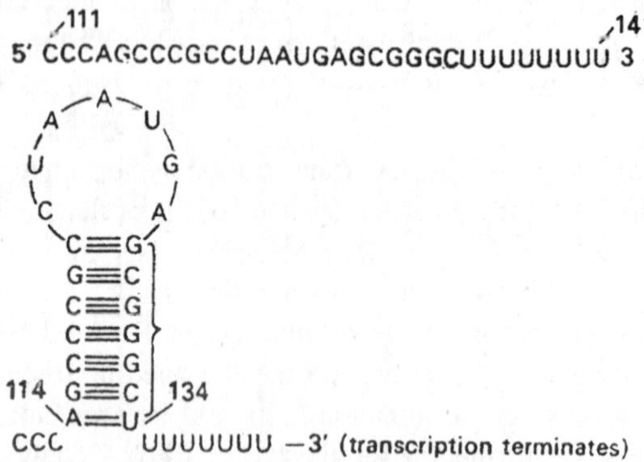

Fig. 11.11. Termination structure.

including 7 adjacent histidines. The run of histidine codons in the mRNA is evidently the sensor for histidyl tRNA concentration.

Two conformations, (*a*) and (*c*), for the histidine leader mRNA that lead to termination at the attenuator helix followed by a run of Us. It also shows a conformation, (*b*), leading to continued transcription into the histidine structural genes. The arrest of the ribosome at a histidine codon allows two preemptor helices to form in this case. *Johnston* and coworkers sequenced the operator DNA of a constitutive mutant, his O 1242, and found a deletion preventing the formation of the attenuator helix—thus the termination conformation could not form. They also found *his O* mutations which cause a *his C* phenotype. These resulted from termination in the leader peptide before the ribosome reached the histidine codons. It is possible that additional areas of the histidine leader sequence participate in conformational control, since it is 151 nucleotides long—an unusually lengthy leader even among transcriptional termination—regulated operons. Identified portions of *his O* regions are shown schematically.

Post-translational Control via Noncovalent Binding of Signal Molecules

The two mechanisms by which prokaryotes generally control the activity of a protein after it has been translated from the corresponding mRNA are (1) non-covalent interactions with signal molecules and (2) covalent modification of the protein itself, for example, phosphorylation of one or more amino acids in the backbone of the protein. This section is concerned with the first alternative, non-covalent signal mechanisms.

The noncovalent binding of a molecule to a protein can have an activating effect or an inhibiting effect on the normal action of the

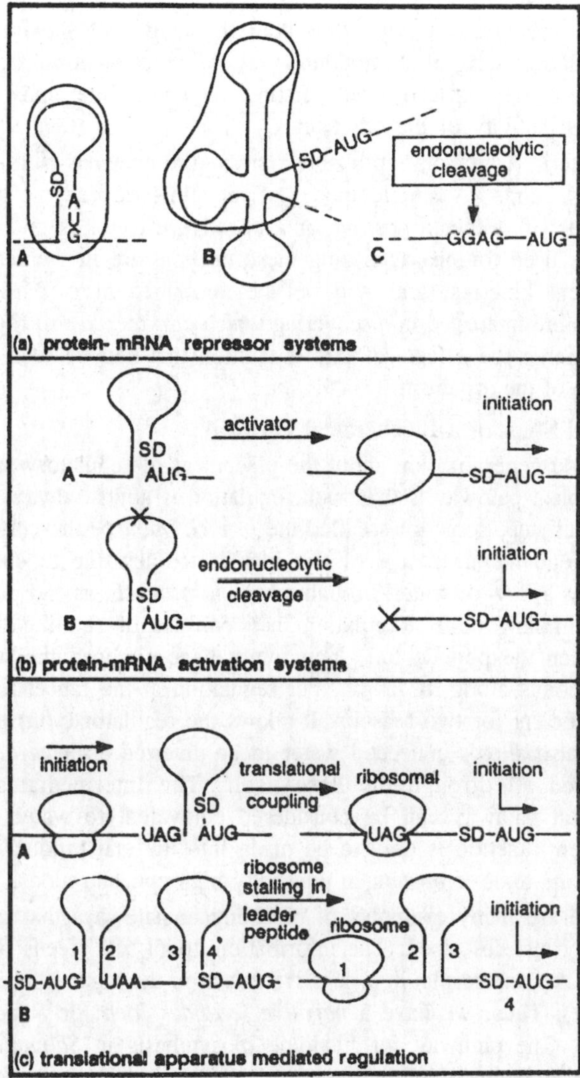

Fig. 11.12. Some models of translational regulation in prokaryotes.

protein. If we consider an enzyme, an activator would increase the rate of reaction catalyzed by the enzyme, whereas an inhibitor would decrease the rate. In many cases, these activations and inhibitions operate via allosteric mechanisms. By definition, *allosteric* regulation is an effect on enzymatic rate caused by binding of a small molecule called an *effector* at a site other than the active sites exist in an enzyme, and binding of a compound (say the reactant or substrate) at one of these sites, instead of at a distinct regulatory site, enhances or inhibits the activity of the other sites.

Allosteric regulation is probably restricted to enzymes with several catalytic subunits. A simple model of an allosteric enzyme and its interaction with either a positive or a negative effector is given. The techniques used for characterizing these enzyme are in the realm of biochemical kinetics, and will not be described here. Molecular geneticists are interested in discovering which enzymes are so regulated and examining the effects of this regulation that can be seen in the phenotype of the organism.

Regulated Steps of a Biochemical Pathway

A good generalization about the placement of regulatory steps in a biochemical pathway is that such regulation is almost always found at the input step, sometimes called the *branch point* or the *committed step*. A good mental image of this idea is to view the biochemical pathway as a flow of water through a hose to produce, as end product, a full swimming pool. Regulation then consists of shutting off the faucet when the pool is full. You might have clamped the hose at multiple points along its length, but regulation at the faucet is more energy efficient for two reasons. It allows the regulator (you) to save work. It also allows unneeded water to be diverted elsewhere rather than stored all through the hose itself. The intermediates in a biochemical pathway can be considered equivalent to water in the hose : their function is only to be made into the end product. They may even be toxic if present in unusually high concentrations.

There are many examples of biochemical pathways that operate in the way just described. The information "pool full" feeds back to the input step and results in a negative effect on the input (shutting off the faucet). Thus, we have a *negative feedback loop*, or a *feedback inhibition*. The pathway for histidine biosynthesis in *Salmonella* is regulated by feedback inhibition as well as by transcriptional control. When histidine is made as the end product, it inhibits the first enzyme in the pathway, phosphoribosyl-ATP synthetase, coded for by the *his G* gene. The pathway to the aromatic amino acids (tryptophan, phenylalanine, and tyrosine) is regulated by feedback inhibition as well, but the design of the inhibition varies in different bacteria. The pathway exhibits *sequential feedback inhibition* in *Bacillus subtilis*. The regulation is sequential because phenylalanine, tyrosine, and tryptophan excesses

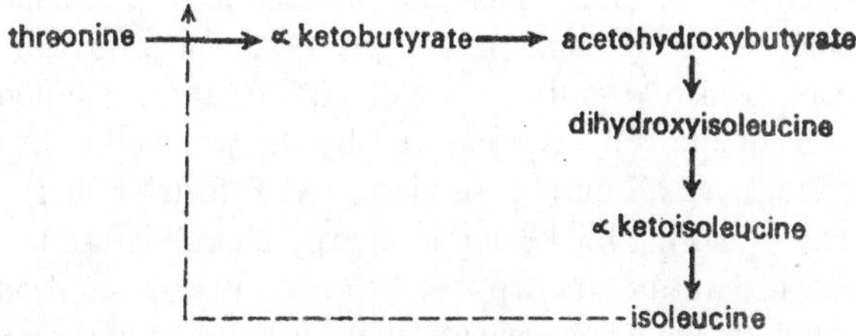

Fig. 11.13. Feedback inhibition in the synthesis of isoleucine in E. coli.

cause buildups of the intermediates chorismate and prephenate. These intermediates in turn inhibit the input step, as well as another step in the pathway.

An alternative regulatory scheme for this pathway, as found in *E. coli*, is an isozymic feedback pattern. There are three isozymes (alternative enzyme catalyzing the same reaction) of the first input enzyme, DAHP synthase. Each of these is feedback inhibited by one of the three major end products. Each end products also shuts off its own input step at or near the branch of the pathway leading to it, via feedback inhibition. There are isozymes of chorismate mutase as well, and the phenylalanine inhibits chorismate mutase P, the isozyme leading towards phenylalanine. In addition, phenylalanine inhibits the next step, prephenate dehydratase, which is in a complex with the chorismate mutase P. Tyrosine inhibits prephenate dehydrogenase; although this enzyme is also complexed with its isozyme of chorismate mutase, the tyrosine does not inhibit the chorismate mutase T activity. Tryptophan inhibits anthranilate synthase, its committed step.

Effects of Noncovalent Interactions with Signal Molecules

One of the most important functions of the signal mechanisms that act after translation is to coordinate the proper allocation of resources within the prokaryotic cell. This coordination results in optimal use of carbon, nitrogen, and energy for the survival of the organism. A mechanism that illustrates the balance of resources is the regulation of the first step in pyrimidine nucleotide biosynthesis, catalyzed by aspartyl transcarbamyolase (ATC). This enzyme is responsible for production of pyrimidine nucleotide precursors from central metabolites and is feedback inhibited by one of the end products, CMP. Even in insufficient amount of CMP, UMP, and TMP are present, the reaction catalyzed by this enzyme will not proceed quickly unless ATP concentration is high. ATP serves as an activator (positive allosteric effector) for the enzyme ATC, whereas CMP serves as an inhibitor negative allosteric effector).

This mechanism tends to facilitate production of a (balanced pool of pyrimidine (C, U, T) and purine (A, G) nucleotides for nucleic acid synthesis. It also keys the influx of nucleotides in the pool to the energy charge of the cell, as measured by the production of adenine nucleotides that have all three phosphates (ATP form) rather than two (ADP) or one (AMP). This keying to energy charge is useful because each nucleotide to be incorporated into nucleic acid must be phosphorylated to the triphosphate level, usually at the expense of ATP.

Another example of metabolic integration via noncovalent signaling can be seen in the inhibition scheme for *E. coli* glutamine synthetase (GS). This enzyme has a three part post-translational regulation scheme. It is stabilized in a more active state by divalent cations, it is inhibited by a variety of compounds, most of which are end products, and it is regulated by covalent modification in response to nitrogen demand. Glutamine produced by GS is a precursor of all the compounds listed below it; one or both of the nitrogens of glutamine are especially important in synthesizing these end products. These six end products serve as separate, independent feedback inhibitor of GS. Their effects are additive; all are required to approach complete inhibition.

Also, glycine and alanine inhibit GS in similar fashion. Although these two amino acids are not made from glutamine, they share the input nitrogen pool with glutamine synthetase, so they are signals relating to the amount of nitrogen already in use by the biosynthetic pathways shown. When the glutamine-dependent and glutamine-independent nitrogen-assimilation pathways all have low levels of products, the GS works fast. Otherwise, the enzyme is slowed by inhibition to an extent. The extent depends on how many important nitrogenous signal compounds are already at high concentrations. Again, a balance (this time between various nitrogen compounds) is achieved by the noncovalent signals. In addition, the GS regulation point switches off nitrogen assimilation, catalyzed by GS and other biosynthetic enzymes, to allow the nitrogen elimination required when nitrogenous compounds are in excess.

The effects of the feedback systems on the phenotype can be summarized by saying that this type of regulation results in vigor by avoiding waste. What is meant can be illustrated by the feedback-resistant type of mutations. In the biosynthesis of histidine in *Salmonella*, for example, the input enzyme, phosphoribosyll ATP synthetase, in feedback inhibited by histidine. Mutants in *his G*, the gene for this input enzyme, sometimes simply abolish the activity of the enzyme.

There is also a class of *his G* mutants that has normal activity for the phosphoribosyl ATP synthetase, but the enzyme is resistant to feedback inhibition by histidine. Such as feedback-resistant mutant wastes energy making excess histidine. One might expect that repression would override this wasted synthesis and turn it off at the transcription level, but in at least one well-characterized feedback-resistant missense mutant of *his G,* repression by histidine also fails to occur. The mechanism of this effect is not yet clarified.

In *Salmonella typhimurium*, the input step for leucine biosynthesis is α-isopropyl malate synthase (IPMS). In wild type cells, this enzyme is strongly inhibited by leucine, which tends to dissociate IPMS into dimers and monomers rather than the uninhibited major form, tetramers. The *Calvos* discovered that a mutant, which can grow in 5, 5, 5,-triluoro-Dl-leucine (unlike wild type), possessed a leucine-feedback-resistant IPMS. This mutant, flr 191, excreted small amounts of leucine. Its growth rate and yield of cells per volume of medium were normal, however. A second flr mutant, flr 19, was resistant to the same leucine analog because of a tenfold increase in the *leu* operon mRNA transcription. This strain also excreted a little leucine but grew normally.

The interesting experiment of the Calvos was to combine these two mutations in the double mutant strains, CV 241. This doubly mutant strain excreted leucine heavily, and, in addition, its yield of cells was reduced by 43%. CV 241 took 84 min for each cell division, whereas the comparable wild type and each of the single flr mutants took only 56 min per division. Clearly, this regulatory failure had serious negative effects on the mutant's ability to survive. We can infer from the two examples described that wasteful production of metabolites, even when they can be deposed of (into the medium), is deleterious to an organism. The system for leucine control is particularly illuminating in that we see how an organism can be protected by having two separate regulatory mechanisms for the same pathway. Redundancy is thus a protective device in some cases.

Post-translational Control via Covalent Modification

The second major protein-level regulatory system in prokaryotes is covalent modification. *Covalent modification* of proteins is catalyzed by specific enzymes. Although the covalent bonds formed are usually more stable than the noncovalent binding discussed in the previous section, these bonds can be broken by specific enzymes, which thus reverse the regulatory effects caused by the modification. Covalent modification is typically used to control bacterium-wide processes responding to environmental signals, rather than to control individual biosynthetic pathways.

Glutamine Synthetase of *E. coli*

Glutamine synthetase (GS) is feedback inhibited by at least eight different nitrogenous compounds. In addition, it is regulated by covalent attachment of adenylates (AMPs) to each of its 12 subunits. The GS is most active when it lacks adenylylation, and when all 12 AMPs are attached (GS) It is slightly active only when MN^{2+} replaces Mg^{2+}. It can still be easily detected because, although GS will not make much glutamine, it still catalyzes a rather rapid glutamyl transferase reaction. The general effect of the covalent modification scheme is that, when there is plenty of α-ketoglutarate and not much glutamine, the enzyme is as GS, unmodified. This form is more active, and furthermore, it resists feedback inhibition by the usual inhibitors except for glycine (not feedback since it is not an end product). When there is little α-ketoglutarate and plenty of glutamine, the GS is adenylylated to form GS. This form of the enzyme is not very active; the more of the 12 subunits that have AMPs attached, the lesser the activity. In addition, this GS is regulated by feedback inhibition, as shown. As you can see, the previous, section contained a description of GS rather than GS, which is not feedback inhibited.

The control of the adenylylation and dedenylylation is rather complex. The same protein, adenylyl transferase (A Tase) adds or removes the AMPs depending on the α-ketoglutarate, ATP, and glutamine concentrations and also on its small regulatory protein, PII. PII also favours the adenylylation or deadenylylation depending on conditions. When α-ketoglutarate and ATP are high but glutamine is low, uridine transferase adds a UMP to PII, another covalent modification, to form PII. PII bound to A Tase favours clipping the AMPs off from GS to regenerate GS. When conditions change, auridylyl-removing enzyme (possibly the same protein as the uridylyl transferase) clips the UMP from the regulatory protein. PII, thus reformed, causes the A Tase to adenylylate GS to for GS.

This covalent modification system is a global system because the influx of nitrogen into metabolism is made to respond to the form in which nitrogen is presented to the cell. Medium with ammonia (NH_3) as nitrogen source results in GS being the predominant form, whereas the slower, more regulated GS predominant when glutamate is the nitrogen source.

Chemotaxis in *E. coli*

E. coli swim by moving their flagella. When the flagella beat counterclockwise (ccw), the bacteria swim smoothly in one direction. When the flagella beat clockwise (cw), the bacteria tumble (change direction drastically). When bacteria are in a homogenous medium, they alternate between cw and ccw beating, generating a pattern of movement called a random walk. *Chemotaxis*, movement in response to a chemical signal, is generated by using more of one or the other direction. For example, a sudden large increase in an attractant chemical or a sudden large decrease in a repellent chemical results in smooth swimming (ccw beat) for several minutes. The response is transient; after a few minutes, the bacteria adapt to the new conditions and return to random walk behaviour. Decrease in attractant or increase in repellent increases cw beat transiently, increasing tumbling for a short period. The combined effect of this system is to bias the random walk so that a bacterium moves toward an attractant and away from a repellent.

The system for chemotaxis consists of flagellar components, signal transduction components, and signal receptor components, all of which are proteins. The genes that are known to code for these proteins. The signal receptors tend to be special transport proteins or permeases for the chemicals eliciting the response. They are often located either in the periplasmic space (between the cell wall and the cell membrane of the bacterium) or in the cell membrane itself. The signal transduction system consists of three known proteins, coded for by the genes *trg, tar,* and *tsr* (equal to *che* D). Mutants in these genes swim normally but lose response to certain chemicals. The *trg*$^-$ mutants, for example, do not respond to ribose or to galactose. The *trg*$^+$ protein probably is the signal component that interacts with the ribose binding protein coded for by *rbs*^{P} and the galactose binding protein coded by *mgl*^{B}.

This concept is supported by the fact that saturation of the ribose receptor will inhibit responses to galactose. The tar$^-$ mutants are unable to respond positively to aspartate or maltose or to respond negatively to hydroxyl ion (OH^-) or divalent metal ions. The tsr$^-$ mutants are not attracted by serine or repelled by hydrogen ions (H^+), fatty acids, indole, or hydrophobic amino acids. Both tar$^+$ and tsr$^+$ code for membrane proteins of about 60,000 daltons molecular weight. The *fla* I gene mentioned in the transcriptional control section as a positive regulatory gene, controls both tar$^+$ and tsr$^+$; fla I$^-$ strains make neither *tsr* nor *tar* gene product.

Changes in the level of covalent modification in this system serve as a switch. A protein methyl transferase, probably coded for by *che* X, transfers methyl groups from S-adenosyl methionine to the *tar* and *tsr* gene products. The role of the methylation and demethylation of these proteins is to allow stimulus transduction and adaptation. Methionine starvation, which prevents methylation of membrane proteins, prevents tumbling (cw rotation); responses to stimuli that either enhance or decrease tumbling frequency cannot occur under these conditions.

During an increase in attractant, methylation of membrane proteins increases. Methylation is decreased (evidently by demthylation) during increases in repellent. A model of methylation changes during responses to attractant concentration changes. The implication of the data is that changes in methylation state occur during response and during adaptation. For example, that normal behaviour or random walk behaviour is characteristic of cells with three different extents of membrane protein methylation. The current idea about these findings is that it is a change in methylation, rather than the extent of methylation itself, that results in the behavioural switch. Mutants that are che X^- exhibit only smooth swimming like *met*– strains starved for the S-adenosyl methionine precursor, methionine. In view of model, you might find this phenotype surprising, since smooth swimming is shown as characteristic of methylated proteins.

Again, it is evidently a change-sensing mechanism that results in starting or inhibiting tumbling; without the protein methyl transferase action, the change is not sensed. The smooth swimming may well be the state of lesser interaction between flagellum and switching molecules. The smooth state would then be thought of as a basal or turned off response, whereas tumbling would be the turned on response. Consistent with this hypothesis is the phenotype of *tar*⁻, *tsr*⁻ double mutants, which have a very low tumbling frequency. Mutants that tumble constantly also exist; they could be altered in the flagellar signal reception proteins such that their flagellar control stem's shape was like the wild type "on" state. The methylation system may work in response to changes in membrane potential that occur when chemotaxic chemical are added to a responsive bacterial culture. The linkage between the system is not yet clarified, however.

Notice that this system, like the GS regulatory scheme using covalent modification, involves whole-organismic responses rather than control of individual pathways.

12

GENE EXPRESSION IN EUKARYOTE

The regulation of gene expression in prokaryotes has been extensively studied. However, the regulation of genes has also been studied in eukaryotic cells. For many protein-coding genes, transcriptional control is the most frequent level of control and regulation at this level is quite strict. Gene control can also occur at a number of post transcriptional levels. The mechanisms and molecules that are involved in the various types of control are unknown.

Potential Sites for Regulation of Enzyme Synthesis

A number of steps are involved in producing a functional mRNA from a protein-coding gene and then having it translated by the ribosomes. That is, the transcription units for mRNAs are often longer than the mature mRNAs. These long pre-mRNAs are processed in the nucleus to produce the final mRNA product. This includes :

1. The addition of a 5' cap, usually before the rest of the RNA is completed. Essentially all RNA chains initiated by RNA polymerase II become 5' capped, so no control operates at this level.
2. The addition of a poly (A) tail at the 3' end.
3. The removal of specific spacer sequences (*introns*) and rejoining of the remaining RNA pieces (*exons*) to produce the mature mRNA.

RNA TRANSCRIPTION

The first level of control is the choice of RNA polymerases : *RNA polymerase I* transcribes rRNA genes, *RNA polymerase II* transcribes protein-coding genes, and *RNA polymerase III* transcribes

5S rRNA and tRNA genes. This specificity is clearly related to the nucleotide sequence of the promoter to which the appropriate RNA polymerase has affinity. Apart from the strongly conserved regions necessary for the initiation of transcription, other sequences in the vicinity are presumably important in regulating access of RNA polymerases to DNA. These latter sequences undoubtedly control the rate at which transcription initiation occurs and, therefore, the amount of pre-RNA produced.

Thus at least for RNA polymerase II, transcription initiation is believed to be the major controlling step in transcriptional control. However, relatively little is known about transcriptional termination and whether control is exerted at that level. For example, not all RNA polymerase II molecules that start a chain complete it.

RNA Processing

While transcriptional control is the predominant level of control, it is not the only one. Some regulation occurs during RNA processing.

Processing of RNA polymerase II-catalyzed transcripts commences by the addition of the 5' cap and 3' poly (A) tail. No control appears to occur with respect to the addition of the 5' cap. The situation regarding the poly (A) tail is less clear, however. Although the precision of transcription termination sites is not understood for RNA polymerase II transcription, the enzyme transcribes past the site for poly (A) addition in a number of well-studied genes. This means that poly (A) addition may need endonucleolytic cleavage and terminal addition of poly (A) to the cleaved 3' end. Poly (A) addition usually precedes the splicing steps of mRNA processing.

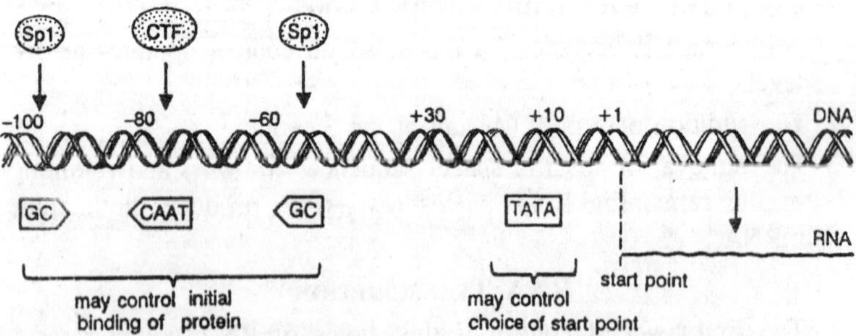

Fig. 12.1. A DNA segment in a eukaryote showing promoter sites and the enhancer site.

Splicing of pre-mRNA to mature mRNA for most transcripts may be an automatic event for some transcription units; however, the primary transcript can give rise to two more mRNAs that encode two or more different proteins. In these cases, differential processing occurs in response to particular regulatory signals.

Another regulatory choice at this level is whether a transcript will processed or discarded in the nucleus. The latter is what is called *nuclear RNA turnover*. For many years it has been known that only a portion of the heterogeneous nuclear RNA is ever processed and found as cytoplasmic mRNA. While all transcripts contain a cap and over 90% of newly capped mRNAs found in polyribosomes contain poly (A), poly (A) is added in the nucleus only to about 25% of all primary transcripts. Presumably the other 75% are discarded with time. What controls these events is unknown.

Cytoplasmic Control

Once the mature mRNA has been produced in the nucleus, at least a proportion of them are transported to the cytoplasm. The mechanism for this is unknown.

Once mRNA have entered the cytoplasm, the fates of the molecules differ. Specifically there is a wide range of half lives for different specific mRNA in the same cells and different half lives for the same mRNA in the same cell under different circumstances. These

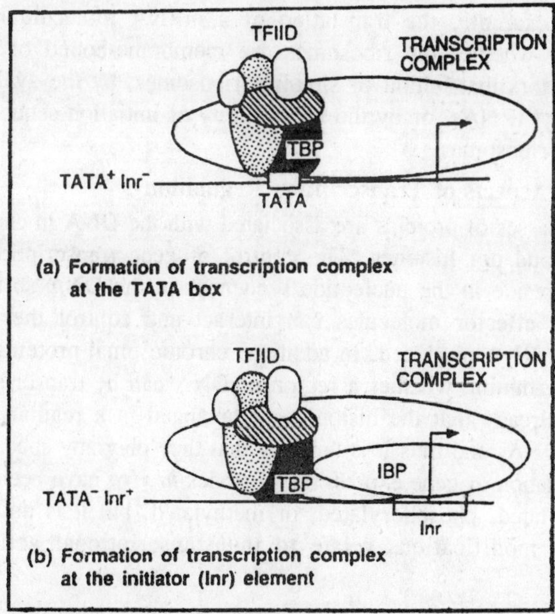

Fig. 12.2. Transcription complexes of TATA box and initiator (Inr) element.

have different consequences to the cell in terms of the relative accumulation of the gene products involved, but the regulatory events operating are here essentially unknown.

Translational Control

There is good evidence, for example, in embryonic development, for differential translation of completed mRNA molecules that are in the cytoplasm. It is also possible that tissue-specific preference for translation of specific mRNAs may occur.

Effector Molecules

Regulation of the transcriptional and post-transcriptional events in the nucleus may be related to the types and concentrations of effector molecules (*e.g.*, inducers, activators, repressors) that are transported into the nucleus from the cytoplasm. Controlling factors that alter the synthesis of these molecules and/or their transport into the nucleus can therefore affect the amount of enzyme ultimately synthesized.

In addition, other subtle control devices may be operating in the cell to regulate the amount of mRNA available for translation, and at present little is known about the regulatory signals for any of this. It is clear also that translational control is operated within eukaryotic cells. For example, the translation of a mRNA molecule may be affected by whether the ribosomes are membrane-bound or not by specific factors that inhibit or simulate ribosomes, by the availability of aminoacyl tRNAs, or by the accessibility of initiation sequences of mRNAs to ribosomes.

Molecular Aspects of Transcription Regulation

Two classes of proteins are associated with the DNA in chromatin : histones and non-histones. The control of gene transcription must ultimately reside in the nucleotide sequences of the DNA so that the appropriate effector molecules can interact and control the amount and type of RNA produced. In addition, chromosomal proteins play a role in determining whether a region of DNA can be transcribed. As discussed already that the histones are arranged in a regular fashion along the DNA, and thus it is unlikely that they play any specific role in the regulation of gene expression. Histones *in vivo* have been shown to be acetylated, phosphorylated, or methylated, but it is not known how these modifications relate to the transcriptional activity of chromatin.

On the other hand, there is much evidence implicating non-histones in the regulation of gene expression in eukaryotes. For example,

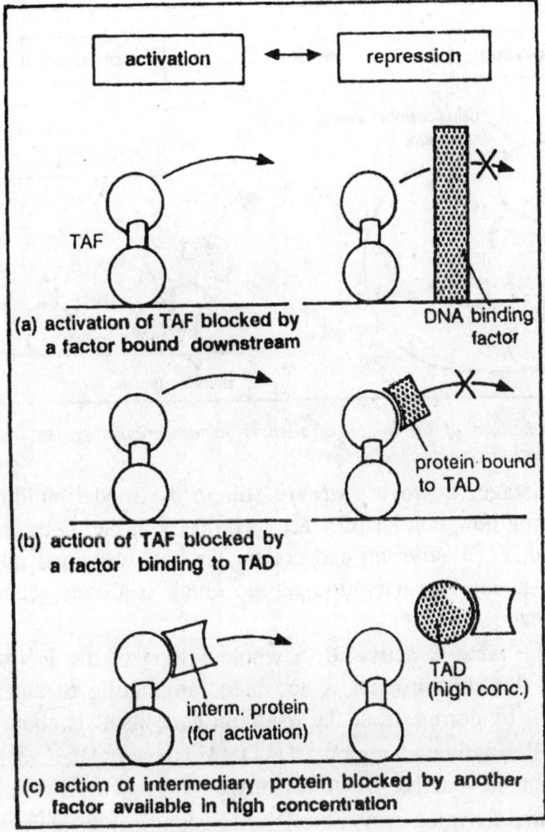

Fig. 12.3. Three ways in which activation (not binding on DNA) by TAF is blocked.

nonhistone proteins have tissue specificity and DNA binding specificity; they are present in higher amounts in transcriptionally active tissues as compared to inactive cell; they are much more diverse than histones; and certain specific classes of nonhistone proteins have, in fact, been linked with the induction of gene activity. In addition, if chromatin from transcriptionally active and inactive tissues is dissociated into DNA, histones, and non-histones, it can be determined by reconstitution experiments which component confers the capacity for transcription. These have shown that the non-histones are indeed the components that determine whether DNA can be transcribed. Thus it is currently believed that nonhistone proteins, presumably in response to specific signals, play a central role in the basic regulation of gene transcription in eukaryotes.

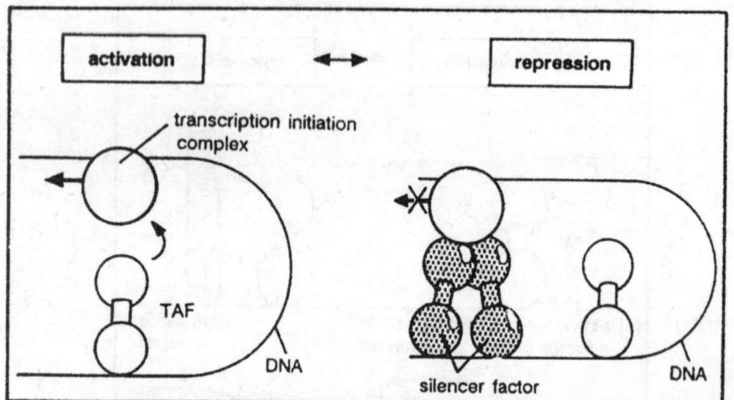

Fig. 12.4. Mechanism of the action of silencer factors, rendering the transcription initiation complex ineffective.

At this stage, however, we are still in the model-building stage for considering how non-histones act specifically at the molecular level. For example, *H. Weintraub* and colleagues have obtained interesting information about the activation of the chick α-globin genes. They concluded that :

1. When the gene is activated, a whole region of the DNA in that area of the chromosome is activated, amounting to about 50 to 100 kbp. by comparison, the α-globin gene itself is about 1 kbp.
2. Within the activated region, the DNA is methylated to a lower level than in inactive DNA regions.
3. Within the activated region the DNA is significantly more sensitive to DNase 1 than is inactive DNA.
4. Within the activated region, specific nonhistone proteins are bound to the chromatin. These are called *high-mobility group (HMG) proteins* because they migrate rapidly in electrophoresis.

In summary, gene activation must involve and unwinding of the highly compact nucleosome secondary and tertiary structure so that RNA polymerase and other regulatory molecules can access the DNA.

Regulation of Gene Expression in Lower Eukaryotes

The demonstration that gene regulation in prokaryotes often involves operons that are controlled in ways analogous to the lactose operon of *E. coli* prompted researchers to investigate whether operons existed in eukaryotes. Indeed much of the early model building concerning the regulation of enzyme synthesis in eukaryotes was influenced by the regulatory models of enzyme synthesis in bacteria. However, enzyme synthesis is regulated in different ways in eukaryotes.

Eukaryotes have many basic similarities to prokaryotes. For example, the processes of DNA replication, transcription, and translation are more or less the same. However, eukaryotes are vastly more complex, with discrete cellular compartments (nuclei, mitochondria, chloroplasts, etc.) that determine the organization of the process. In this regard the lower eukaryotes, and particularly the fungi, have proved to be useful model systems for the study of gene regulation since they are typically eukaryotic in their cellular structure and genetic organization, yet they are microorganisms that can be handled in ways very similar to bacteria. Further, these organisms are simple and live in environments that are subject to rapid changes. Like bacteria, lower eukaryotes must be able to adapt rapidly at the gene expression level when such changes occur.

Yeast and *Neurospora* will be discussed in this section. These organisms have a genome complexity of approximately ten times that of *E. coli*. Early studies concentrated on determining whether operons exist in these fungi. Since they are highly amenable to genetic and biochemical investigations, it is relatively easy to isolate mutants affecting enzyme function and also the regulation of enzyme synthesis. This of course parallels the approach of *Jacob* and *Monod* in their studies of the regulation of the *E. coli* lactose operon. In general the fungal studies showed that, contrary to the findings in prokaryotes, genes with related function are not closely linked but rather tend to be scattered over the chromosomes in the genome. Nonetheless, it has been shown in both lower and higher eukaryotes that there is coordinate synthesis of all the enzymes in a particular biochemical pathway. This presumably involves a regulatory system different from that of prokaryotes. Characteristically, then, the gene products in eukaryotes consist of monocistronic mRNAs and not polycistronic mRNAs.

In some cases, however, evidence was obtained for apparent clustering of genes in fungi, and this raised the possibility of the existence of operons in eukaryotes. Three representative cases will now be discussed, and while the early information was highly suggestive of an operon organization, all the evidence to date runs counter to that possibility.

Galactose Fermentation Genes of Yeast

The first three enzymes for galactose fermentation in yeast are *galactokinase*, *α-D-galactose 1-phosphate uridyl transferease* and *uridine diphosphoglucose 4-epimerase*. The genes for these enzymes, as defined by mutants, are GAL_1, GAL_7, and GAL_{10}, respectively, and these are apparently closely linked on chromosome II in the order GAL_7-GAL_{10}-GAL_1. The three genes are coordinately induced more than 5000-fold by the addition of galactose to the medium.

Regulatory mutants that affected the expression of the GAL genes were studied. Once class of such mutants maps at a locus distinct from the structural genes, and these mutants are characterized by constitutive synthesis of the three GAL enzymes. These mutants are recessive to the wild-type allele. By analogy with the lactose operon of *E. coli*, the locus involved was called *i* and the mutants were designated i^-. A second class of regulatory mutants carry mutations that map to a locus, GAL_4, which is linked to either the *i* locus or the GAL structural genes. These GAL_4 mutants are pleiotropically negative in that they are uninducible by galactose. A third class of regulatory mutants maps immediately adjacent to the GAL_4 locus at the gal_{81} region, and these result in a constitutive production of galactose-fermenting enzymes. The GAL_{81} mutants resemble o^c mutants of the lactose operon in that they behave as cis-dominants in diploids.

From the data *H. Douglas* and *D. Hawthorne* proposed a model for the regulation of expression of the GAL genes. Here the *i* gene produces a repressor that represses the expression of the GAL_4 gene by interacting with the adjacent gal_{81} region if galactose is absent. If galactose is added, the repressor is inactivated and the GAL_4 gene can then be transcribed. Since gal_4 mutants are pleiotropically negative, the GAL_4 product is presumably a positive effector molecule required for the expression of three GAL structural genes. How and where the GAL_4 effector interacts with the GAL gene cluster is not known. In other words, the i^--gal_{81} relationship resembles the repressor-operator

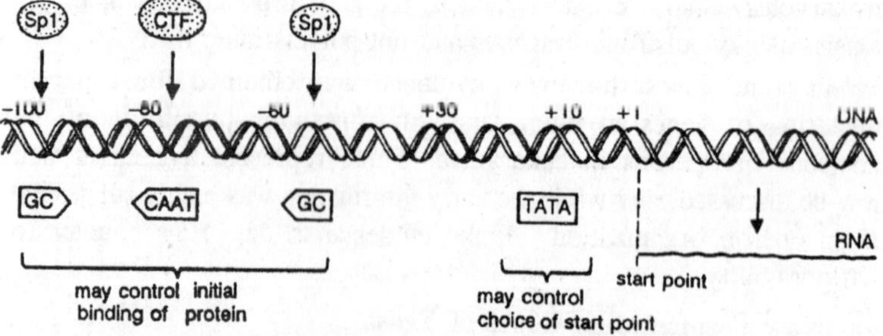

Fig. 12.5. Different consensus sequences (boxes) in the promoter region for thymidine kinase gene.

relationship of bacterial operons. Whether this is formally the case is a question for debate. The existence of i^s (super-repressible) mutations certainly supports the repressor concept in the model. Still unknown, however, is the exact nature of the *i* and GAL_4 gene products at the molecular level, how the *i* product interacts with the gal_{81} region, and whether a polycistronic mRNA is produced from the GAL structural genes.

Genes for Aromatic Amino Acid Biosynthesis is Neurospora

Another potential candidate for an operon in eukaryotes are the genes for the early steps of the pathway for aromatic amino acid (phenylalanine, tyrosine, and tryptophan) biosynthesis in *Neurospora crassa*. These have been studied by *N. Giles*, *M. Case* and their colleagues for many years.

Of particular significance was the discovery that a multienzyme aggregate (molecular weight 230,000 daltons) contained five different enzyme activities. These five activities are coded for by the so-called *arom* gene cluster of five adjacent genes, *aro* 2, *aro* 4, *aro* 5, *aro* 9, and *aro* 1. Genetic studies showed that mutations in a particular gene either affected the individual enzyme activity or caused the loss of two or more of the activities of the complex. These pleiotropic mutants were reminiscent of nonsense mutants in operons of bacteria where polar effects result during the translation of a single polycistronic mRNA. It was suggested, therefore, that the *arom* gene cluster coded for a polycistronic mRNA, and transcription commencing at the *aro*-2 gene. This supported the possibility that the *arom* gene cluster was an operon. However, *F. Gaerner* and *Giles*' group, working independently, have shown that to so-called *arom* gene cluster is actually a *single* structural gene that codes for a single polypeptide of molecular weight 115,000 daltons. This dimerizes to produce an enzyme that has the five enzyme activities just discussed. The separate polypeptides that were found in early investigations have been shown to be artifacts of the cellular fractionation techniques where the pentafunctional

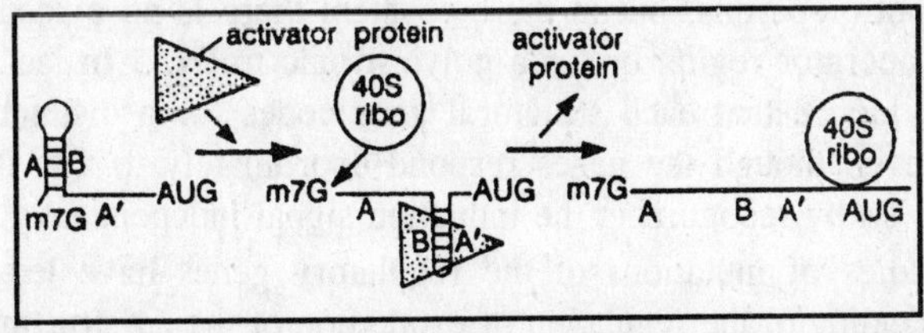

Fig. 12.6. Hypothetical mechanism for activation of translation in mammalian systems.

polypeptide is cleaved by endogenous protease activity. Thus the *arom* system is not an operon but a fusion of five ancestral genes into one. This cluster gene has one promoter region. This not the only example of a multifunctional polypeptide in eukaryotes; a number of other example are known, particularly in the lower eukaryotes. The point it illustrates is that when one breaks open the cell and examines the contents, the results one obtains do not necessarily reflect the situation in vivo.

Regulation of Quinic Acid Metabolism in Neurospora

N. Giles, M. Case, and their colleagues have also studied the regulation of expression of the *qa* genes of *Neurospora*, genes that are involved in the metabolism of quinic acid (QA) as a carbon source. From genetic and DNA-sequencing experiments, the *qa* genes have been shown to be clustered and to involve five structural genes and two regulatory genes.

Transcription of the structural genes is induced 300-to 1000-fold by the addition of QA to the medium. The structural genes *qa*-3, qa-2 and qa-4 code for the enzymes for the first three steps in the catabolism of QA (i.e., quinic acid dehydrogenase, catabolic dehydrogenase, and dehydroshikimic acid dehydrase, respectively). Note that both this catabolic pathway and the aromatic amino acid biosynthetic pathway just described have a dehydroquinase involved. There is no overlap between the pathways, however, since the multifunctional polypeptide of the biosynthetic pathway serves to channel the intermediates effectively.

The other two structural genes, *qa-x* and *qa-y*, were identified by analysis of the DNA sequence of the qa region. It is known that these two structural genes are transcribed and that they respond to QA induction, but the functions of their protein products are currently unknown. At the *qa-y* end of the *qa* structural gene cluster are two regulatory genes *qa*-1S and *qa*-1F. A regulatory gene adjacent to a cluster of structural genes is reminiscent of the organization of the prokaryotic operons, but in the *qa* system there is no evidence either for an operator region or for a polycistronic mRNA. In fact, there is solid evidence that each structural gene codes for a distinct mRNA. Thus, even though the genes respond coordinately to QA induction, they do so by recognizing the induction signal independently.

Studies of mutations of the regulatory genes have led to some understanding of the regulation of expression of the *qa* structural genes. The regulatory gene mutations were studied both in the normal haploid cells and also in heterokaryons involving nuclei with different genetic constitutions in order to study dominance relationships. Mutants of the *qa*-1S gene are either constitutive and recessive (qa-1S⁻alleles). Mutants of the qa-1F gene (qa-1F⁻ alleles) are noninducible and recessive even in the presence of a (constitutive) qa-1S^c allele. The following model has been proposed: qa-1F codes for an activator protein needed for transcription of itself and for all the structural genes except qa-x. qa-1S codes for a repressor protein that interacts with QA, in which state it has no effect on qa-1F transcription. In the absence of QA, however, the qa-1S protein blocks transcription of the qa-1F activator protein so that most of the structural genes are repressed. The exact role of qa-x in the qa system remains to be defined.

Thus the qa gene cluster is not like the arom gene cluster, since the qa genes clone for individual mRNAs and distinct protein products. The absence of a polycistronic mRNA or of an operator-like region in the qa gene system also distinguishes the qa gene cluster from prokaryotic operons.

In summary, in most instances genes with related functions in eukaryotes tend to be unlinked, although coordinately regulated. In lower eukaryotes in particular, where more extensive genetics has been possible, there are a number of examples of clustered genes but no definitive example of a bacterial-like operon. At least in some instances the gene cluster may actually be a single gene that codes for a multifunctional polypeptide.

Regulation of Gene Expression in Higher Eukaryotes

Higher eukaryotes are characterized by the differentiation of cells into tissues, organs, etc., that have specific functions. In this they differ markedly from the comparatively undifferentiated lower eukaryotes. The following section will concentrate on animal systems, in particular the vertebrates.

With the great complexity of cell specialization in higher eukaryotes come different problems in terms of the regulation of gene expression. For example, the specialized cells in these organisms are not subjected to drastic changes in the environment as in the case for lower eukaryotes and prokaryotes, this results in the fact that animals have homeostatic mechanisms that maintain relatively constant extracellular and intracellular environments. This is mediated by the blood, which has a fairly constant composition that is maintained by a variety of mechanisms. In vertebrates, for example, this is controlled by hormones. Thus animal cells are generally not exposed to large changes in the concentration of metabolites or substrates and therefore is less need for rapid changes in the rates of enzyme synthesis. Characteristically, then such changes are less frequent and of less magnitude than in lower eukaryotes or bacterial cells. For example, the enzyme *ornithine decarboxylase*, one of the most rapidly responding enzymes, exhibits a maximum increase of only 10 to 20 fold in four hours when induced. Contrast this with the 1000-fold induction of *β-galactosidase* within minutes in *E. coli*.

Before discussing the role of hormones in the regulation of enzyme synthesis, it must be stated that there are indeed enzyme induction and repression mechanisms operating in higher eukaryotes that are similar to those in prokaryotes. Because of the low number of regulatory mutants available in animal cells, there are relatively few systems that have been investigated in this regard. By contrast, the action of hormones on gene expression have been studied extensively, and in the following section, some of the information that has been obtained will be considered.

Regulation of Enzyme Synthesis by Hormones

A *hormone* may be defined as an effector molecule produced in low concentrations by one cell that evokes a physiological response in another cell. In vertebrates, a large number of classes of molecules have been shown to have hormonal activity, including polypeptides, amino acids, fatty acid derivatives, and steroids. Some of the hormones act directly on the cell's genome whereas others act at the cell surface, thereby activating membrane-bound adenyl cyclase to produce cyclic AMP (cAMP ; 3' 5'-cyclic adenosine monophosphate) from ATP. The cAMP acts as a "second messenger" to evoke the intracellular effects observed following hormonal release.

Hormones act on specific target tissues that posses receptors capable of recognizing and binding to that particular hormone. For most of the polypeptide hormones, the receptors are on the cell surface, whereas the receptors for steroid hormones are in the cytoplasm.

Model for Steroid Hormone Action

Steroid hormones are biosynthetically derived from sterols, which occur only in eukaryotic cells. In general, each class of steroid hormones mediates its biological response by binding to an intracellular receptor protein that is confined to target tissues. The interaction of the hormone with its receptor protein brings about a change in the structure of the protein such that there is increased affinity of the steroid-receptor complex for DNA. While steroid-receptors complexes bind to all

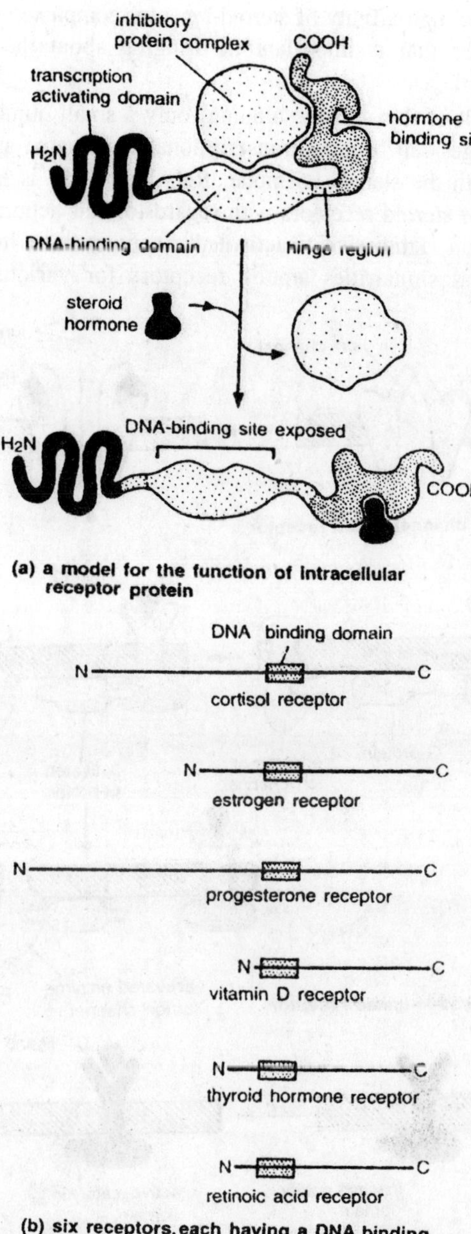

Fig. 12.7. The intracellular receptor superfamily.

DNAs, it is the high affinity of steroid-receptor complexes with *specific DNA* sequences that is important in bringing about the changes in gene expression.

As a result of this specific binding, only a small number of genes within the target cell become transcriptionally activated as a result of interaction with the steroid hormone. Relatively little is known about the structure of steroid receptors with regard to their action in affecting gene expression. Studies of unactivated receptors have indicated that there are great similarities among receptors for various classes of

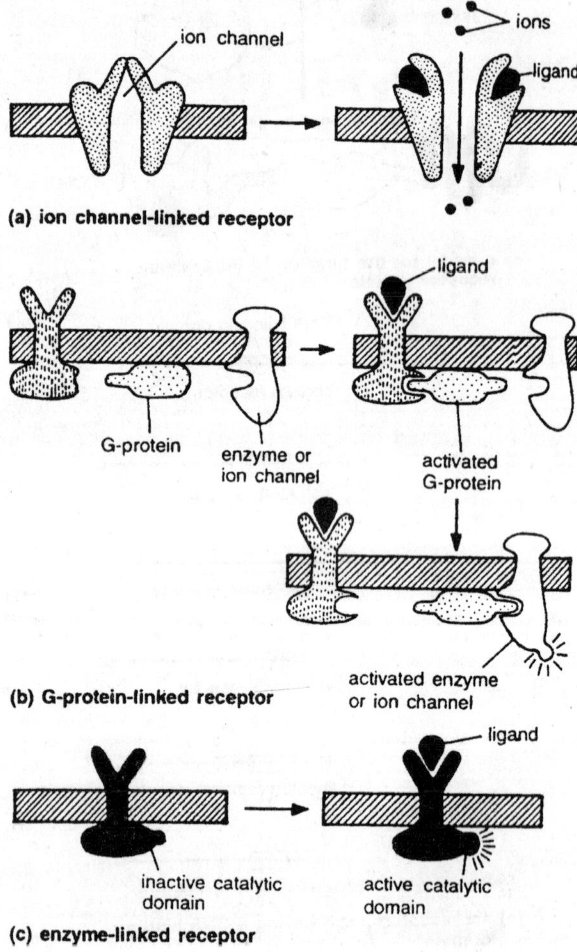

Fig. 12.8. Three classes of cell surface receptors.

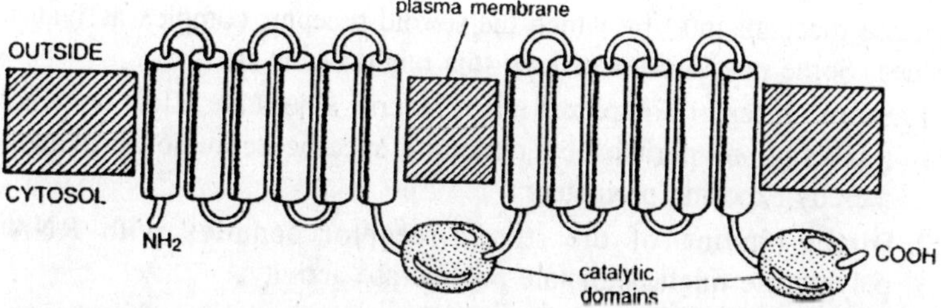

Fig. 12.9. Structure of adenyl cyclase enzyme, showing two clusters of six transmembrane segments, separating two similar cytoplasmic catalytic domains.

steroids. In all cases studied to date, the unactivated receptor exists as a multimer with a molecular weight of 200,000–300,000 daltons, and with sedimentation values of 8-10S. The activated form of steroid receptors always have lower S values than the unactivated forms. In the typical case, activation of an 8-10S receptor leads to the production of a 3-4S activated form, suggesting that the most stable activated form of steroid receptors in a monomeric structure. It should be pointed out, however, that there is evidence that a multimeric from is involved in gene activation. There is no question that a great deal of progress has been made in recent years in establishing the structure of steroid receptors and the nature of their interaction with the steroid molecule itself. Further, in some cases information is coming to hand about the nature of the DNA sequences involved in high-affinity binding of the steroid-receptor complexes in the nucleus. However, much more information is needed before we will have an understanding of the

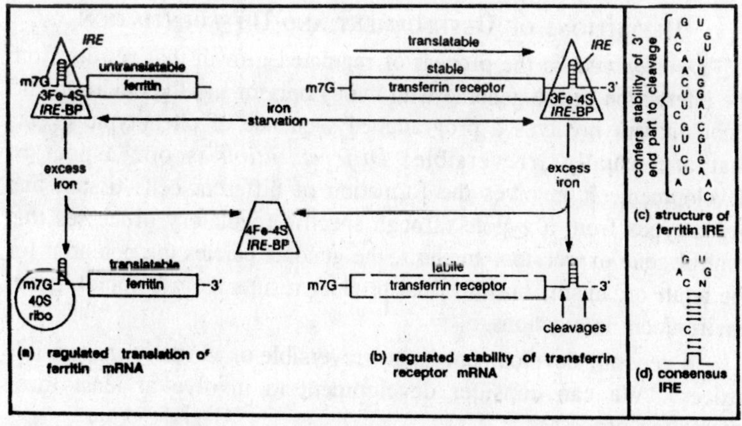

Fig. 12.10. Post-transcriptional regulation of two genes involved in the uptake and detoxification of iron in mammalian cells.

precise mechanism(s) by which the steroid-receptor complex activates genes. Some possibilities of how this might occur are :

1. Steroid-receptor complexes might reverse a negative effect on gene transcription, perhaps by removing specific repressor proteins, thereby exposing promoters.
2. Direct binding of the steroid-receptor complex with RNA polymerase might stimulate polymerase activity.
3. Steroid-receptor complexes might after the conformation of the chromatin (e.g., by removing nuclear proteins) in such a ways as to facilitate RNA polymerase binding to a promoter.

In summary, hormones act to integrate metabolism in higher eukaryotes. In some cases (e.g., in the liver) the coordination of metabolic activities involves the combined actions of several hormones. It is generally accepted that hormones act at the transcriptional level.

Long-term Genetic Regulation in Higher Eukaryotes

The examples of the previous section all show short-term regulation of gene expression in higher eukaryotes, that is, where adjustments are made in cellular activity in response to environmental changes (e.g., hormone release). There are, however, two properties of higher eukaryotes (and some lower eukaryotes) that reflect the long-term regulation of gene expression : development and differentiation. These processes are really outside the area of genetics and in the areas of developmental biology and embryology, and therefore only a very general discussion of them will be given in this text.

Definitions of Development and Differentiation

Development is the process of regulated growth that results from the interaction of the genome with the cytoplasm and the environment. Development involves a programmed sequence of phenotypic events that are usually irreversible. *Differentiation* is one aspect of development. It involves the formation of different cell, tissue, and organ types from a zygote through specific regulatory processes that control gene expression. In short, the genome carries the potential for the adult organism, but the final product results from complex gene-environment interactions.

In general, development is an irreversible or virtually irreversible process. We can consider development to involve at least three interacting processes :

1. The replication of the genetic material.

2. The growth of the organism as a result of cellular metabolic activity.
3. Cellular differentiation by which genetically identical cells diverge in their structure and function to give rise to organized tissues, which in turn associate to form organs.

Differentiation is the formation of different types of cells and tissues form a zygote by the specific regulation of gene activity in temporal and spatial ways.

General Aspects of Development and Differentiation

A number of general attributes of development and differentiation can be related to the genetic concepts that have been presented in the text.

Nuclear DNA remains constant

Early models for gene involvement in development included one where there was a programmed loss of nuclear DNA as the organism developed, or rather that the development processes that occurred were the result of losses of particular genes in an ordered sequence. That is not true. Rather, cells of differentiated tissues contain the same genomic content of DNA as the fertilized egg (although some differentiated cells may be polyploid). On elegant experiment that showed this to be the case was performed by *J. Gurdon*. He transplanted a nucleus from the gut cell of a tadpole of *Xenopus laevis*, the South African clawed toad, into an unfertilized egg of that organism from which the nucleus had been removed. The result was that the egg, once stimulated, developed into a normal adult toad. Thus the differentiated cells of the tadpole exhibited *totipotency*; that is, they contained all of the genetic information required for the egg to develop and differentiate into an adult organism.

DNA is transcribed in a programmed way

All of the available evidence shows that development and differentiation involve a detailed program of transcription of the DNA, which occurs in response to specific activator and repressor molecules. Two lines of evidence to support this will be considered here.

1. As mentioned previously, it is possible to quantify the extent to which RNA isolated from a cell hybridizes to the nuclear DNA. In general, the experiments involve RNA and DNA molecules that are radioactively labelled with different isotopes. A refinement of this technique is competitive DNA : RNA hybridization where

unlabelled RNA is first allowed to hybridize with the DNA before the radioactive RNA is added. If the RNAs are from the same tissue, the unlabelled RNA should effectively block all of the DNA sites to which the labelled RNA can bind, and this would be detected as 100% competition when the radioactivity is measured. If the RNAs are from two different tissues, however, the effect on the amount of radioactive RNA that will bind to the DNA will depend on how many of the RNA species were synthesized in common by both the tissues. One can do this experiment using mRNAs isolated from different tissues (e.g., lung, liver, kidney, muscle) of the same organism. Results of such an experiment show that there is limited competition between the mRNAs of the tissues in the hybridization and leads to the conclusion that differentiated cells reflect differences in the gene transcription activity. This correlates well with other studies showing different spectrums of enzymes and relative differences in enzyme amounts and activities in different tissues of the same organism. These differences must, of course, reflect differential gene activity.

2. In certain insects as *Drosophila*, the chromosomes of the larval salivary gland cells undergo *polytenization*. That is, the chromosomes replicate up to a thousand times but without there being cell division. The replicated chromosomes remain together as the *polytene chromosomes*, which show characteristic banding patterns. The bands are thought to represent the coding sequences of genes, while the function of the interband region is not known. The DNA is continuous along the length of these chromosomes. In *Drosophila*, there are three larval stages, each separated by a molting event. The last larval stage is followed by pupation. This is an interesting model system, therefore, for studying gene activity (since the genes are visible, if indeed the bands are genes) during development. In fact, specific bands "puff" in a particular pattern related to the time of larval development. The puffs are localized loosenings of the compact polytene chromosome that occurs so that RNA polymerase can initiate transcription. Indeed it can be shown that RNA is being actively synthesized in the puffs, and thus the puffs are visual evidence of gene activity. More importantly, the puffing patterns are reproducible, and, as has been stated, they are tissue and developmental stage specific.

Gene-cytoplasm Interactions

The gene activities of a cell are affected by the cytoplasm. Thus when certain genes are turned on during differentiation, particular proteins are synthesized, some of which have regulatory role in maintaining the differentiated state of the cell. For example, in Gurdon's transplantation experiment discussed earlier, we made the point that the nucleus carried all the genetic information necessary for the egg to develop into an adult. However, the fact that the egg cell behaved as an egg cell and not as a tadpole gut cell is an example of how the activity of the nuclear genome is controlled by the cytoplasmic state.

13

PROTEIN SYNTHESIS

Proteins are the molecules responsible for catalyzing most intracellular chemical reactions (enzymes), for regulating gene expression (regulatory proteins), and for determining many features of the structues of cells, tissues, and viruses (structural proteins). A protein is composed of one or more chains of amino acids that are covalently joined. The chains of amion acids are called *polypeptides*. The 20 different amino acids commonly found in natural polypeptides can be in any number and any order. Because the number of amino acids in a polypeptide molecule usually ranges from 100 to 1000, the number of different protein molecules that is possible is enormous.

Each amino acid contains a carbon atom (the α carbon) to which is attached one carboxyl group (—COOH), one amino group (—NH_2), and a side chain commonly called an R group. The R group are generally chains or rings of carbon atoms bearing various chemical groups. The simplest side chains are those of glycine (—H) and of alanine (CH_3). Polypeptide chains are formed when the carboxyl group of one amino acid joins with the amion grioup of a second amino acid; the resulting chemical bond is an ordinary covalent bond called a *peptide bond*. Thus, the basic unit of a protein is a polypeptide chain in which α-carbon atoms alternate with peptide units to for a backbone having an ordered array of side chains.

The two ends of every polypeptide molecule are distinct. One end has a free —NH_2 group and is called the *amino terminus*; the other end has a free -COOH group and is the *carboxyl terminus*. Polypeptides are synthesized by adding individual amion acids to the carboxyl end of the growing chain. Conventionally the amion acids of a polypeptide chain are numbered starting with the amino acid at the amion end.

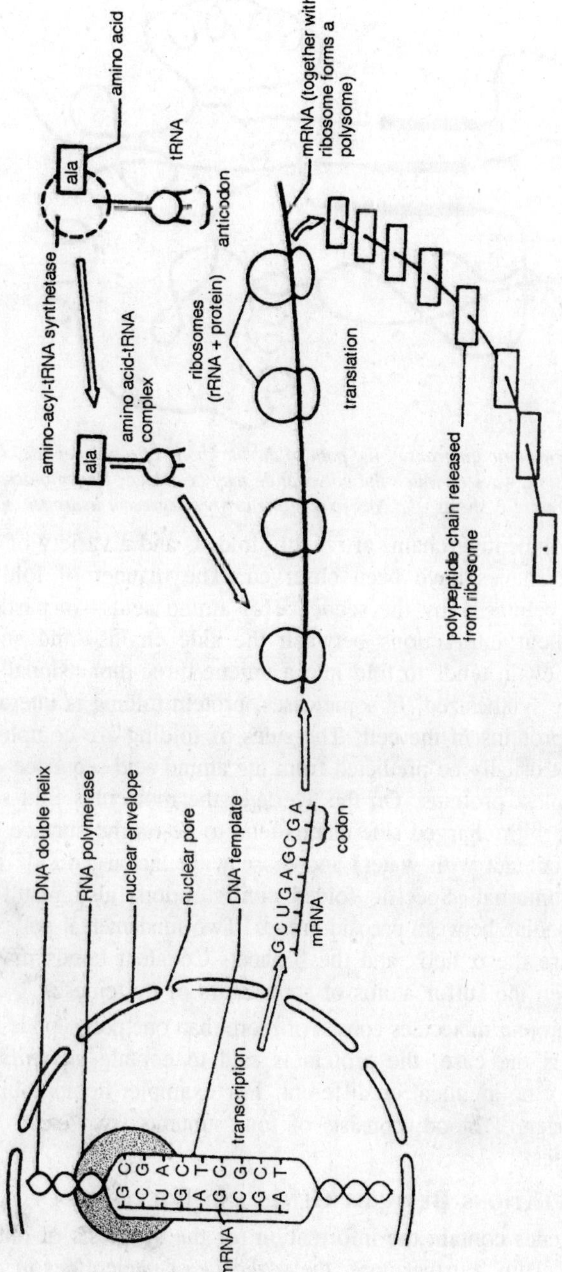

Fig. 13.1. Simplified summary diagram of the major structures and processes involved in protein synthesis in the cell.

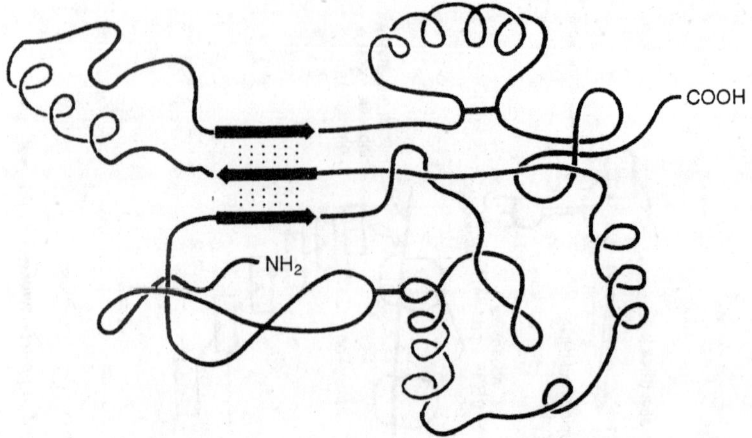

Fig. 13.2. A schematic diagram of the path of the backbone of a polypeptide, showing possible ways in which the polypeptide may be folded. Heavy black arrows represent β sheets; the dots joining the arrows represent hydrogen bonds.

Most polypeptide chains are highly folded, and a variety of three-dimensional shapes have been observed. The manner of folding is determined primarily by the sequence of amino acids—in particular, by noncovalent interactions between the side chains—and so each polypeptide chain tends to fold into a unique three-dimensional shape as it is being synthesized. In some cases, protein folding is interactions with other proteins in the cell. The rules of folding are complex and shape cannot usually be predicted from the amino acid sequence except for the simplest proteins. On the average. the molecules fold so that amino acids with charged side chains tend to be on the surface of the protein (in contact with water) and those with uncharged side chains tend to be internal. Specific folded configurations also result from hydrogen-bonding between prepaid groups. Two fundamental polypeptide structures are the α helix and the β sheet. Covalent bonds may also form between the sulfur atoms of some pairs of cysteines.

Many protein molecules consist of more than one polypeptide chain. When this is the case, the protein is said to contain *subunits*. The subunits may be identical or different. For example, hemoglobin, the oxygen carrier of blood, consists of four subunits, two each of two different types.

Relations Between Genes and Polypeptides

Most genes contain the information for the synthesis of only one polypeptide chain, Furthermore, the *sequence* of nucleotides in a gene determines the *sequence* of amino acids in polypeptide. This point was first proved by studies of the tryptophan synthetase gene *trpA* in *E. coli*, a gene in which many mutations had been obtained and accurately mapped. The effects of numerous mutations on the amino acid sequence of the enzyme were determined by directly analyzing the amino acid sequences of the wildtype and mutant enzymes. Each mutation was found to result in a single amino acid substituting for the wildtype amino acid in the enzyme; more importantly, *the order of the mutations in the genetic map was the same as the order of the affected amino acids in the polypeptide chain.* This attribute of genes and polypeptdes is called *colinearity*, which means that the sequnce of base pairs in DNA determines the sequence of amino in the polypeptide in a colinear or point-to-point manner. Colinearity is universally found in prokaryotes. However, we will see later that in eukaryotes noninformational DNA sequences interrupt the continuity of most genes, the order but not the spacing between the mutations correlates with amino acid substitution.

Transcription

The first step in gene expression is the synthesis of an RNA molecule copied from the segment of DNA that constitutes the gene. The basic features of the production of RNA are described in this section.

General Features of RNA Synthesis

The essential chemical characteristics of the enzymatic synthesis of RNA resemble those of DNA synthesis.

1. The precursors is the synthesis of RNA are the four ribonucleoside 5'-triphosphates—namely, adenosine triphosphate (ATP), guanosine triphosphate (GTP), cytidine triphosphate (CTP. and uridine triphosphate (UTP). They differ from the DNA precursors only in that the sugar is ribose rather than deoxyribose and the base uracil (U) replaces thymine(T).
2. In the formation of RNA, a sugar-phosphate bond is formed between the 3'-hydroxyl group of one nucleo-tide and the 5'-triphosphate of a second nucleotide. This is the same chemical reaction as that which occurs in the synthesis of DNA, but the enzyme is different.

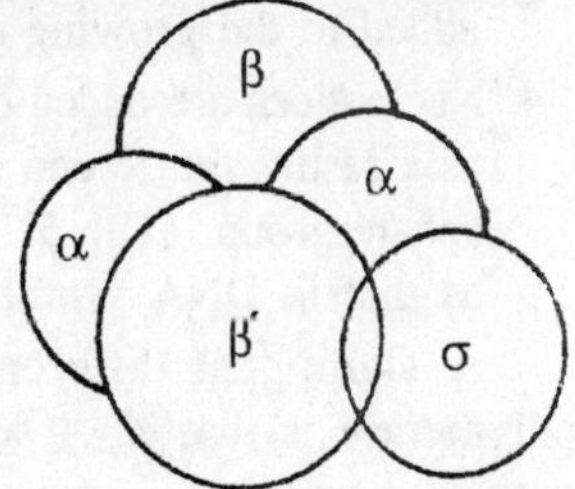

Fig. 13.3. A model of the structure of prokaryotic RNA polymerase showing association of five polypeptides (α₂ββ σ)

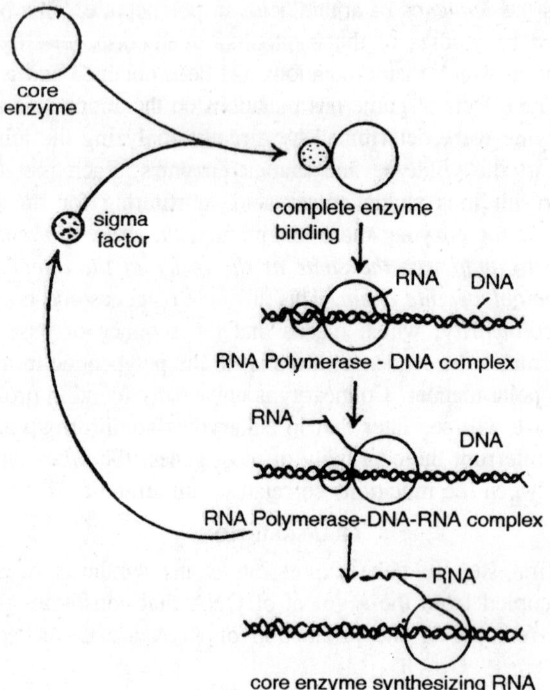

Fig. 13.4. Role of sigma factor and core enzyme of RNA polymerase during transcription.

3. The sequence of bases in an RNA molecule is determined by the base sequence of the DNA template. Each base added to the growing end of the RNA chain is chosen for its ability to base-pair with the DNA template strand: thus, the bases C, T, G, and A in a DNA strand cause G, A, C, and U, respectively, to be added to the growing end of an RNA molecule.
4. Nucleotides are added only to the 3'-OH end of the growing chain: as a result, the 5' end of a growing RNA molecule bears a triphosphate group. (The 5' → 3' direction of chain growth is the same as that in DNA synthesis.)

A significant difference between DNA polymerase and RNA polymerase is that *RNA polymerase is able to initiate chain growth without a primer.*

An important feature of RNA synthesis is the following:

Introns and Exons

In 1977 biologists were surprised to discover that the DNA of a eukaryotic gene is longer than its corresponding mRNA. It should be the same length because the mRNA is a direct copy. It was discovered that immediately after the mRNA is made, certain sections of the molecule are cut out, before it is used in translation. The sections of the gene that code for these unused pieces of RNA are called *introns*. The remaining sections of the gene are the code for the protein and are called *exons*. The size and arrangement of introns is very variable and characteristic for a particular gene. In prokaryotes there are no introns.

One possible function for introns has come with the discovery that the same mRNA may have different introns removed in different cells. The gene therefore has alternative introns and can code for different, though similar, proteins. This increases its potential use.

An example is the calcitonin gene. Two different forms of mRNA can be produced by this gene, depending on which introns are removed. One is produced in the thyroid gland and codes for the protein calcitonin, which had 32 amino acids. Calcitonin is a hormone which acts to lower calcium levels in the blood. The other is produced in the hypothalamus and codes for a protein with 37 amino acids which is similar to calcitonin and is called CGRP (calcitonin gene-related peptide). This is a powerful vasodilator agent. It is also released from nerve endings in some parts of the peripheral nervous system.

Protein Synthesis in Prokaryotes

There are three major steps in protein synthesis: initiaion, elongation, and termination. These will be discussed in turn.

Initiation

Initiator tRNA. The first amino acid in the syntheis of all bacterial polypeptides is N-formyl-methionine (fmet), which is a modified methionine amino acid in which the α-amino group is "blocked" and therefore cannot participate in peptide bond formation. The formyl group is added to the methionine after the amino acid has become attached to a specific tRNA, called tRNA. fmet. This reaciton is catalyzed by the enzyme *transformylase*. In many cases the fmet that starts a polypeptide chain is subsequently removed by enzymatic action.

Biochemical analysis has shown that at least two species of tRNA that can be charged with methionine are present in all prokaryotic organisms: one is that tRNA involved with initiaton, and the other species is responsible for the insertion of methionine elsewhere in the polypeptide chain. The latter rRNA is designed rRNA.met. Both of these tRNAs in bacteria are aminoacylated by the same enzyme, but only tRNA, fmet is a substrate for the transformylase-catalyzed reaction.

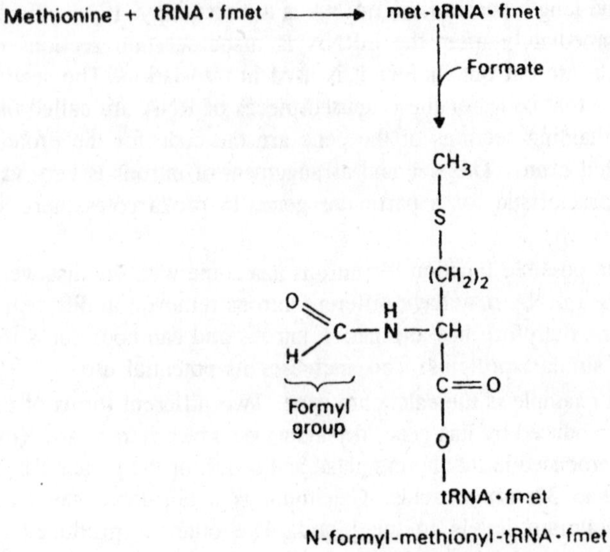

Methionine + tRNA · fmet ⟶ met-tRNA · fmet

↓ Formate

CH_3
|
S
|
$(CH_2)_2$
|
O=C(H)—N(H)—CH
|
C=O
|
O
|
tRNA · fmet

Formyl group

N-formyl-methionyl-tRNA · fmet

Fig. 13.5. Synthesis of N-formylmethionyl-tRNA.

Both tRNAs read AUG (the only methionine codon), but in addition tRNA.fmet can recognize GUG and UUG codons. RNA-sequencing studies have shown that both molecules havae an anticodon that is complementary to AUG. The two tRNA molecules do differ in some other properties. For example, the binding of fmet-TRNA.fmet to ribosomes is catalyzed by an initiator factor whereas the binding of met-tRNA.met is catalyzed by elongation factors. The two tRNAs apparantly bind to the ribosome at different sites. Thus it is clear that fmet.tRNA.fmet must have a strucutre that is specific for its role in initiation.

Ribosome binding sites

In bacteria, the first step in initiation is the formation of a complex between the 30S ribosomal subunit, fmet-TrNA, and an mRNA molecule. The 50S subunit is added later to form the active 70S ribosome (monosome). The mRNA may contain information for one to several distinct polypeptide chains. For each of the segments coding for a polypeptide, there is a specific nucleotide sequence for orienting the mRNA correctly and in the right reading frame on the ribosome. These sequences are called the *ribosome binding sites*.

Most of the ribosome binding sites have apurine-rich sequence about 8 to 12 bases upstream from the AUG start codon. This equence

Message origin	Ribosome binding site sequence
E. coli lac Z	UUC ACA CAG GAA ACA GCU AUG ACC AUG AUU
E. coli trp B	AUA UUA AGG AAA GGA ACA AUG ACA ACA UUA
E. coli RNA polymerase β	AGC GAG CUG AGG AAC CCU AUG GUU UAC UCC
Phage λ cro	AUG UAC UAA GGA GGU UGU AUG GAA CAA CGC

Fig. 13.6. Some prokaryotic ribosome binding sites. The initiation codon, AUG, is boxed. The larger boxed regions indicate the regions of contiguous complementarity (including allowable G-U base pairs) to the 3' end of 16S rRNA.

and other bases in this region are complementary to a pyrimidine-rich region, including at least CCUCC at the 3' end of 16S rRNA. The mRNA region that binds in this way is called the *Shine-Dalgarno sequence* after the discoverers of this relationship. Thus it appears that the formation of complementary base pairs between mRNA and 15 S rRNA in the 30S ribosomal subunit allows the ribosomes to locate and bind to the initator regions in the mRNA.

Initiation factors and initiation

In addition to mRNa, fmet-tRNA, and ribosomal subunits, three protein initiaion factors (IF-1, IF-2, and IF-3) and GTP are required for the initiation process to occur. First the properties of the initiation factors are discussed and then the scheme proposed for the initiation processes in protein synthesis is presented.

1. *IF-3.* The IF-3 factor weighs 23,000 daltons and functions in binding mRNA to the 30S subunit. It also acts as a dissociation factor for separating the 30S and 50S subunits after polypeptide synthesis is complete. Like all the IFs, IF-3 is found bound to free 30S subunits and can be released by washing the subunits in 0.5 M ammonium chloride.

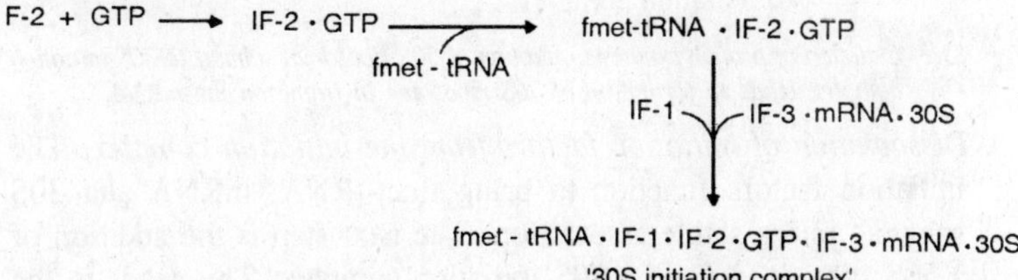

Fig. 13.7. Initiation of protein synthesis: steps in the formation of the 30S initiation complex.

Experiments with radioacative If-3 have shown that it is capable of binding to both 30S subunits and to mRNA molecules. In an in vitro protein-synthesizing system, IF-3 enhances the binding of fmet-tRNA to mRNA.30S subunit complexes. It is attractive to suppose that IF-3 recognizes mRNAs by the AUG or GUG initiation codons, but there is no solid evidence on this point.

In summary, the initiation reaction in which IF-3 is involved is:

IF-3 + mRNA + 30S subunit → (IF-3.mRNA.30S) complex

2. *IF-2*. The 80,000 dalton IF-2 protein in involved with the binding of the initiator tRNA to the IF-3.mRNA.30S complex. The high-energy molecule GTP is used in this reaction. In vitro experiments have shown that IF-2 and GTP will bind to form a complex that is stabilized when it is turn forms a complex with fmet-tRNA. This latter complex then binds with the IF-3.mRNA.30S complex and the IF-1 protein factor (9000 daltons) to forms the 30S initiation complex.

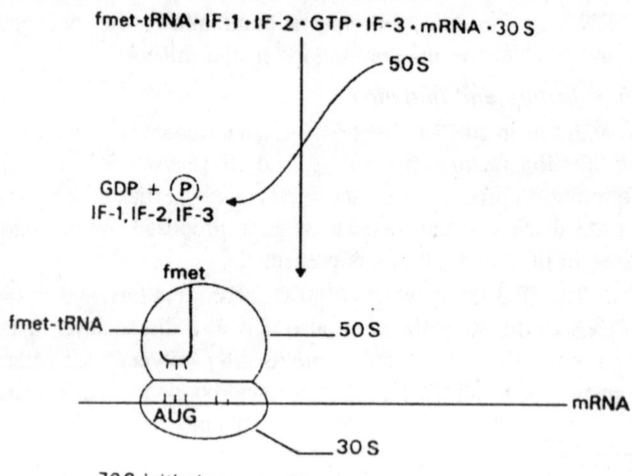

Fig. 13.8. Initiation of protein synthesis: addition of 50S ribosomal subunit to 30S initiation complex leads to formation of 70S ribosome in frame on the mRNA.

3. *Dissociation of initiation factors from the initiation complex*. The initiation factors function to bring fmet-tRNA, mRNA and 30S subunits into a stable association. The next step is the addition of a 50S subunit to form a 70S initiation complex. This leads to the hydrolysis of GTP to GDP + P and the release of three intiation factors. The factors can then be used for further initiaton reactions on the same or different mRNA.

Elongation

The 70S ribosome has two sites for binding aminoacyl-tRNA. In protein synthesis, charged tRNA binds first to a site called the A (aminoacyl) site. Then the amino acid it carries becomes joined to the growing polypeptide chain carried by the tRNA at the site called the P (peptidyl) site by the formation of a peptide bond.

It is not known whether the fmet-tRNA enters the A site and then moves to the P site or whether it enters the P site directly. Before

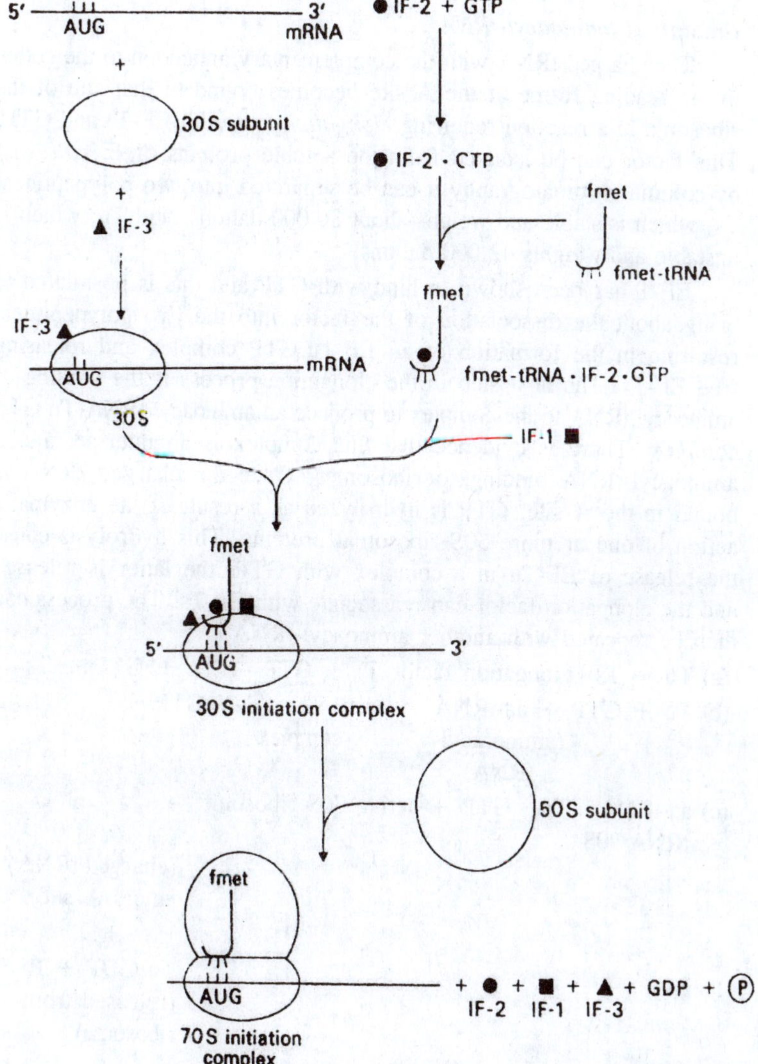

Fig. 13.9. Summary of the steps in the initiation of protein synthesis in prokaryotes.

further protein synthesis can occur, however, the fmet-tRNA must become located in the P site hydrogen-bonded to the start codon on the mRNA. Once this has occurred a cyclic sequence of events commences in which one amino acid at a time is added to the growing polypeptide chain. This is called *elongation*.

Binding of aminoacyl-tRNA

The charged tRNA with the complementary anticodon to the codon in the reading frame of the A site becomes bound to that site of the ribosome in a reaction requiring *elongation factor* T (EE-T) and GTP. This factor can be isolated from the soluble proteins of *E. coli*, and by column chromatography it can be separated into two polypeptides: Ts, which is stable and weighs about 30,000 daltons, and Tu, which is unstable and weights 42,000 daltons.

EF-T has been shown to bind with GTP and this is postulated to bring about the dissociation of the factor into the two polypeptides, resulting in the formation of an EF-Tu.GTP complex and releasing free EF-Ts. The next step in the elongation process is the binding of aminoacyl-tRNA to the complex to produce an aminoacyl-tRNA.Tu.GTP complex. There is evidence that this complex is an intermediate in aminoacyl-tRNA binding to ribosomes. Once the charged tRNA is bound in the A site, GTP is hydrolyzed as a result of the enzymatic action of one or more 50S ribosomal proteins. This hydrolysis cause the release of EF-Tu in a complex with GTP. the latter is released and the elongation factor can reassociate with EF-Ts. The process can then be repeated with another aminoacyl-tRNA.

(a) Tu + Ts (Elongation factor T) + GTP → Tu + GTP + Ts

(b) Tu + GTP + aa-tRNA → aa-tRNA · Tu · GTP
(amino acyl-tRNA) complex

(c) aa-tRNA · Tu · GTP + active 70S ribosome → a a tRNA·70S
(charged tRNA) enters A site)
+
Tu·GDP + P_i
(released from ribosome)

(d) Tu · GDP + Ts → Tu · Ts

Experiments with an analog of GTP that cannot be hydrolyzed have shown that GTP hydrolysis is required for release of EF-Tu from the ribosome but it is not needed for aminoacyl-tRNA binding to the ribosome. Other experiments have shown that binding of fmet-tRNA to the ribosome does not require EF-Tu.

Peptide bond formation

At the beginning of the this stage, a tRNA carrying the growing polypeptide chain is located in the P site, and an aminoacyl-tRNA is

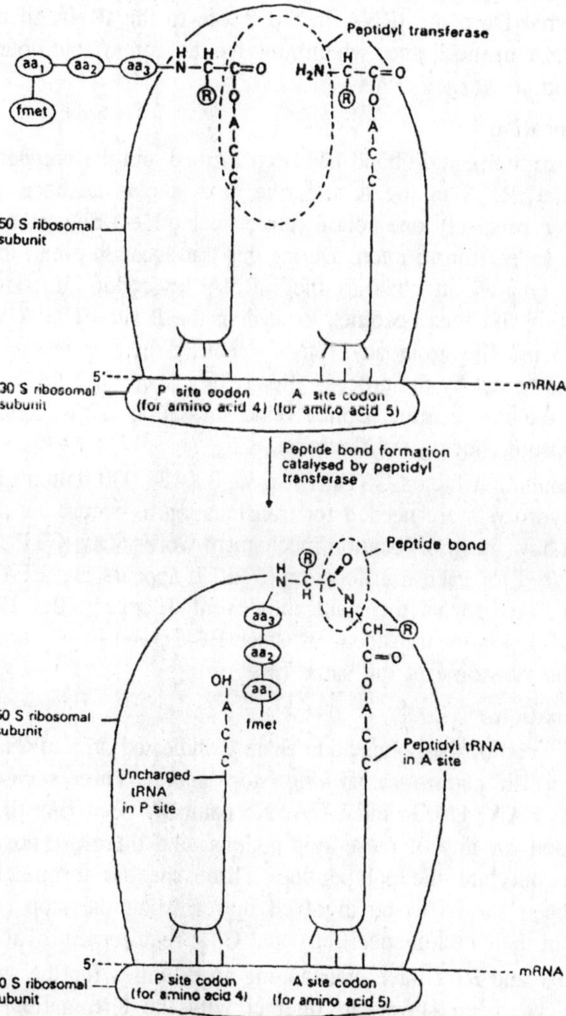

Fig. 13.10. Diagrammatic representation of peptide bond formation on ribosomes catalyzed by peptidyl transferase.

located in the A site. Thee tRNAs are maintained in positions conducive for peptide bond formation of the hydrogen bonds between the respective codons and anticodons and by the tertiary structure of the ribosome. The peptide bond is formed with the aid of the enzyme, *peptidyl transferase*, which is a ribosomal protein of the 50S subunit. The end result of the reaction is that the polypeptide chain is one amino acid longer, and the growing polypeptide chain has been transferred from the tRNA in the P site to the tRNA in the A site. The tRNA in the P site, which now has no amino acid bound to it, is called an *uncharged tRNA*.

Translocation

Once the peptide bond has been formed and the polypeptide chain is on the tRNA in the A site, the next step is advancement of the ribosome precisely one codon (three nucleotides) down the mRNA, a process called *translocation*. During this translocation event, the peptidyl-tRNA remains attached to the mRNA by codon-anticodon–pairing properties and thus becomes located in the P site. The A site is then vacant, and the aminoacyl-tRNA specified by the new codon there becomes bound by the process already described. The uncharged tRNA left in the P site after peptide bond formation is also released from the ribosome during translocation.

Elongation factor G (EF-F), a 72,000–84,000 dalton protein, and GTP hydrolysis are needed for translocation to occur, but it is not yet known how the translocation mechanism works. One GTP molecule is hydrolyzed for each translocation event. It appears that EF-G molecule is hydrolyzed for each translocation event. It appears that EF-G leaves the ribosome after translocation, since EF-Tu and EF-G cannot interact with the ribosome at the same time.

Termination

The end of the polypeptide chain is indicated on a mRNA molecule by a specific *chain-terminationg* (stop) *codon*. Three such codons are known: UAA, UAG, and UGA. No naturally occurring tRNA has an anticodon for any of these stop codons, and therefore no amino acid can be put into the polypeptide. Three specific termination factors have been shown to be involved in regarding the stop codon. they differ in their codon specificity and GTP requirement (Table 13.1).

RF1 and RF2 have overlapping specificities for the stop codons. They have been shown to interact with the termination codons by interaction at the A site. The RF3 factor apparently plays a stimulatory role in RF1 and RF2 activity. There is some evidence for a GTP requirement in the RF3 factor's activty. In any event, chain termination, as mediated by these factors, involves the cleavage of the carboxyl group of the C-terminal end of the polypeptide chain from the tRNa in the P site. This results in the release of the polypeptide and the now uncharged tRNA. The ribosome will then move along the mRNA until a new initiation sequence is encountered (as it may be in polycistronic mRNAs), or it will dissociate from the mRNA. If none is found, when the ribosome is released from the mRNA, IF-3 functions to keep the two subunits apart. Thus, when a new 70S initiation complex is formed, the two subunits are drawn randomly from the free pools of 30S and 50S subunits.

Table 13.1. Properties of prokaryotic termination factors.

Termination factor	*Molecular weight (daltons)*	*Stop codons recognized*	*GTP requirement*
RF1	44,000	UAA and UAG	No
RF2	47,000	UAA and UGA	No
RF3	46,000	None	Yes

While the polypeptide chain is being synthesized, the primary sequence of amino acids directs the three-dimensional shape. In other words, the elongating chain begins to assume its final shape as it is being made. Indeed, some enzyme activity can be detected on ribosomes that have not yet completed the synthesis of an enzymatic polypeptide.

Polysomes

Efficient translation of an mRNA molecule cannot be achieved by a single ribosome moving along it. The amount of space a ribosome

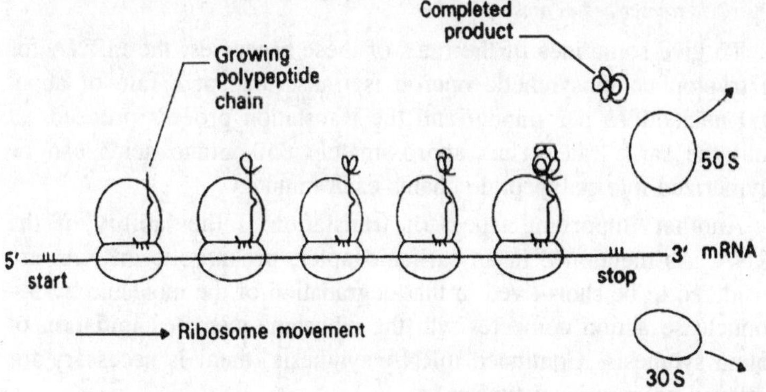

Fig. 13.11. Diagrammatic representation of a polysome engaged in protein synthesis.

takes up on a mRNA is relatively small, and thus several ribosomes can work on the mRNA at once. The association of a number of ribosomes on a single mRNA chain is called a *polyribosome* or *polysome*, and this allows several polypeptide chains to be made from each mRNA. The length of the polypeptide chain on a given ribosome will be directly proportional to how far the ribosome has moved along the mRNA from the 5' end of the molecule. The existence of polysomes explains why a cell needs so little mRNA, while at the same time it contains so much more protein.

Relationship of Transcription and Translation

In bacteria the mRNA typically becomes associated wth ribosomes while synthesis of the mRNA molecule is continuing. This is possible owing to the lack of a nuclear membranes so that as the 5' end of the growing mRNA molecule is displaced form the DNA as the double helix reforms, the ribosome binding site becomes available. Ribosomes then load on to the mRNA in rapid sequence, the first being close behind the RNA polymerase.

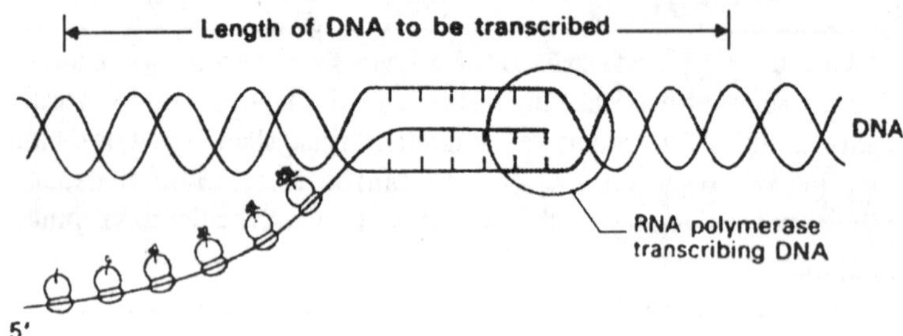

Fig. 13.12. Schematic of the possible translation of an mRNA while it is still being transcribed in prokaryotes.

To give some idea of the rates of these processes, the mRNA for the tryptophan biosynthetic operon is transcribed at a rate of about 1000 nucleotides per minute and the translation process proceeds at about the same rate. Thus approximately 350 amino acids can be polymerized into polypeptide chains each minute.

Another important aspect of translation is the lability of the mRNA. As mentioned in an earlier chapter, prokaryotic mRNAs are considered to be short-lived in that degradation of the molecule by 5'–exonuclease action competes wth the ribosome-mediated initiation of protein synthesis. Continued mRNA synthesis, then, is necessary for continued polypeptide synthesis.

Protein Synthesis in Eukaryotes

The steps and mechanisms of protein synthesis are similar in both eukaryotes and prokaryotes. As previously discussed, the ribosomes are different, and this is also the case with the soluble protein factors.

Initiation

Initiation of protein synthesis involves the binding of mRNA to the ribosomes. In eukaryotes it has now been established that the 5'-cap structure is necessary to get efficient binding but the 3' poly(A) sequence apparently is not needed. There is good evidence, however, that the poly(A) sequence stablizes the mRNA during the translation process. The precise mechanism whereby the eukaryotic message binds to the ribosome is not known, although it is most likely that RNA–RNA and RNA–protein interactions are involved. To date the nucleotide sequence has been determined to the 5'-side of the AUG start codon for a number of eukaryotic mRNAs and there are very few common features. Strikingly, while the last 50 nucleotides of *E. coli* 16S rRNA and eukaryotic 18S rRNA are highly homologous and similiar secondary structure models can be built, the eukaryotic rRNA does *not* have the CCUCC sequence characteristic of prokaryotic rRNA. Also, there is no Shine-Dalgarno sequence in the eukaryotic mRNa. Thus it appears that, unlike the case in prokaryotes, there is not a fixed nucleotide sequence that plays a role in ribosome binding to the message.

As in prokaryotes the initiation codon is AUG, and a special initiator methionyl-tRNa recognizes that signal in the message. Unlike its prokaryotic counterpart the methonine carried by the tRNA does not become formylated, since the appropriate enzyme system does not exist in eukaryotic cells. The initiator tRNAcan be distinguished from the met-tRNA that reads AUG codons elsewhere in the message, however, by the fact that it can be formylated in vitro in the presence of an *E. coli* extract. Hence in eukaryotes it is also appropriate to define the two methionine-accepting tRNAs as tRNA.fmet and tRNA.met.

At least in mammals, there are many more initiation factors (labeled eIFs for eukaryotic initiation factors) than in prokaryotes. In many cases the proteins have not been purified to homogeneity, and thus their absolute roles in protein synthesis are uncertain. Between them all, they carry out the initiation events performed by the three prokaryotic IFs. It remains to be seen to what extent the situation in mammals is generalizable throughout the eukaryotes.

Elongation

As in prokaryotes, there are two elongation factors: eEF-1 (equivalent to prokaryotic EF-T) and eEF-2 (equivalent to prokaryotic EF-G). eEF-1 has been studied in a number of systems, and in general it exists in multiple forms. Purified eEF-1 from rabbit reticulocytes, for example, has a molecular weight of 186,000 and consists of three subunits weighing 62,000 daltons each. In some systems the subunits aggregate to produce molecules of greater than 1 million daltons. Regardng the function of eEF-1, much less is known than in the prokaryotes. The eEF-1 from rabbit reticulocytes, for example, has been shown to bind to aminoacyl-tRNA and to GTP, and thus to facilitate bindng of the aminoactyl-tRNA to the A site in ribosomes. During this step, GTP is hydrolyzed to DGP as a result of GTPase activity of the elongation factor, and an eEF-1.DGP complex is released from the ribosome.

The eukaryotic eEF-2 is similar to prokaryotic EE-G, although the two are not interchangeable in *in vitro* systems. The factor from rabbit reticulocytes has a molecular weight of 96,500–110,000 daltons and, after binding with GTP, binds to the ribosome. This event results in hydrolysis of GTP to GDP, translocation of the ribosome one codon down the message, and release of an eFF-2.GDP complex. All these steps are similar to the events that take place in prokaryotes, one exception being that the eukaryotic factor forms a stable complex with GTP whereas the prokarotic factor does not.

Termination

The same chain termination codons are functional in eukaryotes as in prokaryotes. One release factor has been identified in rabbit reticulocytes, and this has a molecular weight of 115,000 daltons and may be a dimer. This factor recognizes all three chain termination codons and it requires GTP to carry out its function. No stimulatory factor analogous to the porkaryotic RF-3 has been found in eukaryotes.

Protein Synthesis and Cellular Compartmentation in Eukaryotes

In prokaryotes there is no nuclear membrane to separate the transcription process from the translation process. The process of a nuclear membrane and the various modification process peculiar to mRNAs in eukaryotes present many levels at which the regulation of gene expression can be affected. These include transcription itself, the processing of the primary transcript to produce mature mRNA,

RNA-RNA splicing, and the movement of mRNA from the nucleus to the cytoplasm.

In eukaryotic cells, the cytoplasm contains a network of interconnecting channels bounded by membranes called the *endoplasmic reticulum* (ER). The membranes involved are continuous and may in fact connect to the nuclear membrane and the cell membrane. Close examination of the ER reveals that it is differentiated into two types, *smooth* (SER) and *rough* (RER), which are distinguished by the fact that the latter has ribosomes bound to it (hence the rough appearance) whereas the former does not. Thus, ribosomes in the cytoplasm are either membrane bound or free. The membrane-bound ribosomes synthesize proteins that are either secreted from the cell or that are packaged in lysosomes (where the proteins degrade other proteins). Thus, for example, pancreatic cells that secrete enzymes into the intestine are extremely rich in RER. The free ribosomes synthesize all other proteins found in the cell, that is, those in the cytoplasm, nucleus, mitochondria and chloroplasts (if present).

The proteins is to be secreted are made on the ribosomes of the RER and then transferred across the membrane into the channel system. The mRNAs for the secreted proteins must somewhow become associated specifically with the ribosomes of the RER. In 1975, G. Blobel and B. Dobberstein proposed a *signal hypothesis* to explain this. They suggested that there is a unique sequence of codons located to the 3' side of the initiator AUG codon which is present only in mRNAs for proteins that must be transferred across membranes. Translation of these codons results in a specific amino acid sequence at the N-terminal end of the protein. Then they postulated that the special end of the protein facilitates the attachment of the ribosome to the membrane so that the protein can be transferred across it. Once the protein has been completed, they proposed that the ribosome dissociates from the ER.

14

GENE CLONING

In the last several decades considerable advances have been made in the prevention of viral disease by vaccination. In many cases the material of choice for a vaccine is killed or otherwise attenuated form of the virus itself. This approach, although effective, has several disadvantages. First and foremost, the risk that pathogenic variants will escape the inactivation or attenuation processes is of considerable concern. Second, some viruses are impossible to grow under commonly used culture conditions, and in such cases no vaccine of import exists.

The techniques of genetic engineering are of particular utility for the production of viral antigens and, is hoped, viral vaccines. In brief, these techniques permit the production of proteins derived from one organism in a second, often unrelated, organism. Thus it is now possible to produce antigens from several pathogenic viruses in organisms such as *Escherichia coli*, *Bacillus subtilis*, and the yeast *Saccharomyces cerevisiae*. Such an approach eliminates all risk, and even potential risk, of reversion of the virus to its virulent form. Further, innocuous fragments of DNA from viruses that it is impossible to propagate under laboratory conditions can be transferred from primary biological material to a microbial vector and subsequently propagated in the host organism of choice.

In this review we describe the general techniques used for isolating, propagating, and expressing, in microorganisms, those segments of viral genomes coding for major antigens. In our selection of material we have restricted ourselves to a discussion of the better studies cases; however, it should be noted that similar studies have been performed upon other important viruses, in particular those of the herpes and paramyxovirus groups. We do not aim at giving precise experimental details of the now conventional technologies of genetic engineering; these can be found in the original literature, in one of the excellent technical manuals, or in a recent review. Instead, we use the viral antigen as a model for the discussion of the technical aspects of cloning viral nucleic acids and the problems related to expression of foreign proteins in microorganisms.

Expression of Foreign Coding Sequences in Escherichia coli

In this review we concentrate on the techniques that have been used to isolate and identify DNA sequences coding for viral antigens. Techniques for cloning RNA in the form of DNA copy (cDNA cloning) have been extensively reviewed elsewhere as have the techniques for direct cloning of DNA genomes. The isolation of an antigen-coding sequence is an essential step toward the production of the corresponding vaccine. Nevertheless, the efficient expression of this sequence as an antigenic polypeptide is a further hurdle that must be surmounted. For this reason we will briefly discuss strategies that may be employed for expressing a foreign gene in *E. coli*. Other host microorganisms are described in Sections VIII and IX.

Vector Copy Number

The rate of synthesis of a given protein depends, by and large, on the number of copies of the cognate gene that exists in each cell. Thus, rather than integrating cloned genes into the bacterial chromosome, most workers have elected to use high-copy-number episomal vectors based on plasmids such as pBR322 and bacteriophages such as M13 or λ. This strategy has disadvantages in that high copy numbers of certain DNA sequences may confer a selective disadvantage, and populations then suffer from high rates of rearrangement of such DNA sequences. In order to offset this problem, several vectors have been developed whose copy number may be maintained at a low level and amplified when required by, for instance, a temperature shift. Of particular interest are those vectors that have the capacity to integrate into the bacterial chromosome (where they are thought to be stable), from which they may be excised at a later stage and induced to replicate.

Signals for Gene Expression

Gene expression in *E. coli* requires signals that specify transcription into RNA and translation of this message. Much work has been carried out in recent years to identify such signals, and this is described briefly below.

Transcription initiation

Promoters for transcription in *E. coli* comprise at least two sequence elements, the "Pribnow box," a short AT-rich zone lying about 100 nucleotides prior to the first transcribed nucleotide, and a second sequence, the -35 region, lying somewhat upstream. Transcription initiation is, in many cases, subject to further controls, exemplified by the lac (lactose utilization) operon. Here, expression is controlled both positively and negatively. Positive control is effected by a protein activator, catabolic activator protein (CAP), which interacts with DNA sequences upstream of the promoter and enhances RNA polymerase binding. Negative control is effected by the lac repressor, which binds to the promoter region and sterically blocks subsequent binding of the RNA polymerase. Negative control of bacterial promoters has proved to be of enormous utility in the controlled expression of foreign genes.

Translation initiation

It is not known exactly that elements are required for the efficient initiation of translation at a given ATG. Certainly the Shine-Dalgarno (S/D) sequence, a short purine-rich sequence a number of nucleotides upstream of the initiation codon, is of essential importance. This sequence is though to hybridize to the 3' terminus of the bacterial 16 S ribosomal RNA and specify translation initiation at the appropriate ATG. Other factors are involved, such as the mRNA secondary structure and other as yet unidentified sequence motifs. In an extensive analysis of translation initiation sites, Gold et al., (1981) and Scherer et al. (1980) demonstrated that such regions show nonrandom base distributions both upstream of the initiating ATG (e.g., the S/D sequence) and downstream, within the coding sequence. This result has important consequences for the expression of foreign protein-coding sequences in *E. coli*.

Termination of transcription and translation

Transcription termination sites in *E. coli* are, by and large, identified by a G-C rich duster, accompanied in some cases by a stem-loop structure, immediately followed by a poly(T) sequence.

Translation termination is brought about by the simple presence, in phase, of one of the three termination codons, but additional parameters are likely to be involved.

Fused and Unfused Proteins

As mentioned earlier, sequences necessary for efficient translation initiation may lie on both sides of an ATG initiation codon. Thus positioning of a foreign coding sequence immediately beyond the ATG may compromise expression. For this reason, many groups have chosen to fuse foreign genes, in phase, at a site distal to the initiation codon of an endogenous bacterial gene. This strategy may also have unexpected benefits in terms of stability of the hybrid protein, but in certain instances the presence of an extended covalently bound polypeptide of bacterial origin may mask antigenic sites in a hybrid protein produced for vaccine purposes.

Inducible and Noninducible Systems

It has been clearly demonstrated that the expression of certain proteins in E. coil is severely detrimental to the survival of the host. Of particular note is the surface antigen of hepatitis B virus. For this reason a better choice is to employ expression systems in which transcription of the coding sequence in question may be blocked, when necessary, by a repressor protein. Although early work employed uncontrollable ("constitutive") expression systems deriving from, for instance, the β-lactamase gene of plasmid pBR322, we restrict our discussion here to systems subject to regulation in vivo.

The lac system

Transcription from the *lac* promoter of *E. coli* is normally blocked by the presence of the lac repressor, the product of the *lac*I gene. Addition of isopropylthiogalactoside (IPTG) to the culture medium results in repressor inactivation and the induction of transcription and translation of the *lac*Z gene or coding sequences fused in phase with this gene. One note of caution: the presence of the *lac* promoter on a multicopy plasmid is sufficient to out-titrate the available repressor protein, and expression may ensue even in the absence of IPTG. To surmount this problem, strains containing the *lacI*^{q} allele (which generates elevated levels of *lac* repressor) are often used.

As mentioned earlier, expression from the *lac* promoter additionally requires the presence of an activator factor CAP. The activity of this factor is in turn controlled by the metabolic state of the cell and, in particular, by the level of glucose. Thus a mutant *lac* promoter (*lacUV5*) that has lost its dependence on the activator protein.

The trp system

The synthesis of tryptophan in *E. coli* requires the presence of five enzymes encoded by the *trp* operon, which is normally repressed by the *trp* repressor in the presence of tryptophan. Several plasmid-borne systems have been described in which protein coding sequences may be fused in phase with the first (*trp*E) or second (*trp*D) genes of the operon or, indeed, with the leader polypeptide *trp*L. Here, tryptophan starvation or the addition of indoleacrylic acid (IAA) may be used to lift repression and allow production of the relevant protein. In contrast the *lac* system, a single copy of the *trp*R (repressor) gene seems to be sufficient to block transcription from the *trp* promoter present on a multicopy plasmid.

The λ P_L system

The use of the major !eftward promoter of phage λ, P_L, has a number of advantages. Under normal conditions, a single copy of the λ repressor gene (*cI*) present on the host chromosome is able completely to block expression from the P_L promoter located on a multicopy plasmid. Further, by using the *cI*857 allele, a temperature shift from 30 to 42°C can be used to inactivate the mutant repressor and allow transcription to proceed. Note that systems based on the rightward promoter of· phage l are also available, and these have similar properties.

The use of P_L has, in addition, a further advantage. In phage λ the first gene downstream of P_L, the *N* gene, codes for a protein that interacts with the RNA polymerase in such a way as to prevent transcription arrest at most termination signals. Plasmids incorporating both P_L and *N* may thus be used with confidence to express foreign protein-coding sequences, since transcription termination within the foreign gene can be eliminated.

The *cI*857/P_L /N regulation, as it stands, lacks signals for initiating translation of a foreign DNA sequence. For this reason, a number of groups have incorporated sequences from other well-translated genes, such as the MS2 replicase gene or the λ *cII* gene. The DNA segment containing the powerful translation initiation signals associated with the membrane lipoprotein (*lpp*) gene of *E. coli* is a good candidate for combination with the λ P_L system.

Hybrid Expression Systems

Tac

De Boer et al. (1982, 1983) described the construction of hybrid expression control elements. As mentioned earlier, transcription initiation signals in *E. coli* comprise at least two recognition elements, the –10 (Pribnow box) and –35 sequences. The *trp* promoter has been shown to possess on optimal (consensus) –35 transcription recognition region but a poor sequence at −10. In contrast, *lacUV35* comprises an optimal −10 sequence and a poor -35 region. Thus two different fusions, *tac*I and *tac*II, were constructed in which the optimal elements of both promoters were combined. Both constructs retain their sensitivity to repression by the *lac*I protein (and thus inducibility by IPTG) characteristic of the lac promoter. When combined with a synthetic transcription translation region, these promoters were shown to direct the expression of a foreign gene 11 and 7 times (*tac*I and *tac*II) more efficiently than the parental *lacUV5* block. Similar constructs have been described by Russell and Bennett (1982).

Rac

In a further construction, De Boer et al. (1982) fused the upstream (-35) control elements of the powerful ribosomal RNA (*rrn*) genes with downstream sequences from *lac*. This system, which is as active as the *tac* constructs, may take advantage of a termination override mechanism associated with *rrn* promoters.

Other systems

Hybrids broadly similar to those described above have been constructed by Russell and Bennett (1982), who exchanged sequence elements between *trp*, *lac*, and the tetracycline resistance (*tet*) gene from pBR322. Note must be made of the synthetic consensus promoter and translation-initiation sequences which, in combination, may supplant the other systems described earlier.

As a broad guideline, the transcription of a foreign DNA sequence mediated by a powerful promoter is, in many systems, a matter of routine. In contrast, the efficient translation of the messenger so produced, and the concomitant accumulation of the product, is a matter of delicacy. Particular factors that affect the yield are the precise sequence of the translation-initiation region (both before and after the initiating ATG), the secondary structure of the messenger RNA around the initiation codon, and the stability of the protein product. No general rules have so far been found that can adequately portray these parameters.

In the following sections we describe the isolation and cloning of viral genomes and, in particular, how the systems described above have been adapted to the production of viral antigens.

Hepatitis B Virus

Owing to its considerable clinical interest, hepatitis B virus (HBV) was one of the first main targets for recombinant DNA technology.

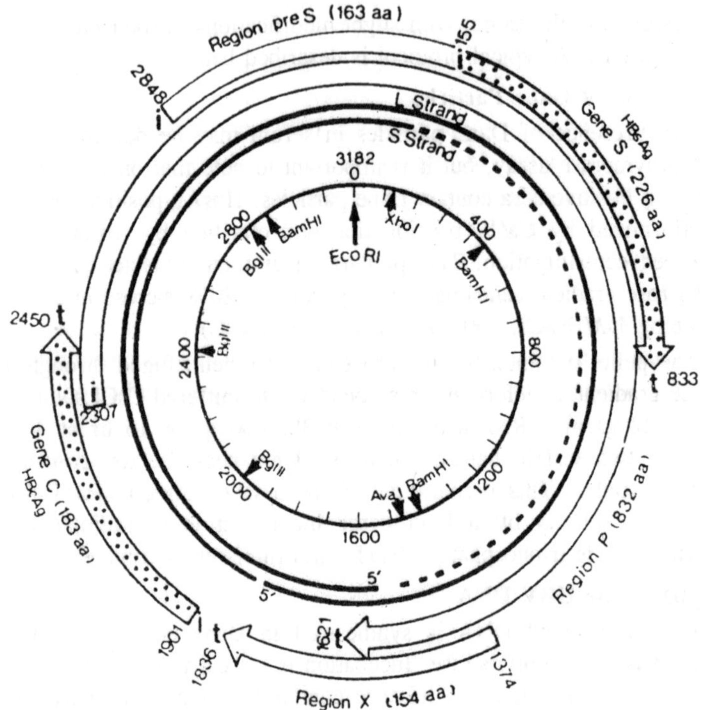

Fig. 14.1. Genetic organization of the hepatitis B virus (HBV) genome.

The HBsAg gene was located by comparison of the DNA sequence with partial amino acid data. The HBcAg gene was identified within a region of DNA that programmed, in *E. coli*, the synthesis of serologically active HBcAg. Neither the coding regions of the HBcAg gene nor those of the HBsAg gene appear to contain intervening sequences.

It has been suggested that the largest open reading frame, which occupies more than 75% of the genome and overlaps the HBc gene and encompasses the HBs gene, encodes the viral DNA polymerase. A fourth open translational coding frame, the so-called X gene, has been identified within the viral sequence, but no function has so far been associated with this gene; the expression of the X protein in *E. coli* may permit specific antibodies to be raised against this protein with the potential of detecting protein X in human serum.

Expression of HBV Surface Antigen in E. coli

As mentioned earlier, the inability of HBV to grow in tissue culture cells has hampered the development of a conventional vaccine.

This has led to attempts to find other preparations that might produce active immunity, including the use of inactivated HBsAg purified from plasma of asymptomatic human carriers. The safety, immunogenicity, and high protective efficiency of such preparation have been demonstrated. This type of subunit vaccine, however, suffers to considerable disadvantages: large quantities of human pooled plasma, with high titers of HBsAg, are necessary, and the manufacturing process is expensive. Containment facilities and strict safety testing of the vaccine are required, including a test for residual infectivity of HBV in susceptible chimpanzees. Thus the use of genetic engineering technology has become an attractive approach for producing HBsAg in quantity.

A first attempt to express HBV antigens in *E. coli* was described by Burrel *et al.* (1979) in a so-called shotgun cloning experiment. Complete double-stranded HBV DNA (generated *in vitro*) was digested with different restriction enzymes, 3'-oligo(dC) sequences were added to the fragments using polynucleotide terminal transferase, and the fragments were annealed to plasmid pBR322 to which oligo(dG) sequences had been attached at the *Pst*I site. These DNA preparations were then used to transform competent cells of *E. coli*. Transformants were subsequently screened by colony hybridization with radiolabeled Dane-particle DNA.

Colonies carrying recombinant plasmids were screened for the production of HBV antigens. The technique developed by Broome and Gilbert (1978) was applied using both anti-HBc and anti-HBs (human and hyperimmune animal sera) antibodies. This experiment revealed only a small number of faint positive colonies.

After identification to the nucleotide sequence of the HBV genome corresponding to the surface antigen, it became possible to perform precise constructions in order to express the HBs antigen. McKay *et al.* (1981) inserted different fragments of the HBV genome containing all or nearly all of the coding sequence for the surface antigen into the PstI site within the β-lactamase gene of plasmid pBR322. These constructions were designed to result in the synthesis of HBs peptides, either as a fusion with the N terminus of the β-lactamase protein or by reinitiation of translation after termination at a TGA stop codon lying upstream of the natural HBs initiation codon.

In all clones tested, solid-phase radioimmunoassay, gel electrophoretic analysis of proteins made in minicells or an assay using microtiter wells coated with anti-HbsAG all gave weak, variable,

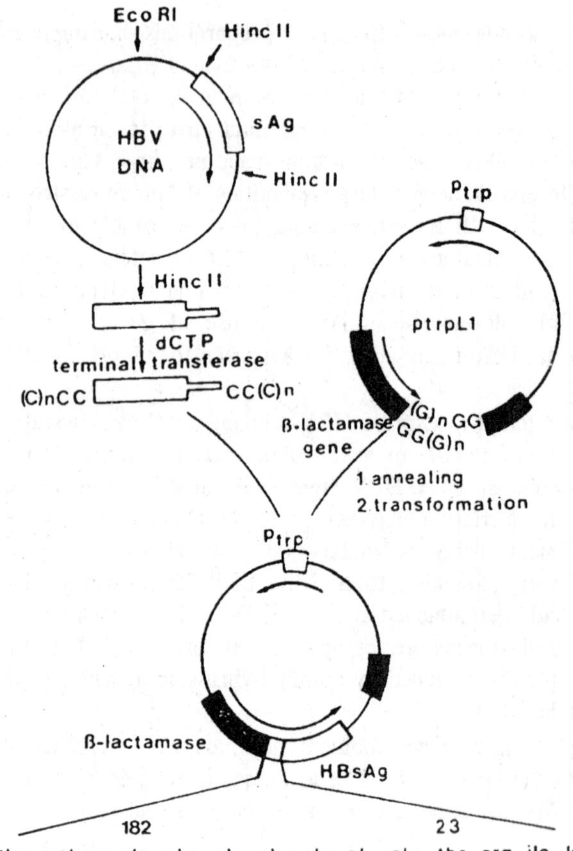

Fig. 14.2. Construction of plasmids expressing a β-lactamase: HBsAg fusion.

and, in general, inconclusive results. Nevertheless, some extracts prepared from these bacteria induced a low titer of anti-HBV antibodies in rabbits, but only after a long period of repeated booster injections.

Edman *et al.*, (1981) also integrated the HBs gene into the *Pst*I site of a plasmid-borne β-lactamase gene but used a system in which the β-lactamase gene was under the control of the *trp* promoter. A 744 base pair HindII fragment of HBV DNA containing the coding sequence for 204 amino acids of HbsAg was purified by polyacrylamide gel electrophoresis, tailed with d(C) using terminal transferase, and hybridized with dG-tailed plasmid *ptrp*L1 prior to transformation of *E. coli* HB101.

The DNA sequence of this HBV DNA fragment predicted that, when the HBsAg sequence was in the proper orientation and the same reading phase as β-lactamase, a 43-kilodalton polypeptide would be produced. This hybrid protein should contain 183 amino acids of the pre-β-lactamase (21 kilodaltons), 5-10 glycine residues from the G/C tail (ca. 600 daltons), and 204 amino acids (22.6 kilodaltons) deriving from HbsAg. If the β-lactamase "Signal" presequence were cleaved, the fusion protein would have a predicted molecular weight of 41,000. Colonies containing recombinant plasmids were examined, and a number were shown to produce the 41-kilodaltons polypeptide. After immunoprecipitation with anti-HBs IgG, it was found that the levels of 41-kilodalton polypeptide produced in *E. coli* varied widely among different experiments and in some cases was barely detectable.

Charnay et al. (1980) described recombinant λ phage derivatives in which part of the HBsAg gene was fused to the β-galactosidase gene. The resulting fusion polypeptide (138 kilodaltons) comprised almost the entire β-galactosidase polypeptide followed by HBsAg lacking its first 29 amino acid residues. This polypeptide could be precipitated with anti-HBsAg antibodies, but reacted poorly in radioimmunoassays. Note that the structure of this fused polypeptide differs considerably from the construction of Edman et al. (1981) and McKay *et al*. (1981), in which few foreign amino acids were fused to HBsAg.

Numerous explanations for the poor expression of HBsAg in *E. coli* have been proposed, but it is widely believed that the problems is due to a deleterious effect of the HBsAg protein on the growth and metabolism in *E. coli*. This phenomenon is well illustrated in an experiment performed at Smith-Kline-RIT. Here, an attempt was made to express in E. coli the surface antigen derived from the two main serotypes, adW2 and ayW2. When these sequences were fused downstream of the *lac*Z initiation codon present on vector pPCψ3, induction of the *lac* promoter with IPTG resulted in marked retardation of cell growth. Furthermore, cultivation in the presence of inducer resulted in extensive loss of the recombinant plasmid. These results indicate that production of a polypeptide containing HBsAg may be lethal to *E. coli*. To analyze this possibility further, a set of recombinant plasmids was constructed in which all or part of the HBsAg gene was fused to the 5' end of either the *lac*Z or *trp*E genes. These plasmids were fused to transform strains containing either normal or substantially elevated levels of the cognate repressor protein (lacI$^+$, *lac*I^q, or *trp*R$^+$) or, alternatively, strains lacking an active repressor (Δ*lac* or *trp*R$^-$). It was observed that strains lacking repressor were transformed 10- to 100-fold less efficiently than strains having normal or elevated levels of repressor. None of the phenomena descried above was observed when analogous plasmids were used in which HBV sequences were present in an orientation unsuitable for expression. Such experiments demonstrate that expression of HBsAg sequences is deleterious to the *E. coli* cell. It should be noted that removing either the N-terminal or both the N- and C-terminal sequences from the constructs did not enhance expression.

Expression of HBV Core Antigen in E. coli

In contrast to the surface antigen, the hepatitis B core antigen can be expressed in large amounts in *E. coli*, where it seems to be relatively stable. The expression of HBc antigenic determinants in *E. coli* was first described by Burrel et al. (1979). Clear, positive, and reproducible results were obtained. Nucleotide sequence analysis showed that the HBV core antigen was synthesized by "translation restart" phenomenon on the mRNA.

Fragments carrying the core antigen gene have been placed under the control of the lac promoter of *E. coli*. Several of the recombinants direct high levels of synthesis of the antigen the exact level depending on the precise structure of the plasmid and the mRNAs synthesized. Note that bacterially synthesized antigen is a satisfactory diagnostic reagent for the presence, in human sera, of antibodies directed against the core antigen.

The *trp* promoter very efficiently enhances the expression of genes linked to its coding sequence. Edman et al. (1981) placed the core antigen under the control of the *trp* promoter in plasmid *ptrp*L1. Here the HBV DNA contains two potential initiation codons of HbcAg. These authors focused their efforts on a construction involving the first met codon, but there is now strong evidence that the second *met* codon is the natural initiation codon used *in vivo*. Thus, the core antigen resulting from the construction of Edman et al., (1981) carries an additional 29 amino acids at its N terminus.

A 1005 base pair (bp) *Hha*I fragment of HBV DNA containing the entire HbcAg gene was purified by gel electrophoresis; in this fragment the first *met* codon for HbcAg lies only 15 bp from one extremity. After treatment with *Hpa*II methylase to prevent cleavage of the fragment during subsequent restriction enzyme digestion, the fragment

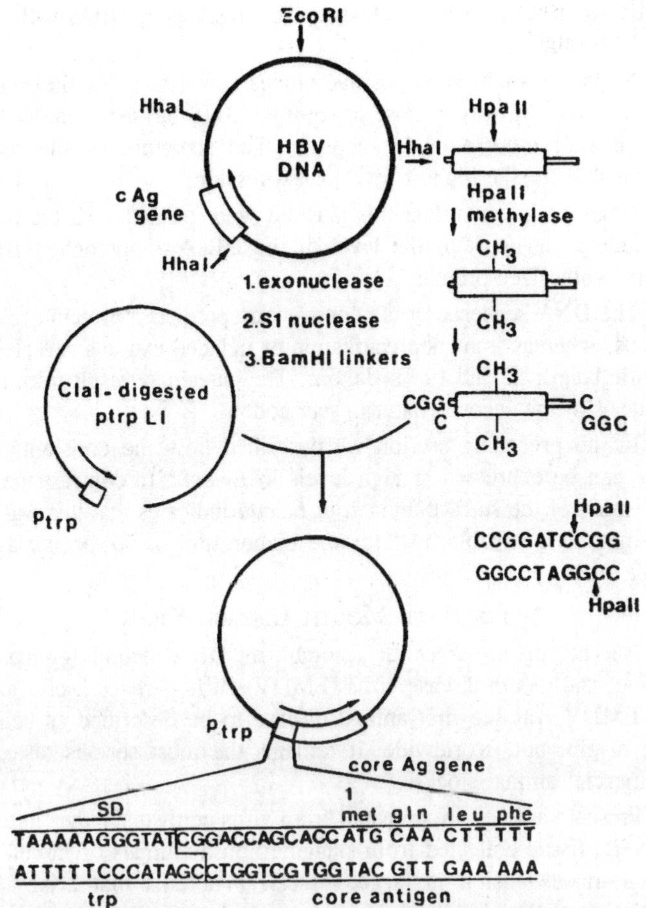

Fig. 14.3. Construction of plasmids expressing a trp:HBcAg fusion.

was subsequently treated with T4 polymerase in the absence of dNTPs (permitting 3' →5'- nuclease activity) for 30 sec at 37°C. Under these conditions, between 0 and 10 nucleotides were removed from each 3' end. The digested fragments were then treated with S1 nuclease, and dodecameric *Bam*H1 linkers (CCGGATCCGG, which contain the *Hpa*II recognition sequence, CCGG) were ligated to the termini using T4 ligase. After digestion of the linkers with *Hpa*II, the linker fragments were removed and the DNA was ligated with *Cla*I-digested and calf intestinal alkaline phosphatase-treated ptrpL1. The ligation mixture was used to transform *E. coli* strain HB101 and transformants were screened for recombinant plasmid DNA. These were then tested for the production of HBcAg using a double-antibody radioimmunoassay (RIA) with human anti-HBcAgIgG.

Sequence analysis of positive clones confirmed that the initiation codon of HBcAg was in close proximity (13–16 bp) the Shine-Dalgarno sequence of the *trp* leader peptide. The structure of the plasmid responsible for the highest level of expression.

When cells containing this plasmid were treated with the inducer 3-β-indolylacrylic acid, the level of the HBcAg approached 10% of newly synthesized protein.

The DNA sequence of the construction predicts a protein of 22,000 daltons, whereas immunoprecipitation of induced extracts revealed two specific bands: 22 and 19 kilodaltons. The latter may result from restart synthesis at the second (natural) *met* codon.

In the preceding section we described how the core antigen of HBV can be expressed at high levels in *E. coli*. In contrast the poor expression of the surface antigen in *E. coli* indicates that this organism may not be a suitable host for the elaboration on a vaccine against HBV.

Foot-and-Mouth Disease Virus

Another prime target for recombinant DNA technology has been the foot-and-mouth disease virus (FMDV). It is perhaps useful to note that FMDV was the first animal disease to be described as being of viral origin, but, worldwide, it remains the most serious disease of commercial animal stocks.

Present vaccines are prepared from virus cultivated either in tongue epithelial tissue collected from slaughterhouse animals or in cell lines grown in suspension in large vessels. Problems that arise in the production of the viral antigens are commonly related to certain strains of the virus and especially to the multiplicity of serotypes, which change regularly. Although much has been achieved with available vaccines, there still is a need for new methods of producing the viral antigens. The technology of genetic engineering could be considered to be a promising alternative to traditional methods.

FMDV is a picornavirus, consisting of a linear single-stranded RNA about 8000 nucleotides in length. It is polyadenylated at its 3' end, has a poly(C) tract close to the 5' end and a protein covalently attached to the 5' terminus. Translation is initiated to the 3' side of the 150-nucleotide poly(C) tract and yields a single polyprotein, which is subsequently processed to generate the structural proteins of the

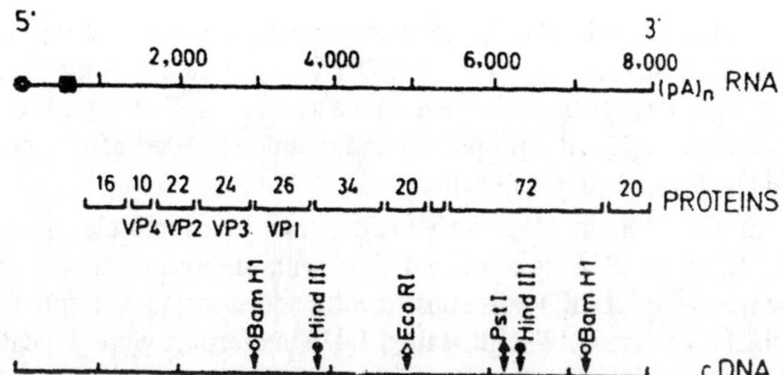

Fig. 14.4. Organization of the FMDV genome and location of restriction sites on the cDNA.

virus: VP1, VP2, VP3, and VP4. It has been shown that the immunizing activity of FMDV particles is associated primarily with VP1, this protein alone can elicit seroneuralizing antibodies.

Purification of FMDV RNA

Virus, usually isolated during an epizootic outbreak, is typically isolated a few times from single plaques and passaged about 50 times in cell culture. Note that the identity of passaged virus must always be confirmed by complement-binding assays with the original antibodies.

Grubman *et al.* (1979) infected roller cultures of baby hamster kidney (BHK) cells with purified virus (10 PFU/cell) and harvested both intracellular and extracellular virus 3–6 hr after infection. Intracellular virus was released by swelling the cells in reticulocyte stranded buffer + heparin and Dounce homogenization. Nucleic were removed by low speed centrifugation, and virus was purified by centrifugation through a 10–50% (w/v) linear sucrose gradient at 60,000 g for 17 hr at 4°. Extracellular virus was concentrated by polyethylene glycol (PEG) precipitation and similarly purified by sucrose gradient centrifugation as described above.

Purified virus in buffer was digested for 60 min at 37° with 500 μg of proteinase K per milliliter in the presence of 1% sodium dodecyl sulfate (SDS), and the viral RNA was further purified by centrifugation on a 15–30% (w/v) linear sucrose gradient containing 0.5% SDS (typically 46,000 g for 17.5 hr at 23°C).

Cloning of FMDV cDNA

Kupper *et al.*, (1981) prepared single-stranded cDNA by primary AMV reverse transcriptase with oligo(dT) hybridized to the polyadenylate tract present at the 3' end of the virion RNA. The RNA–DNA hybrid so formed was heat denatured (3 min at 100°C) and quickly cooled. The single-stranded cDNA was rendered double-stranded by treatment with *E. coli* DNA polymerase I in the presence of the four deoxyribonucleoside triphosphates for 2 hr at 15°C. The reaction was arrested with phenol and double stranded cDNA purified by Sephadex G-150 gel filtration.

To insert the double-stranded cDNA molecules into plasmid vector pBR322, the cDNA was treated first with single-strand specific S1 nuclease and oligo(dC) tails subsequently added to the 3' termini using terminal transferase. The dC-tailed DNA molecules were annealed to pBR322 to which 3'-oligo(dG) sequences has been added at the pstI site and used to transform competent cells of *E. coli* HB101. Tetracycline-resistant transformants were transferred to microtiter plates, and replicas of each plate were used for hybridization with ^{32}P-labeled fragmented viral RNA. Colonies displaying strong hybridization with the probe were selected for analysis. A similar protocol was developed by Botthryod et al. (1981).

Characterization of the VP1 Gene

Restriction enzyme analysis of the cDNA inserts generated a restriction map of the FMDV genome aligned relative to the 3' end of the RNA. Approximate correlations between the different viral proteins and the viral genome were deduced from the order of translation of the different polypeptides and from their apparent molecular weights. This correlation has then been used to predict the position of viral genes within specific restriction fragments of the FMDV genome. Although approximate, this analysis did locate the VP1 gene. Clones predicted to contain the VP1 gene were sequenced and compared with the known sequence of NH2-terminal 40 amino acids of VP1; this allowed the unambiguous identification of the VP1 coding region.

The use of ^{32}P-labeled fragments of FMDV RNA as probes to align the restriction map with the physical map was reported by Boothroyd et al. (1981). The sequence of the complete VP1 gene of FMDV is of particular interest since VP1 proteins harbors the main antigenic determinant of the virion. Moreover, changes in its amino acid sequence are responsible for the high antigenic variability of FMDV. The complete nucleotide sequences of the VP1 gene corresponding to different serotypes have been reported.

A rapid method for selective cloning of VP1 gene from different viral serotypes was necessary to characterize the new amino acid changes found after each new outbreak. Such a method was set up by

Yansura et al., (1983). Purified RNA from the new strain is annealed to three synthetic deoxyoligonucleotide primers, each 10 nucleotides long, which hybridize to RNA sequences approximately 200 nucleotides 3' to the VP1 gene. These sequences were chosen because of homology noted in this area when comparison are made of FMD virus types A12, 01, A27, and C3.

In the presence of reverse transcriptase enzyme and deoxynucleotide triphosphates, the RNA is copied into DNA (cDNA) beginning with the primers. After denaturation, this is converted into double-stranded cDNA (ds-dDNA) using the Klenow fragment of DNA polymerase and treated with nuclease S1 to digest single-stranded regions. Double-stranded cDNA longer than about 2000 base pairs is isolated by gel electrophoresis, tailed with dCTP using terminal transferase, and annealed to linearized pBR322 carrying poly(dG) tracts at the PstI site. This material is then used to transform *E. coli*. Plasmids with incorporated ds-cDAN are isolated, and those containing inserts larger than about 1500 base pairs are analyzed by filter hybridization to radiolabeled DNA from the VP1 clone of virus type A12. Plasmids containing the VP1 gene may then be analyzed by restriction mapping and the nucleotide sequence of the VP1 variant gene determined using standard procedures.

Expression of the FMDV VP1 Gene in E. coli

The strong leftward promoter (P_L) of phage λ is a good candidate to promote expression of eukaryotic genes in *E. coli*; the activity of this promoter can be controlled by using host strains that carry the gene for a temperature-sensitive λ repressor. A vector (pPL_C24) carrying the P_L promoter plus the SD region of the MS2 replicase gene was constructed by Remaut et al. (1981). This vector, which has been successfully used to express the gene for human β interferon carries a unique restriction site for *Bam*Hi at a position corresponding to amino acid 99 of the MS2 replicase gene and a unique site for HindIII 10 nucleotides downstream. A *Bam*HI-*Hind*III fragment of the FMDV cDNA which carries nearly all the VP1 coding sequence (the BamHI site is located at the ninth amino acid) was inserted in the pPLc24 vector. A fusion polypeptide of 396 amino acids comprising the N-terminal 99 amino acids of MS2 replicase, 284 amino acids of FMDV, and 13 amino acids coded by the vector was predicted to b synthesized from the recombinant plasmid. Such a "tribrid" protein has been identified in induced *E. coli* extracts, at a yield of about 1000 molecules per cell.

Fusion of the VP1 gene (corresponding) to the VP1 Protein from serine 7, *Pst*I site, to glutamine 211, *Pvu*II site) with the *trp*E protein has been reported by Kleid *et al.* (1981). The yield of fused protein obtained in this expression system was about 17% of the total protein. Moreover, the fusion protein appears to be stable in the cell extract and can be used as a effective vaccine.

Enormous efforts to localize precisely the major antigenic site on the VP1 protein have been attempted. Examination of VP1 amino acid sequences from various types and subtypes revealed that there exists a particular region in the protein that is highly variable. Studies indicated that this region of the VP1 protein is exposed on the virus surface and possesses a major antigenic site. A 13,000 –dalton peptide that contains the variable region, isolated from CNBr-treated protein, has been shown to induce neutralizing antibodies and protect swine from FMD.

Synthetic peptides from this region and linked to keyhole limpet hemocyanin carrier have been shown to induce neutralizing antibodies in guinea pigs and rabbits. Such peptides have been proposed as the basis of synthetic vaccines indeed, a single inoculation of a synthetic peptide corresponding to amino acids 141–160 can elicit neutralizing antibodies, with a titer greater than that obtained with the purified capsid protein VP1, irrespective of whether this is produce by disruption of virus particles or by expression in *E. coli*. Thus the prospects for a totally synthetic vaccine against FMD are encouraging.

In an alternative approach, Yansura *et al.*, (1983) elected to express this main antigenic site of the VP1 protein in *E. coli* in a form of a fused protein.

The coding sequence for amino acids 130–157 of the VP1 protein was recovered as an 83 bp *Alu*I-*Rsa*I fragment and ligated to plasmid pUC9 which had.been cleaved previously with SmaI. The pUC9 plasmid, when used to transform *E. coli* JM83 to amplicillin resistance, codes for the amino-terminal portion (*lac*Z') of β-galactosidase. This complements the carboxyterminal β-galactosidase protein expressed in the *E. coli* JM83 host and the reconstituted β-galactosidase can be detected by a suitable chromogenic substrate (e.g., X-gal). Introduction of the 83 base pair VP1 gene fragment into the SmaI site introduced a termination codon in the same reading frame with the initial ATG codon; thus derivatives containing the 83 base pair insert were easily identified by the absence of staining with X-gal.

In pUC9 the *Sma*I site in the *lac*Z' gene is flanked by *Pst*I or *Eco*RI sites, and one derivative containing the 83 base pair insert was treated with *Pst*I and *Eco*RI to recover the formerly blunt-ended fragment, now flanked by these sites. This fragment was introduced into the expression plasmid pNCV to give the derivative pFMB-42. In plasmid pNCV the *trp*P promoter lies upstream of a coding sequence generated by fusing the leader polypeptide (*trp*L) to the carboxyterminus of the trpE gene (the *trp* LE gene product, which has 190 amino acids, is extremely useful as part of a fusion protein, since it appears to be both insoluble inside the bacteria and resistant to proteolytic degradation). A *Pst*I site had been introduced into the *trp* LE fusion at the *trp* E termination codon of pNCV. When linked to the PstI site of the modified VP1 gene fragment described above, a fused gene was created with the codons of the LE protein and the antigenic region of the VP1 protein in the same reading frame. The pNCV derivative, pFMB-42, was used to transform *E. coli*. *Escherichia coli*/pFMB-42, when grown under depressed conditions, was found to express high levels of the desired fusion protein. Like the *trp*LE gene product, the fusion protein was found to be stable in *E. coli* and gave rise to insoluble intracellular inclusions visible by phase contrast microscopy. Levels of up to 20% of total cellular protein were obtained by this method. The product could be recovered from disrupted cells by centrifugation and the fusion protein be purified by a modification of a method similar to that described for FMD virus VP1 protein.

We have described how a single major epitope, either as a chemically synthesized oligopeptide or expressed in bacteria as a fusion protein, can elicit the production of seroneutralizing antibodies. Thus the prospects for a new FMDV vaccine are encouraging.

Poliovirus: Purification of Polio RNA and Cloning of Polio cDNA

Although the development of both killed and live vaccines has been spectacularly successful in controlling poliomyelitis in industrialized nations, polio remains a serious threat in much of the world. In spite of intensive research, the precise mechanisms by which poliovirus replicates and causes cytopathic effects remain unknown. The molecular basis for viral attenuation has also not been solved. Poliovirus, like FMDV, is a member of picornavirus group. A variety of experimental results suggest that the poliovirus genome is also translated into a single polyprotein from which the functional viral proteins are derived by proteolysis.

The RNA molecule contained in the Mahoney strain is 7433 nucleotides long, polyadenylated at the 3' terminus, and covalently linked to a small protein (VPg) at the 5' terminus. An open reading frame of 2207 consecutive triplets spans over 89% of the nucleotide sequence and codes for the viral polyprotein NCVPOO. Twelve viral polypeptides have been mapped by amino acid sequence analysis and were found to proteolytic cleavage products of the polyprotein, cleavage occurring predominantly at Gln-Gly pairs.

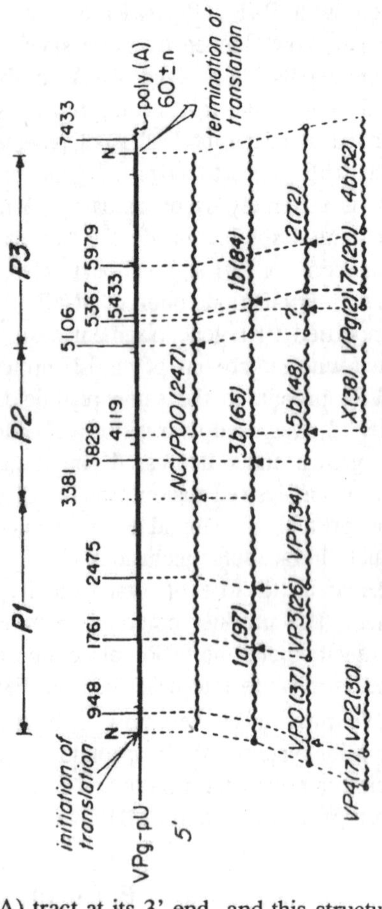

Poliovirus RNA may be purified by the technique of Lee et al. (1979). In brief, the method involves the release of virus particles from infected heLa cells by disruption in hypotonic medium. Virions are recovered by high speed centrifugation (78,000 g, 3 hr), and viral RNA is purified by serial phenol/chloroform extraction in the presence of 0.2% SDS followed by precipitation.

Fig. 14.5. Gene organization of poliovirus RNA.

Polio RNA contains a poly(A) tract at its 3' end, and this structural element has permitted copying the poliovirus RNA genome into cDNA using an oligo(dT) primer and reverse transcriptase. Cloning of the genome as a cDNA copy is usually a prerequisite for determining the primary structure of poliovirus genome; however, restrictions imposed by recombinant DNA guidelines led Kitamura et al., (1981) to sequence polio cDNA by a modification of Sanger's chain termination method without recourse to cloning.

Polio cDNA was synthesized *in vitro*, and chains of 7000–7400 deoxyribonucleotides were selected by zonal centrifugation. In turn, polio RNA was digested exhaustively with RNase T1 or with RNase A and large oligonucleotide products separated by two-dimensional gel electrophoresis. These were eluted from the gel, the 3'-phosphate group

Apart from Australia and New Zealand, and certain countries such as Britain and Japan that have succeeded in eradicating the disease, rabies is still present in most parts of the world. Infection of man, often through being bitten by a rabid animal, is usually fatal without treatment. Nevertheless, rabies is unusual in that postexposure vaccination, if administered soon, can prevent development of the disease.

Infection with rabies virus is often followed by a long incubation period. During this time the virus travels through peripheral nerves to the spinal chord and brain, and at this stage the behavioral symptoms become apparent. Subsequent appearance of the virus in other organs, such as the salivary glands, facilitates it transmission.

The virus consists of a rod- or bullet-shaped virion enclosed in a lipid bilayer through which the glycoprotein (G; hemagglutinin) protrudes. Other proteins contained in the virus include the transcriptase (L), the matrix proteins (NS, M2), and the nucleoprotein (N). The viral RNA, about 10 kb of single-stranded (negative-strand) RNA, exists as a helical ribonucleoprotein complex with N protein.

After infection, mRNA transcription mediated by L proceeds directly upon the viral negative strand, and the five mRNAs, corresponding to the five viral proteins, are polyadenylated by a virus-dependent function. The generation, in addition, of full-length plus-strand transcripts permits the subsequent production of viral RNA for encapsidation in the virion.

The glycoprotein spike of rabies virus is the only protein that traverses the envelope, and for this reason it is the sole antigen capable of binding neutralizing antibodies and of eliciting their production. Existing vaccines, however, rely on the use of whole virus inactivated by treatment with β-propiolactone or a similar agent, but the high price of such vaccines has restricted their use. Several groups have therefore attempted to use recombinant DNA techniques to produce a more cost-efficient rabies vaccine.

Preparation of Rabies Glycoprotein Messenger RNA

As mentioned above, the G protein is the only viral antigen capable of eliciting the production of neutralizing antibodies; thus work has concentrated on this protein.

Wunner et al. (1980) infected monolayer cultures of rodent cells (BHK21-S3) in roller bottles with rabies virus (ERA strain) at a multiplicity of infection of 25 and recovered the infected cells after 20 hr. Washed cells were treated for 30 min with 2.5% sodium dodecyl sulfate (SDS) and 200 μg of Pronase per milliliter. Total RNA was extracted with phenol and precipitated with ethanol prior to DNase digestion and reextraction with phenol.

This preparation was heat denatured and subjected to oligo(dT)-cellulose chromatography. RNA bound in 300 mM salt was eluted with distilled water and concentrated by ethanol precipitation.

Wunner et al. (1980) were able to demonstrate that this material, after injection into amphibian oocytes, was capable of programming the synthesis of proteins reacting with antiglycoprotein and antinucleoprotein antisera.

Since G, without its carbohydrate groups has a molecular weight of about 65,000–70,000, the mRNA should be about 1.8 kb in length and sediment at 18 S. N is slightly smaller (ca. M_r 60,000), and the corresponding mRNA should sediment slower, at about 16 S. Accordingly, Wunner et al. (1980) were able partially to separate the two messenger species. After sucrose gradient centrifugation they injected each fraction into *Xenopus laevis* oocytes and analyzed the newly synthesized proteins with monoclonal antibodies raised against G or N. By this means they were able to isolate an mRNA fraction enriched for glycoprotein coding sequences.

Cloning of Glycoprotein Coding Sequences in E. coli

Commencing with this enriched material, Anilionis et al. (1981) prepared cDNA by using AMV reverse transcriptase and an oligo(dT) primer hybridized to the polyadenylate tails. A second DNA strand was synthesized using *E. coli* DNA polymerase. After single-strand specific S1 nuclease treatment, the double-stranded cDNA was tailed (3') with dCMP residues using terminal transferase, and large (≥ 1 kb) tailed cDNA was further purified by sucrose gradient centrifugation. This material was annealed with pBR322 plasmid, which had been previously tailed at its PstI site with dGMP residues, then transformed into *E. coli* strain χ1776.

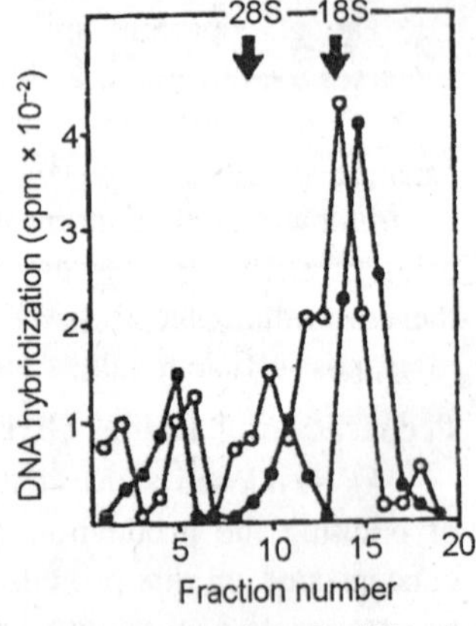

Fig. 14.6. Detection of mRNAs for glycoprotein and nucleocapsid by hybridization to cloned DNAs.

Tetracycline-resistant colonies emerging from this procedure were examined for the presence of rabies sequences by hybridization with radiolabeled virions RNA. Positive clones were further examined by hybridization to sucrose gradient-fractionated mRNA. Two hybridization profiles were obtained, one corresponding to an 18 S mRNA, presumably the G protein messenger, and the other to a 16S mRNA, the N protein messenger.

Sequence analysis of the G protein clones revealed a contiguous open reading frame comprising 1572 nucleotides capable of coding for a 524 amino acid protein (Anilionis et al., 1981). Comparison of the deducted amino acid sequence with the partial amino acid sequence of mature G revealed a stretch of 19 additional residues preceding the first amino acid sequence of mature G, probably representing a signal sequence for the partially extracellular protein. Further, a domain of 22 hydrophobic amino acids exists near the C terminus of the encoded polypeptide, and this is likely to represent a transmembrane "anchor" zone for the membrane-bound glycoprotein.

Similarly, the G protein coding sequence from the related CVS strain of rabies virus has been cloned and characterized.

Mapping of Antigenic Determinants

Dietzschold et al. (1982) reported that native G protein could be cleaved by cyanogen bromide into seven peptide bands, and they separated these fragments by polyacrylamide gel electrophoresis in the presence of SDS. Only two peptides reacted with antiglycoprotein antiserum, whereas three peptides induced the formation of virus-neutralizing antibodies after injection into mice. Thus at least three antigenic determinants are present in the glycoprotein polypeptide.

In a subsequent study, Dietzschold et al., (1983) examined antigenic variants of rabies virus that had been selected for the ability to grow in the presence of a monoclonal virus-neutralizing antibody. In one case the non-pathogenic mutant virus isolated by this procedure was shown to have an altered amino acid (Arg → Ile) at position 333; this alteration falls outside, but adjacent to, the third region identified by analysis of cyanogen bromide peptides.

Expression of Rabies Glycoprotein Sequences in E. coli

Expression in a Bacteriophage vector

Bacteriophage M13 and its derivatives can be usefully employed for the expression of foreign genes in *E. coli*. Although a virus, M13 infection does not kill the host. Instead, the viral genome proliferates in the cytoplasm as a double-stranded circular episomal form.

Concomitantly, phage particles containing circular single-stranded DNA are exported from the cell surface.

Derivatives of M13 have been constructed in which the *lac* promoter (*lac*P) has been integrated into a nonessential region of the phase genome. The inducible (by isopropyl thiogalactoside, IPTG) *lac* promoter is adjacent to the N-terminal section of the *E. coli lacZ* gene encoding β-balactosidase and a region containing multiple restriction recognition sequences. Thus the transcription signals for efficient and controlled gene expression are followed by suitable sites for the integration of exogenous DNA segments.

It should be noted that in a related rhabdovirus, vesicular stomatitis virus (VSV), expression of the glycoprotein N-terminal hydrophobic signal peptide in *E. coli* has been shown to be detrimental to the host. In contrast, expression of the C-terminal hydrophobic domain had no deleterious effect. Nevertheless, this result stresses the importance of using an inducible system for the expression of foreign proteins in bacteria.

Subfragments of the rabies glycoprotein coding sequence lacking the N-terminal signal sequence region have been cloned adjacent to the *lacP-lacZ* fragment in such a way as to permit the synthesis of a hybrid β-balactosidase-rabies glycoprotein polypeptide. Here advantage was taken of a unique *Hin*dIII recognition sequence lying 9 amino acids downstream from the first amino acid of the mature glycoprotein coding sequence. These 9 amino acids were replace, in most cases, by 11 amino acids from the N terminus of the *E. coli* β-balactosidase.

Infected bacterial cultures were induced with IPTG, radiolabeled with [^{35}S]methionine, and subjected to immunoprecipitation with polyclonal antiserum raised against the rabies virus glycoprotein. The bacterially synthesized glycoprotein, and subfragments thereof, were efficiently recognized by the antiserum. Surprisingly, a number of other proteins of bacterial origin were also precipitated. This is probably due to cross-reaction between *E. coli* proteins and bacterial proteins present in the adjuvant used to prepare the antiglycoprotein antiserum.

Expression in a plasmid vector

Yelverton et al., (1983) described a cDNA clone corresponding to the G protein of a different laboratory strain of rabies, the CVS strain. At the amino acid level the two proteins share about 90% homology. In order to express the gene coding for G-CVS in bacteria, a different approach was adopted. Commencing with a cloned full-length cDNA copy of the G messenger, oligonucleotide mutagenesis was used to introduce a bacterial translation initiation signal at a point immediately before the first amino acid of the mature glycoprotein coding sequence, thus "deleting" the hydrophobic signal peptide. At the C terminus, the hydrophobic trans membrane domain was removed by introducing a TAG translation stop codon at a *Bgl*II recognition site 14 amino acids prior to the hydrophobic zone. This coding sequence was placed under the control of an inducible (by tryptophan starvation) promoter derived form the trp operon of *E. coli*, and directed the synthesis, in *E. coli*, of a polypeptide cross-reacting with antiserum raised against the authentic rabies-CVS glycoprotein. We have seen that peptides corresponding to rabies glycoprotein can be produced in a bacterial host. This eliminated the costly production and containment procedures necessary for vaccines based upon inactivated rabies virus.

As mentioned earlier, rabies is unusual in that postinfection vaccination is effective in preventing development of the disease. Nevertheless, it has not yet been demonstrated that purified authentic glycoprotein can protect when administered after exposure, although it does elicit the production of high titers of neutralizing antibodies.

The process for a bacterial vaccine for preexposure use are much brighter, for purified authentic glycoprotein gives a level of protection comparable to that obtained with inactivated virus vaccine. Levels of antibodies produced, and the protection afforded, nevertheless depend greatly upon the presentation of the antigenic determinants.

Expression of Viral Antigens in Bacillus subtilis

Gene Expression Vectors Based on Plasmids

The ability to express efficiently heterologous genes and the variety of molecular techniques now available to accomplish this have made *E. coli* the organism of choice for cloning and expressing viral genes. A foreign gene product manufactured by *E. coli* must, however, undergo costly purification procedures to free it from containing endotoxins, which are known to exert a pyrogenic effect on man. Biologically safe organisms, such as *B. subtilis* and *Saccharomyces cerevisiae*, have become promising alternative hosts for molecular cloning of genes whose products are destined for vaccines, foods, or beverages.

Bacillus subtilis, a nonpathogenic bacterium free of endotoxins, is not a normal resident of the gut, as is *E. coli*. Many of the *Bacillus* spp. are important producers of enzymes and antibiotics purified from highly developed, large-scale fermentation processes. Secretion of many extracellular proteins into the surrounding growth medium is an additional feature that makes *B. subtilis* a suitable candidate as the host microorganism for cloned genes. Furthermore, *B. subtilis* is the most widely studied and thoroughly mapped gram-positive microorganism. It is capable of genetic transformation at relatively high frequencies, sometimes as high as 10%. Since the discovery that *Staphylococcus aureus* plasmids can replicate and express genetic information in *B. subtilis*, recombinant DNA technology on *Bacillus* sp. has advanced rapidly. Several plasmid and bacteriophage cloning vehicles have been successfully employed.

To express efficiently foreign genes cloned in *B. subtilis*, transcriptional and/or translational signals that differ from those of *E. coli* are required. For this reason, so-called "promoter-probe" vectors are needed to detect genetic regulatory signals that allow expression of heterologous genes in the host of interest. Plasmid vectors suitable for cloning in *E. coli* fragments of DNA that carry transcriptional promoter of termination signals are well characterized. Detection in these systems is based on expression of genes that encode β-galactosidase or confer antibiotic resistance to host cells. More recently, three promoter-probe systems for *B. subtilis* have been constructed. All are based on activating the expression of genes that are not normally expressed in *B. subtilis*. This "insertional activation" is accomplished by ligating a fragment of DNA that promotes gene expression into a cleavage site upstream from the structural gene. The latter must encode a gene product that is readily detected and its activity easily quantified. Of the three systems for *B. subtilis*, two are similar in that they are based upon expression of chloramphenicol acetyltransferase (CAT) genes originating from either *Bacillus pumilus* or the *E. coli* transposable genetic element Tn9. The other approach is a method whereby fragments of DNA that promote expression of a foreign gene in *B. subtilis* are detected by a change of colour of bacterial colonies. This sensitive chromogenic assay is based upon the expression, in *B. subtilis*, of the *xyl*E gene that encodes for catechol 2,3-oxygenase (C230) of the TOL plasmid from *Pseudomonas putida* mt-2. A brief description of all three systems follows.

Methods for isolating fragment of DNA that promote expression of CAT genes in B. Subtilis

In the system devised by Lovett and co-workers, a 2.2 kb *Eco*RI-generated fragment of DNA carrying the CAT gene responsible for chloramphenicol resistance (Cm^r) of *B. pumilus* NCIB 8600 is transferred to the neomycin-resistance (Nm^r, 10 μg/ml) plasmid pUB110. The new plasmid, pPL531, confer high-level Cm^r (200 μg/ml) to *B. subtilis*

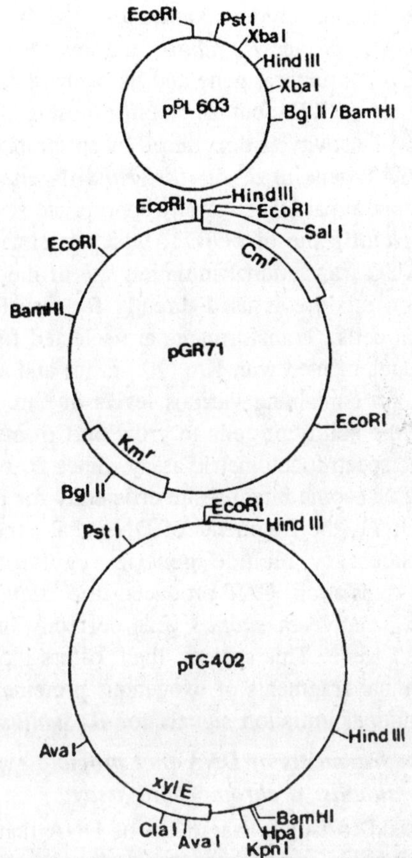

Fig. 14.7. Promoter-probe plasmid vectors for Bacillus subtillis.

host cells. Sequential digestion with *Pst*I, ligation, digestion with *Bam*HI and *Bgl*II, followed by religation leads the promoter-probe plasmid pPL603. Because the CAT gene of pPL603 no longer carries its promoter, host cells demonstrate only low-level Cm^r (5 μg/ml). Insertion of *Eco*RI or *Eco*RI fragments of DNA into the unique *Eco*RI site upstream from the CAT gene can lead to high-level Cm^r. The authors demonstrated that select fragment of DNA from *B. subtilis*, *Bacillus licheniformis*, plasmids pUB110 and pPL10, and bacteriophages SP02 and ϕ105 promote expression of the *B. pumilus* CAT gene in *B. subtilis*. It was not determined whether the genetic regulatory signals exerted their control at the transcriptional or translational level.

As similar approach exploits the CAT gene from Tn9. The promoter-probe plasmid is the 8.4 kb pGR71, an E. coli/*B. subtilis* cointegrate. The plasmid confers kanamycin (Km^r)/neomycin (Nm^r) resistance to both *E. coli* and *B. subtilis* and low-level Cm^r (5 μg/ml) in *E. coli*. The CAT structural gene and its putative ribosome binding site are present on pGR71, but *B. subtilis* host cells are Cm^s and demonstrate no CAT activity as determined by spectrophotometric assay. To activate the CAT gene in *B. subtilis*, *Hin*dIII-generated fragments of *B. subtilis* chromosomal DNA (or any appropriate source) are ligated into the unique *Hin*dIII site of pGR71, which is situated 33–38 base pairs 5' to the AUG translational initiation site of the CAT structural gene. The ligation mixture is used directly for transformation of *B. subtilis* competent cells. Transformants are selected first on complete medium plates supplemented with Km (10 μg/ml) and are then replica-plated onto medium containing various levels of Cm. CAT activities are estimated by the ability of cells to grow on Cm medium or, more accurately, by the spectrophotometric assay. Since *B. subtilis* is unable to use the *E. coli* ribosome binding site efficiently for the Tn9-derived CAT gene on pGR71, the fragments of DNA that generate Cm^r in *B. subtilis* must necessary include genetic regulatory signals for transcription and translation. CAT produced in *B. subtilis* results from a fusion between a native *B. subtilis* gene correctly in reading phase with the Tn9 CAT gene. This system, then, offers the opportunity to select amino-terminal fragments of exogenous proteins that also carry functional translational initiation signals for *B. subtilis*.

Method to isolate fragments of DNA that promote expression of the XylE gene in B. subtilis: a chromogenic assay

A novel method to isolate fragments of DNA that promote gene expression in *B. subtilis*, and perhaps many other microorganisms, has been descried. The system is based upon production of catechol 2,3-oxygenase encoded by the *P. putida* TOL plasmid gene *xyl*E. The gene is transferred as a *Bam*Hi/*Xho*I fragment of TOL to the *B. subtilis*/*E. coli* cointegrate plasmid pH33. This results in the promoter-probe vector ptG402. Although *xyl*E is functionally expressed in *E. coli*, C230 is not detected in *B. subtilis* unless a fragment of DNA capable of promoting gene expression is ligated into a cleavage site (*Bam*HI, *Hpa*I, *Kpn*I) on pTG402 upstream from *xyl*E. The *Bam*HI site is amenable to ligation with DNA fragments generated by digestion with *Bam*HI, *Bcl*I, *Bgl*II, or *Sau3*A, whereas any blunt-ended fragment can be ligated into the *Hpa*I site. A unique feature of this system is the ease of detection: colonies of cells that express *xyl*E become yellow within seconds after Cm selection plates are sprayed with a 0.5 M solution of catechol. The latter, a colourless substrate, is converted by C230 to the yellow product, 2-hydroxymuconic semialdehyde. The C230 produced in *B. subtilis* is identical to that produced in *P. putida* and *E. coli* by criteria of molecular weight and enzymatic activity as determined in a simple spectrophotometric assay. Strong complementarity between the ribosome binding site for *xyl*E and *B. subtilis* 16 S rRNA suggests that *xyl*E mRNA translation in *B. subtilis* most likely commences at the same initiation codon as that recognized by *P. putida* and *E. coli*. This system, therefore permits the detection of transcriptional promoters; it does not necessitate the synthesis of fusion polypeptides. The authors have demonstrated that fragments of DNA from *B. subtilis*, *B. licheniformis*, *B. pumilus*, *E. coli*, plasmid pUB110 (unpublished), and bacteriophage ϕ29 (unpublished) all have the capacity to promote *xyl*E gene expression in *B. subtilis*. The assay is rapid and inexpensive, does not require special indicator plates but offers the advantages of a genetic indicator test and can be used for the development of plasmid gene expression vectors.

Expression of HBV Core Antigen in B. subtilis

Because the construction of gene expression vectors for *B. subtilis* is a recent development, expression of viral antigens in the microorganism has yet to be fully explored. Success in expressing the genes for HBV core antigen and the major antigen of foot-and-mouth disease virus (FMDV) has, however, been realized. To express the HBV core antigen gene, the following method was employed.

The plasmid vector pKH80 was made by joining pBD9 to pBR322 by ligating the two plasmids at their respective and unique EcoRI sites. Plasmid pkH80 replicates in *B. subtilis* wherein genes to render host cells resistant to Km (10 μg/ml) and erythromycin (EMr, 10 μg/ml) are expressed. The Em^r gene encodes a protein of 29,000 daltons, the synthesis of which is inducible by subinhibitory concentrations (0.05 μg/ml), of erythromycin. The entire sequence of the EM^r gene has been determined. In *E. coli*, cells with pKH80 are resistant to ampicillin (Ap^r, 50 μg/ml), tetracycline (Tc^r, 15 μg/ml), Km (10 μg/ml), and Em (at high concentrations such as 60 μg/ml, Em partially inhibits *E. coli* unless pkH80 is present).

The HBV core antigen was isolated as a PstI fragment from pHBV139A. An average of 100 bp was trimmed from each end of the fragment by using exonuclease *Bal*31, then *Bam*HI and *Hin*dIII linkers were ligated, at random, to the extremities. Digestion with *Bam*HI and purification of the DNA fragment led to the isolation of the HBV core antigen gene flanked by a *Hin*dIII linker 5' to the gene, a *Bam*HI linker 5' to the *Hin*dIII linker, and a single *Bam*HI linker 3' to the HBV core gene. This *Bam*HI fragment was amplified by cloning it in the *Bam*HI site of pBR322. The *Bam*HI-generated fragment was ligated in the unique *Bcl*I site of pKH80; the *Bcl*I cleavage site lies within the EM^r gene. This construct maintains the proper reading frame for expressing the HBV core antigen under the control of genetic regulatory signals for the inducible EM^r gene on the plasmid vector. Ribosomes translating the mRNA of the EM^r gene should arrive in phase at a termination codon (UAG) that lies 4–6 bp 5' to the AUG initiation codon of the HBV core antigen gene. Translation would then start de novo from the transcript of the viral DNA sequence.

Although *B. subtilis* strains that carry this hybrid plasmid gave positive results when assayed for HBV core antigen by solid-phase radioimmunoassay, the concentration of HBV core protein was determined to be $\leq$ 0.1% of total cellular protein. The stop-start type of construction is presumably inefficient in this case, but expression might be increased by incorporating an efficient ribosome binding site for *B. subtilis* upstream from the restart site.

Expression of the Major FMDV Antigen in B. Subtilis

To express the VP1 antigen for FMDV in *B. subtilis*, Hardy et al. (1981) utilized the VP1 gene from *E. coli* plasmid pPLVP1. The latter was digested with *Bam*HI then joined to pBD9 previously digested with *Bel*I. In one of the two possible orientations, the pPLVP1 and pBD9 hybrid plasmid (pEVP1) should result in a DNA sequence encoding a fusion polypeptide comprising 73 amino acids of the Em^r gene on pBD9, 284 amino acids of FMDV, and 13 amino acids of the pPL c24 moiety of pPLVP1. The plasmid was introduced into *B. subtilis* by transformation, then cells were treated with subinhibitory concentrations of Em to induce expression of the fused gene. Proteins from an extract of these cells were resolved on a SDS-polyacrylamide (10%) gel. The polypeptides in the gel were transferred to nitrocellulose filters, then a radioimmunoassay was used to detect the expected 41,000-dalton fusion polypeptide that binds VP1-specific antibodies. Approximately 1% of the total cellular protein was calculated to be VP1 protein in its hybrid form.

Expression of Viral Antigens in Saccharomyces cerevisiae

Plasmid Vectors for Expression of Genes in Yeast

The yeast *Saccharomyces cerevisiae* has been developed as a host organism for cloned genes. Because it is a eukaryotic organisms, yeast is a suitable host for the expression of genes from higher organisms: posttranscriptional and posttranslational processes closely resemble those of higher cells, the organism has complex membrane systems, and the ability to secrete and glycosylate proteins has been observed. Cloning vectors for yeast are normally constructed as yeast/*E. coli* cointegrate plasmids. They consist of an origin of replication and antibiotic resistance gene(s) that function in *E. coli*, the 2-μm DNA sequence needed for replication of the plasmid in yeast, and a yeast gene that encodes a product that allows selection by complementation in yeast and *E. coli* recipient cells that carry the corresponding mutation. For details on cloning procedures in yeast, Beggs (1981) provides an excellent review.

Expression of HBV Surface Antigen in Yeast

As previously noted, the HBV surface antigen gene cloned in *E. coli* results in very inefficient production of HBsAg. However, in the yeast systems described below, not only is a substantial level of HBsAg produced, but it is synthesized in the form of particles or aggregates similar in size and shape to those found hepatoma cell lines or in sera from HBV-infected individuals.

In the method of Valenzuela et al. (1982), the HBsAg gene was expressed in yeast by using a vector that utilizes the 5'-flanking region of the yeast alcohol dehydrogenase I (ADH I) gene as a promoter to transcribe HBsAg coding sequences. The HBsAg gene was isolated on a 835 bp *TacI-Hpa*I fragment from pHVB-3200. The fragment, which carries 26 bp preceding the AUG methionine codon of the mature HBsAg, was ligated to *Eco*RI linkers and cloned in the EcoRI site of pBR322. The EcoRI-generated fragment was then purified by preparative gel electrophoresis and ligated to the yeast/*E. coli* hybrid plasmid, pMA-56, which was previously digested with EcoRI treated with bacterial alkaline phosphatase. The resulting plasmid was introduced into yeast cells, and HBsAg was detected by radioimmunoassay from cell extracts of mid-log phase cultures. From the assay data, it was calculated that 10-25 ng of HBsAg were made per milliliter of yeast culture.

An alternative method to express HBsAg in yeast employs the promoter from the gene P-60, which encodes a 60-kilodalton acid phosphatase. The expression of the P-60 gene is controlled by the level of inorganic phosphate (P_i) in the growth medium; i.e., low levels of P_i (1.5 g/liter in Burkholder minimal medium) induce the synthesis of P60. Cells are grown to a density of 4×10^6 cells/ml in high P_i

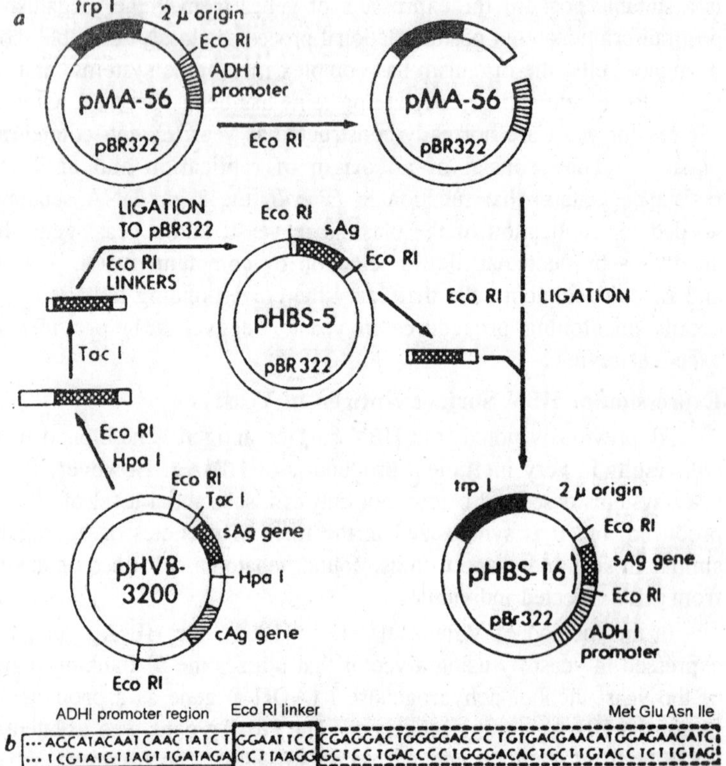

Fig. 14.8. (a) Construction of plasmids for the expression of HBV surface antigen in yeast. (b) Nucleotide sequence of the ADH1-HBsAg gene junction present in pHBS-16.

(1.5 g KH_2PO_4/liter), then a sample is removed, centrifuged, and suspended in low- P_i minimal medium for induction. Control samples are treated similarly, except that resuspension is high-P_i minimal medium.

The plasmid vector pAM82, used to express HBsAG, was constructed from pAT77, which carries the sequence of the NH_2-terminal coding region of P-60 and its upstream control region. The P-60 polypeptide coding sequence was completely removed by digesting pAT77 with *Sal*I. The plasmid was then treated with exonuclease BAL-31 under conditions adjusted to remove 85–130 bp of DNA, and ligated with XboI linkers.

HBV DNA was removed from pHBV4, a derivative of pACYC177 that carries the entire HBV (subtype *adr*) genome in double-stranded form as a *Xho*I fragment. pHBV4 was digested with *Xho*I to yield a fragment of DNA with a *Xho*I terminus located 27 bp upstream of the methionine codon fro the mature HBsAG. The *Xho*I-generated fragment was ligated in the *Xho*I site of pAM82.

The recombinant plasmid was propagated in *E. coli* χ1776, then transformed into a yeast recipient strain AH22 (a *leu*2 *his*4 *can*1 *cir*$^+$) or its acid phosphatase constitutive derivative, AH22 *pho*80. Cells were grown in liquid medium, then induction was performed in phosphate-free medium. Radioimmunoassays for HBsAg from cell extracts demonstrated that induction was necessary for high-level synthesis of HBsAg. Induced cells were estimated to produce 250 ng of HBsAG per milliliter of yeast culture. Uninduced cells produced 10^3-fold less HBsAg.

HBsAg produced in yeast by either cloning method can be purified by equilibrium sedimentation through a discontinuous CsCI gradient, where it bands at 1.2 g/cm^3. Velocity sedimentation in sucrose gradients demonstrates that HBsAg from yeast sediments at ~50S. Both values are in agreement with those of HBsAg from the human hepatoma cell line PLC/PRF/5 and suggest that HBsAg is synthesized in yeast in the form of particles. Examination of yeast HBsAG purified by sedimentation or by precipitation with anti-HBsAG antibodies established that the yeast particles are more variable in size and smaller than Dane particles. The particles are free of DNA and do not contain significant quantities of higher molecular weight glycoprotein, which suggests that glycosylation is not required for particle formation. Because the HBsAg produced in yeast was demonstrated to be sufficiently immunogenic, this suggests that glycosylation is also not required for immunogenicity. Taken together, the data clearly suggest that the HBsAg particle produced in yeast bears valuable potential for use as a vaccine.

Conclusion

In this review we have discussed in some depth how viral antigen scan be cloned and expressed in *E. coli* and other microorganisms. For the most part we have not approached in similar depth question of how suitable such microbially synthesized antigen are likely to be for vaccine purposes.

However, the production of viral protein via recombinant DNA technology has so far yielded extremely promising results, and we await developments in this field with exception.